Chemical Thermodynamics
of Materials

Chemical Thermodynamics of Materials

Macroscopic and Microscopic Aspects

Svein Stølen
Department of Chemistry, University of Oslo, Norway

Tor Grande
Department of Materials Technology, Norwegian University of Science and Technology, Norway

with a chapter on *Thermodynamics and Materials Modelling* by
Neil L. Allan
School of Chemistry, Bristol University, UK

John Wiley & Sons, Ltd

Other Wiley Editorial Offices

John Wiley & Sons Inc., 111 River Street, Hoboken, NJ 07030, USA

Jossey-Bass, 989 Market Street, San Francisco, CA 94103-1741, USA

Wiley-VCH Verlag GmbH, Boschstr. 12, D-69469 Weinheim, Germany

John Wiley & Sons Australia Ltd, 33 Park Road, Milton, Queensland 4064, Australia

John Wiley & Sons (Asia) Pte Ltd, 2 Clementi Loop #02-01, Jin Xing Distripark, Singapore 129809

John Wiley & Sons Canada Ltd, 22 Worcester Road, Etobicoke, Ontario, Canada M9W 1L1

Wiley also publishes its books in a variety of electronic formats. Some content that
appears in print may not be available in electronic books.

Library of Congress Cataloging-in-Publication Data

Stølen, Svein.
    Chemical thermodynamics of materials : macroscopic and microscopic
aspects / Svein Stølen, Tor Grande.
       p. cm.
Includes bibliographical references and index.
    ISBN 0-471-49230-2 (cloth : alk. paper)
    1. Thermodynamics. I. Grande, Tor. II. Title.
QD504 .S76 2003
541'.369--dc22

                                                    2003021826

British Library Cataloguing in Publication Data

A catalogue record for this book is available from the British Library

ISBN 10: 0-471-49230-2 (H/B)
ISBN 13: 978-0-471-49230-6 (H /B)

Typeset in 10/12 pt Times by Ian Kingston Editorial Services, Nottingham, UK
Printed and bound in Great Britain by Antony Rowe Ltd, Chippenham, Wiltshire
This book is printed on acid-free paper responsibly manufactured from sustainable forestry
in which at least two trees are planted for each one used for paper production.

Contents

Preface

Why write yet another book on the thermodynamics of materials? The traditional approach to such a text has been to focus on the phenomenology and mathematical concepts of thermodynamics, while the use of examples demonstrating the thermodynamic behaviour of materials has been less emphasized. Moreover, the few examples given have usually been taken from one particular type of materials (metals, for example). We have tried to write a comprehensive text on the chemical thermodynamics of materials with the focus on cases from a variety of important classes of materials, while the mathematical derivations have deliberately been kept rather simple. The aim has been both to treat thermodynamics macroscopically and also to consider the microscopic origins of the trends in the energetic properties of materials that have been considered. The examples are chosen to cover a broad range of materials and at the same time important topics in current solid state sciences.

The first three chapters of the book are devoted to basic thermodynamic theory and give the necessary background for a thermodynamic treatment of phase diagrams and phase stability in general. The link between thermodynamics and phase diagrams is covered in Chapter 4, and Chapter 5 gives the thermodynamic treatment of phase stability. While the initial chapters neglect the effects of surfaces, a separate chapter is devoted to surfaces, interfaces and adsorption. The three next chapters on trends in enthalpy of formation of various materials, on heat capacity and entropy of simple and complex materials, and on atomistic solution models, are more microscopically focused. A special feature is the chapter on trends in the enthalpy of formation of different materials; the enthalpy of formation is the most central parameter for most thermodynamic analysis, but it is still neglected in most thermodynamic treatments. The enthalpy of formation is also one of the focuses in a chapter on experimental methods for obtaining thermodynamic data. Another special feature is the final chapter on thermodynamic and materials modelling, contributed by Professor Neil Allan, University of Bristol, UK – this is a topic not treated in other books on chemical thermodynamics of materials.

The present text should be suitable for advanced undergraduates or graduate students in solid state chemistry or physics, materials science or mineralogy. Obviously we have assumed that the readers of this text have some prior knowledge of chemistry and chemical thermodynamics, and it would be advantageous for students to have already taken courses in physical chemistry and preferably also in basic solid state chemistry or physics. The book may also be thought of as a source of information and theory for solid state scientists in general.

We are grateful to Neil Allan not only for writing Chapter 11 but also for reading, commenting on and discussing the remaining chapters. His effort has clearly improved the quality of the book. Ole Bjørn Karlsen, University of Oslo, has also largely contributed through discussions on phase diagrams and through making some of the more complex illustrations. He has also provided the pictures used on the front cover. Moreover, Professor Mari-Ann Einarsrud, Norwegian University of Science and Technology, gave us useful comments on the chapter on surfaces and interfaces.

One of the authors (TG) would like to acknowledge Professor Kenneth R. Poeppelmeier, Northwestern University, for his hospitality and friendship during his sabbatical leave during the spring semester 2002. One of the authors (S^2) would like to express his gratitude to Professor Fredrik Grønvold for being an inspiring teacher, a good friend and always giving from his great knowledge of thermodynamics.

Svein Stølen
Tor Grande
Oslo, October 2003

1
Thermodynamic foundations

1.1 Basic concepts

Thermodynamic systems

A thermodynamic description of a process needs a well-defined **system**. A thermo-dynamic system contains everything of thermodynamic interest for a particular chemical process within a boundary. The **boundary** is either a real or hypothetical enclosure or surface that confines the system and separates it from its **surroundings**. In order to describe the thermodynamic behaviour of a physical system, the interaction between the system and its surroundings must be understood. Thermodynamic systems are thus classified into three main types according to the way they interact with the surroundings: **isolated systems** do not exchange energy or matter with their surroundings; **closed systems** exchange energy with the surroundings but not matter; and **open systems** exchange both energy and matter with their surroundings.

The system may be homogeneous or heterogeneous. An exact definition is difficult, but it is convenient to define a **homogeneous system** as one whose properties are the same in all parts, or at least their spatial variation is continuous. A **heterogeneous system** consists of two or more distinct homogeneous regions or **phases**, which are separated from one another by surfaces of discontinuity. The boundaries between phases are not strictly abrupt, but rather regions in which the properties change abruptly from the properties of one homogeneous phase to those of the other. For example, Portland cement consists of a mixture of the phases β-Ca_2SiO_4, Ca_3SiO_5, $Ca_3Al_2O_6$ and $Ca_4Al_2Fe_2O_{10}$. The different homogeneous phases are readily distinguished from each

Chemical Thermodynamics of Materials by Svein Stølen and Tor Grande
© 2004 John Wiley & Sons, Ltd ISBN 0 471 492320 2

other macroscopically and the thermodynamics of the system can be treated based on the sum of the thermodynamics of each single homogeneous phase.

In colloids, on the other hand, the different phases are not easily distinguished macroscopically due to the small particle size that characterizes these systems. So although a colloid also is a heterogeneous system, the effect of the surface thermodynamics must be taken into consideration in addition to the thermodynamics of each homogeneous phase. In the following, when we speak about heterogeneous systems, it must be understood (if not stated otherwise) that the system is one in which each homogeneous phase is spatially sufficiently large to neglect surface energy contributions. The contributions from surfaces become important in systems where the dimensions of the homogeneous regions are about 1 μm or less in size. The thermodynamics of surfaces will be considered in Chapter 6.

A homogeneous system – solid, liquid or gas – is called a **solution** if the composition of the system can be varied. The **components** of the solution are the substances of fixed composition that can be mixed in varying amounts to form the solution. The choice of the components is often arbitrary and depends on the purpose of the problem that is considered. The solid solution $LaCr_{1-y}Fe_yO_3$ can be treated as a quasi-binary system with $LaCrO_3$ and $LaFeO_3$ as components. Alternatively, the compound may be regarded as forming from La_2O_3, Fe_2O_3 and Cr_2O_3 or from the elements La, Fe, Cr and O_2 (g). In La_2O_3 or $LaCrO_3$, for example, the elements are present in a definite ratio, and independent variation is not allowed. La_2O_3 can thus be treated as a single component system. We will come back to this important topic in discussing the Gibbs phase rule in Chapter 4.

Thermodynamic variables

In thermodynamics the state of a system is specified in terms of macroscopic **state variables** such as volume, V, temperature, T, pressure, p, and the number of moles of the chemical constituents i, n_i. The laws of thermodynamics are founded on the concepts of internal energy (U), and entropy (S), which are functions of the state variables. Thermodynamic variables are categorized as intensive or extensive. Variables that are proportional to the size of the system (e.g. volume and internal energy) are called **extensive variables**, whereas variables that specify a property that is independent of the size of the system (e.g. temperature and pressure) are called **intensive variables**.

A **state function** is a property of a system that has a value that depends on the conditions (state) of the system and not on how the system has arrived at those conditions (the thermal history of the system). For example, the temperature in a room at a given time does not depend on whether the room was heated up to that temperature or cooled down to it. The difference in any state function is identical for every process that takes the system from the same given initial state to the same given final state: it is independent of the path or process connecting the two states. Whereas the internal energy of a system is a state function, work and heat are not. Work and heat are not associated with one given state of the system, but are defined only in a transformation of the system. Hence the work performed and the heat adsorbed by the system between the initial and final states depend on the choice of the transformation path linking these two states.

Thermodynamic processes and equilibrium

The state of a physical system evolves irreversibly towards a time-independent state in which we see no further macroscopic physical or chemical changes. This is the state of **thermodynamic equilibrium**, characterized for example by a uniform temperature throughout the system but also by other features. A **non-equilibrium state** can be defined as a state where irreversible processes drive the system towards the state of equilibrium. The rates at which the system is driven towards equilibrium range from extremely fast to extremely slow. In the latter case the isolated system may appear to have reached equilibrium. Such a system, which fulfils the characteristics of an equilibrium system but is not the true equilibrium state, is called a **metastable** state. Carbon in the form of diamond is stable for extremely long periods of time at ambient pressure and temperature, but transforms to the more stable form, graphite, if given energy sufficient to climb the activation energy barrier. Buckminsterfullerene, C_{60}, and the related C_{70} and carbon nanotubes, are other metastable modifications of carbon. The enthalpies of three modifications of carbon relative to graphite are given in Figure 1.1 [1, 2].

Glasses are a particular type of material that is neither stable nor metastable. Glasses are usually prepared by rapid cooling of liquids. Below the melting point the liquid become supercooled and is therefore metastable with respect to the equilibrium crystalline solid state. At the glass transition the supercooled liquid transforms to a glass. The properties of the glass depend on the quenching rate (thermal history) and do not fulfil the requirements of an equilibrium phase. Glasses represent **non-ergodic** states, which means that they are not able to explore their entire phase space, and glasses are thus not in internal equilibrium. Both stable states (such as liquids above the melting temperature) and metastable states (such as supercooled liquids between the melting and glass transition temperatures) are in internal equilibrium and thus **ergodic**. Frozen-in degrees of freedom are frequently present, even in crystalline compounds. Glassy crystals exhibit translational periodicity of the molecular

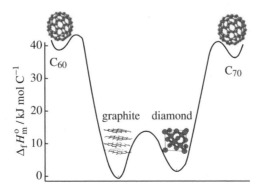

Figure 1.1 Standard enthalpy of formation per mol C of C_{60} [1], C_{70} [2] and diamond relative to graphite at 298 K and 1 bar.

centre of mass, whereas the molecular orientation is frozen either in completely random directions or randomly among a preferred set of orientations. Strictly spoken, only ergodic states can be treated in terms of classical thermodynamics.

1.2 The first law of thermodynamics

Conservation of energy

The first law of thermodynamics may be expressed as:

> Whenever any process occurs, the sum of all changes in energy, taken over all the systems participating in the process, is zero.

The important consequence of the first law is that energy is always conserved. This law governs the transfer of energy from one place to another, in one form or another: as heat energy, mechanical energy, electrical energy, radiation energy, etc. The energy contained within a thermodynamic system is termed the **internal energy** or simply the **energy** of the system, U. In all processes, reversible or irreversible, the change in internal energy must be in accord with the first law of thermodynamics.

Work is done when an object is moved against an opposing force. It is equivalent to a change in height of a body in a gravimetric field. The energy of a system is its capacity to do work. When work is done on an otherwise isolated system, its capacity to do work is increased, and hence the energy of the system is increased. When the system does work its energy is reduced because it can do less work than before. When the energy of a system changes as a result of temperature differences between the system and its surroundings, the energy has been transferred as **heat**. Not all boundaries permit transfer of heat, even when there is a temperature difference between the system and its surroundings. A boundary that does not allow heat transfer is called **adiabatic**. Processes that release energy as heat are called **exothermic**, whereas processes that absorb energy as heat are called **endothermic**.

The mathematical expression of the first law is

$$\sum \mathrm{d}U = \sum \mathrm{d}q + \sum \mathrm{d}w = 0 \qquad (1.1)$$

where U, q and w are the internal energy, the heat and the work, and each summation covers all systems participating in the process. Applications of the first law involve merely accounting processes. Whenever any process occurs, the net energy taken up by the given system will be exactly equal to the energy lost by the surroundings and vice versa, i.e. simply **the principle of conservation of energy**.

In the present book we are primarily concerned with the work arising from a change in volume. In the simplest example, work is done when a gas expands and drives back the surrounding atmosphere. The work done when a system expands its volume by an infinitesimal small amount $\mathrm{d}V$ against a constant external pressure is

$$\mathrm{d}w = -p_{\mathrm{ext}} \mathrm{d}V \qquad (1.2)$$

Table 1.1 Conjugate pairs of variables in work terms for the fundamental equation for the internal energy U. Here f is force of elongation, l is length in the direction of the force, σ is surface tension, A_S is surface area, Φ_i is the electric potential of the phase containing species i, q_i is the contribution of species i to the electric charge of a phase, E is electric field strength, p is the electric dipole moment of the system, B is magnetic field strength (magnetic flux density), and m is the magnetic moment of the system. The dots indicate scalar products of vectors.

Type of work	Intensive variable	Extensive variable	Differential work in dU
Mechanical			
Pressure–volume	$-p$	V	$-p\mathrm{d}V$
Elastic	f	l	$f\mathrm{d}l$
Surface	σ	A_S	$\sigma\mathrm{d}A_S$
Electromagnetic			
Charge transfer	Φ_i	q_i	$\Phi_i\mathrm{d}q_i$
Electric polarization	E	p	$E\cdot\mathrm{d}p$
Magnetic polarization	B	m	$B\cdot\mathrm{d}m$

The negative sign shows that the internal energy of the system doing the work decreases.

In general, dw is written in the form (intensive variable)·d(extensive variable) or as a product of a force times a displacement of some kind. Several types of work terms may be involved in a single thermodynamic system, and electrical, mechanical, magnetic and gravitational fields are of special importance in certain applications of materials. A number of types of work that may be involved in a thermodynamic system are summed up in Table 1.1. The last column gives the form of work in the equation for the internal energy.

Heat capacity and definition of enthalpy

In general, the change in internal energy or simply the energy of a system U may now be written as

$$\mathrm{d}U = \mathrm{d}q + \mathrm{d}w_{pV} + \mathrm{d}w_{\text{non-e}} \tag{1.3}$$

where $\mathrm{d}w_{pV}$ and $\mathrm{d}w_{\text{non-e}}$ are the expansion (or pV) work and the additional non-expansion (or non-pV) work, respectively. A system kept at constant volume cannot do expansion work; hence in this case $\mathrm{d}w_{pV} = 0$. If the system also does not do any other kind of work, then $\mathrm{d}w_{\text{non-e}} = 0$. So here the first law yields

$$\mathrm{d}U = \mathrm{d}q_V \tag{1.4}$$

where the subscript denotes a change at constant volume. For a measurable change, the increase in the internal energy of a substance is

$$\Delta U = q_V \tag{1.5}$$

The temperature dependence of the internal energy is given by the **heat capacity at constant volume** at a given temperature, formally defined by

$$C_V = \left(\frac{\partial U}{\partial T} \right)_V \tag{1.6}$$

For a constant-volume system, an infinitesimal change in temperature gives an infinitesimal change in internal energy and the constant of proportionality is the heat capacity at constant volume

$$dU = C_V \, dT \tag{1.7}$$

The change in internal energy is equal to the heat supplied only when the system is confined to a constant volume. When the system is free to change its volume, some of the energy supplied as heat is returned to the surroundings as expansion work. Work due to the expansion of a system against a constant external pressure, p_{ext}, gives the following change in internal energy:

$$dU = dq + dw = dq - p_{ext} \, dV \tag{1.8}$$

For processes taking place at constant pressure it is convenient to introduce the **enthalpy** function, H, defined as

$$H = U + pV \tag{1.9}$$

Differentiation gives

$$dH = d(U + pV) = dq + dw + V dp + p dV \tag{1.10}$$

When only work against a constant external pressure is done:

$$dw = -p_{ext} \, dV \tag{1.11}$$

and eq. (1.10) becomes

$$dH = dq + V dp \tag{1.12}$$

Since $dp = 0$ (constant pressure),

$$dH = dq_p \tag{1.13}$$

and

$$\Delta H = q_p \tag{1.14}$$

The enthalpy of a substance increases when its temperature is raised. The temperature dependence of the enthalpy is given by the **heat capacity at constant pressure** at a given temperature, formally defined by

$$C_p = \left(\frac{\partial H}{\partial T}\right)_p \tag{1.15}$$

Hence, for a constant pressure system, an infinitesimal change in temperature gives an infinitesimal change in enthalpy and the constant of proportionality is the heat capacity at constant pressure.

$$dH = C_p dT \tag{1.16}$$

The heat capacity at constant volume and constant pressure at a given temperature are related through

$$C_p - C_V = \frac{\alpha^2 VT}{\kappa_T} \tag{1.17}$$

where α and κ_T are the **isobaric expansivity** and the **isothermal compressibility** respectively, defined by

$$\alpha = \frac{1}{V}\left(\frac{\partial V}{\partial T}\right)_p \tag{1.18}$$

and

$$\kappa_T = -\frac{1}{V}\left(\frac{\partial V}{\partial p}\right)_T \tag{1.19}$$

Typical values of the isobaric expansivity and the isothermal compressibility are given in Table 1.2. The difference between the heat capacities at constant volume and constant pressure is generally negligible for solids at low temperatures where the thermal expansivity becomes very small, but the difference increases with temperature; see for example the data for Al_2O_3 in Figure 1.2.

Since the heat absorbed or released by a system at constant pressure is equal to its change in enthalpy, enthalpy is often called heat content. If a phase transformation (i.e. melting or transformation to another solid polymorph) takes place within

Table 1.2 The isobaric expansivity and isothermal compressibility of selected compounds at 300 K.

Compound	$\alpha/10^{-5}\ K^{-1}$	$\kappa_T/10^{-12}\ Pa$
MgO	3.12	6.17
Al_2O_3	1.62	3.97
MnO	3.46	6.80
Fe_3O_4	3.56	4.52
NaCl	11.8	41.7
C (diamond)	0.54	1.70
C (graphite)	2.49	17.9
Al	6.9	13.2

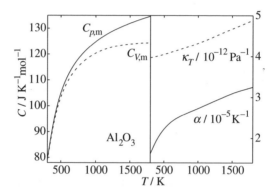

Figure 1.2 Molar heat capacity at constant pressure and at constant volume, isobaric expansivity and isothermal compressibility of Al_2O_3 as a function of temperature.

the system, heat may be adsorbed or released without a change in temperature. At constant pressure the heat merely transforms a portion of the substance (e.g. from solid to liquid – ice–water). Such a change is called a **first-order phase transition** and will be defined formally in Chapter 2. The standard enthalpy of aluminium relative to 0 K is given as a function of temperature in Figure 1.3. The standard enthalpy of fusion and in particular the standard enthalpy of vaporization contribute significantly to the total enthalpy increment.

Reference and standard states

Thermodynamics deals with processes and reactions and is rarely concerned with the absolute values of the internal energy or enthalpy of a system, for example, only with the changes in these quantities. Hence the energy changes must be well defined. It is often convenient to choose a reference state as an arbitrary zero. Often the reference state of a condensed element/compound is chosen to be at a pressure of 1 bar and in the most stable polymorph of that element/compound at the

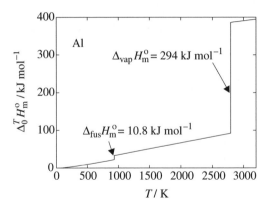

Figure 1.3 Standard enthalpy of aluminium relative to 0 K. The standard enthalpy of fusion ($\Delta_{fus}H_m^o$) is significantly smaller than the standard enthalpy of vaporization ($\Delta_{vap}H_m^o$).

temperature at which the reaction or process is taking place. This reference state is called a **standard state** due to its large practical importance. The term standard state and the symbol ^{o} are reserved for $p = 1$ bar. The term **reference state** will be used for states obtained from standard states by a change of pressure. It is important to note that the standard state chosen should be specified explicitly, since it is indeed possible to choose different standard states. The standard state may even be a **virtual state**, one that cannot be obtained physically.

Let us give an example of a standard state that not involves the most stable polymorph of the compound at the temperature at which the system is considered. Cubic zirconia, ZrO_2, is a fast-ion conductor stable only above 2300 °C. Cubic zirconia can, however, be stabilized to lower temperatures by forming a solid solution with for example Y_2O_3 or CaO. The composition–temperature stability field of this important phase is marked by Css in the ZrO_2–$CaZrO_3$ phase diagram shown in Figure 1.4 (phase diagrams are treated formally in Chapter 4). In order to describe the thermodynamics of this solid solution phase at, for example, 1500 °C, it is convenient to define the metastable cubic high-temperature modification of zirconia as the standard state instead of the tetragonal modification that is stable at 1500 °C. The standard state of pure ZrO_2 (used as a component of the solid solution) and the investigated solid solution thus take the same crystal structure.

The standard state for gases is discussed in Chapter 2.

Enthalpy of physical transformations and chemical reactions

The enthalpy that accompanies a change of physical state at standard conditions is called the standard **enthalpy of transition** and is denoted $\Delta_{trs}H^o$. Enthalpy changes accompanying chemical reactions at standard conditions are in general termed standard **enthalpies of reaction** and denoted $\Delta_r H^o$. Two simple examples are given in Table 1.3. In general, from the first law, the standard enthalpy of a reaction is given by

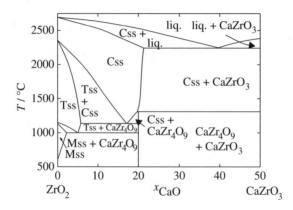

Figure 1.4 The ZrO_2–$CaZrO_3$ phase diagram. Mss, Tss and Css denote monoclinic, tetragonal and cubic solid solutions.

Table 1.3 Examples of a physical transformation and a chemical reaction and their respective enthalpy changes. Here $\Delta_{fus}H_m^o$ denotes the standard molar enthalpy of fusion.

Reaction	Enthalpy change
Al (s) = Al (liq)	$\Delta_{trs}H_m^o = \Delta_{fus}H_m^o = 10789$ J mol^{-1} at T_{fus}
$3SiO_2$ (s) + $2N_2$ (g) = Si_3N_4 (s) + $3O_2$ (g)	$\Delta_r H^o = 1987.8$ kJ mol^{-1} at 298.15 K

$$\Delta_r H^o = \sum_j v_j H_m^o (j) - \sum_i v_i H_m^o (i) \tag{1.20}$$

where the sum is over the standard molar enthalpy of the reactants i and products j (v_i and v_j are the stoichiometric coefficients of reactants and products in the chemical reaction).

Of particular importance is the standard molar **enthalpy of formation**, $\Delta_f H_m^o$, which corresponds to the standard reaction enthalpy for the formation of one mole of a compound from its elements in their standard states. The standard enthalpies of formation of three different modifications of Al_2SiO_5 are given as examples in Table 1.4 [3]. Compounds like these, which are formed by combination of electropositive and electronegative elements, generally have large negative enthalpies of formation due to the formation of strong covalent or ionic bonds. In contrast, the difference in enthalpy of formation between the different modifications is small. This is more easily seen by consideration of the enthalpies of formation of these ternary oxides from their binary constituent oxides, often termed the standard molar **enthalpy of formation from oxides**, $\Delta_{f,ox}H_m^o$, which correspond to $\Delta_r H_m^o$ for the reaction

$$SiO_2 \text{ (s)} + Al_2O_3 \text{ (s)} = Al_2SiO_5 \text{ (s)} \tag{1.21}$$

Table 1.4 The enthalpy of formation of the three polymorphs of Al_2SiO_5, kyanite, andalusite and sillimanite at 298.15 K [3].

Reaction	$\Delta_f H_m^o$ / kJ mol^{-1}
2 Al (s) + Si (s) + 5/2 O_2 (g) = Al_2SiO_5 (kyanite)	−2596.0
2 Al (s) + Si (s) + 5/2 O_2 (g) = Al_2SiO_5 (andalusite)	−2591.7
2 Al (s) + Si (s) + 5/2 O_2 (g) = Al_2SiO_5 (sillimanite)	−2587.8

These are derived by subtraction of the standard molar enthalpy of formation of the binary oxides, since standard enthalpies of individual reactions can be combined to obtain the standard enthalpy of another reaction. Thus,

$$\Delta_{f,ox} H_m^o (Al_2 SiO_5) = \Delta_f H_m^o (Al_2 SiO_5) - \Delta_f H_m^o (Al_2 O_3)$$
$$- \Delta_f H_m^o (SiO_2) \tag{1.22}$$

This use of the first law of thermodynamics is called **Hess's law**:

The standard enthalpy of an overall reaction is the sum of the standard enthalpies of the individual reactions that can be used to describe the overall reaction of Al_2SiO_5.

Whereas the enthalpy of formation of Al_2SiO_5 from the elements is large and negative, the enthalpy of formation from the binary oxides is much less so. $\Delta_{f,ox} H_m$ is furthermore comparable to the enthalpy of transition between the different polymorphs, as shown for Al_2SiO_5 in Table 1.5 [3]. The enthalpy of fusion is also of similar magnitude.

The temperature dependence of reaction enthalpies can be determined from the heat capacity of the reactants and products. When a substance is heated from T_1 to T_2 at a particular pressure p, assuming no phase transition is taking place, its molar enthalpy change from $\Delta H_m (T_1)$ to $\Delta H_m (T_2)$ is

Table 1.5 The enthalpy of formation of kyanite, andalusite and sillimanite from the binary constituent oxides [3]. The enthalpy of transition between the different polymorphs is also given. All enthalpies are given for $T = 298.15$ K.

Reaction	$\Delta_r H_m^o = \Delta_{f,ox} H_m^o$ / kJ mol^{-1}
Al_2O_3 (s) + SiO_2 (s) = Al_2SiO_5 (kyanite)	−9.6
Al_2O_3 (s) + SiO_2 (s) = Al_2SiO_5 (andalusite)	−5.3
Al_2O_3 (s) + SiO_2 (s) = Al_2SiO_5 (sillimanite)	−1.4
Al_2SiO_5 (kyanite) = Al_2SiO_5 (andalusite)	4.3
Al_2SiO_5 (andalusite) = Al_2SiO_5 (sillimanite)	3.9

$$\Delta H_m(T_2) = \Delta H_m(T_1) + \int_{T_1}^{T_2} C_{p,m} dT \tag{1.23}$$

This equation applies to each substance in a reaction and a change in the standard reaction enthalpy (i.e. p is now $p^o = 1$ bar) going from T_1 to T_2 is given by

$$\Delta_r H^o(T_2) = \Delta_r H^o(T_1) + \int_{T_1}^{T_2} \Delta_r C_{p,m}^o dT \tag{1.24}$$

where $\Delta_r C_{p,m}^o$ is the difference in the standard molar heat capacities at constant pressure of the products and reactants under standard conditions taking the stoichiometric coefficients that appear in the chemical equation into consideration:

$$\Delta_r C_{p,m}^o = \sum_j v_j C_{p,m}^o(j) - \sum_i v_i C_{p,m}^o(i) \tag{1.25}$$

The heat capacity difference is in general small for a reaction involving condensed phases only.

1.3 The second and third laws of thermodynamics

The second law and the definition of entropy

A system can in principle undergo an indefinite number of processes under the constraint that energy is conserved. While the first law of thermodynamics identifies the allowed changes, a new state function, the **entropy** S, is needed to identify the spontaneous changes among the allowed changes. The second law of thermodynamics may be expressed as

The entropy of a system and its surroundings increases in the course of a spontaneous change, $\Delta S_{tot} > 0$.

The law implies that for a reversible process, the sum of all changes in entropy, taken over all the systems participating in the process, ΔS_{tot}, is zero.

Reversible and non-reversible processes

Any change in state of a system in thermal and mechanical contact with its surroundings at a given temperature is accompanied by a change in entropy of the system, dS, and of the surroundings, dS_{sur}:

$$dS + dS_{sur} \geq 0 \tag{1.26}$$

The sum is equal to zero for reversible processes, where the system is always under equilibrium conditions, and larger than zero for irreversible processes. The entropy change of the surroundings is defined as

$$dS_{sur} = -\frac{dq}{T}$$ (1.27)

where dq is the heat supplied to the system during the process. It follows that for any change:

$$dS \geq \frac{dq}{T}$$ (1.28)

which is known as the **Clausius inequality**. If we are looking at an isolated system

$$dS \geq 0$$ (1.29)

Hence, for an isolated system, the entropy of the system alone must increase when a spontaneous process takes place. The second law identifies the spontaneous changes, but in terms of both the system and the surroundings. However, it is possible to consider the specific system only. This is the topic of the next section.

Conditions for equilibrium and the definition of Helmholtz and Gibbs energies

Let us consider a closed system in thermal equilibrium with its surroundings at a given temperature T, where no non-expansion work is possible. Imagine a change in the system and that the energy change is taking place as a heat exchange between the system and the surroundings. The Clausius inequality (eq. 1.28) may then be expressed as

$$dS - \frac{dq}{T} \geq 0$$ (1.30)

If the heat is transferred at constant volume and no non-expansion work is done,

$$dS - \frac{dU}{T} \geq 0$$ (1.31)

The combination of the Clausius inequality (eq. 1.30) and the first law of thermodynamics for a system at constant volume thus gives

$$TdS \geq dU$$ (1.32)

Correspondingly, when heat is transferred at constant pressure (pV work only),

$$TdS \geq dH \tag{1.33}$$

For convenience, two new thermodynamic functions are defined, the **Helmholtz** (A) and **Gibbs** (G) **energies**:

$$A = U - TS \tag{1.34}$$

and

$$G = H - TS \tag{1.35}$$

For an infinitesimal change in the system

$$dA = dU - TdS - SdT \tag{1.36}$$

and

$$dG = dH - TdS - SdT \tag{1.37}$$

At constant temperature eqs. (1.36) and (1.37) reduce to

$$dA = dU - TdS \tag{1.38}$$

and

$$dG = dH - TdS \tag{1.39}$$

Thus for a system at constant temperature and volume, the equilibrium condition is

$$dA_{T,V} = 0 \tag{1.40}$$

In a process at constant T and V in a closed system doing only expansion work it follows from eq. (1.32) that the spontaneous direction of change is in the direction of decreasing A. At equilibrium the value of A is at a minimum.

For a system at constant temperature and pressure, the equilibrium condition is

$$dG_{T,p} = 0 \tag{1.41}$$

In a process at constant T and p in a closed system doing only expansion work it follows from eq. (1.33) that the spontaneous direction of change is in the direction of decreasing G. At equilibrium the value of G is at a minimum.

Equilibrium conditions in terms of internal energy and enthalpy are less applicable since these correspond to systems at constant entropy and volume and at constant entropy and pressure, respectively

$$dU_{S,V} = 0 \tag{1.42}$$

$$dH_{S,p} = 0 \tag{1.43}$$

The Helmholtz and Gibbs energies on the other hand involve constant temperature and volume and constant temperature and pressure, respectively. Most experiments are done at constant T and p, and most simulations at constant T and V. Thus, we have now defined two functions of great practical use. In a spontaneous process at constant p and T or constant p and V, the Gibbs or Helmholtz energies, respectively, of the system decrease. These are, however, only other measures of the second law and imply that the total entropy of the system and the surroundings increases.

Maximum work and maximum non-expansion work

The Helmholtz and Gibbs energies are useful also in that they define the maximum work and the maximum non-expansion work a system can do, respectively. The combination of the Clausius inequality $TdS \geq dq$ and the first law of thermodynamics $dU = dq + dw$ gives

$$dw \geq dU - TdS \tag{1.44}$$

Thus the **maximum work** (the most negative value of dw) that can be done by a system is

$$dw_{max} = dU - TdS \tag{1.45}$$

At constant temperature $dA = dU - TdS$ and

$$w_{max} = \Delta A \tag{1.46}$$

If the entropy of the system decreases some of the energy must escape as heat in order to produce enough entropy in the surroundings to satisfy the second law of thermodynamics. Hence the maximum work is less than $|\Delta U|$. ΔA is the part of the change in internal energy that is free to use for work. Hence the Helmholtz energy is in some older books termed the (isothermal) work content.

The total amount of work is conveniently separated into expansion (or pV) work and non-expansion work.

$$dw = dw_{non-e} - pdV \tag{1.47}$$

For a system at constant pressure it can be shown that

$$dw_{\text{non-e,max}} = dH - TdS \tag{1.48}$$

At constant temperature $dG = dH - TdS$ and

$$w_{\text{non-e,max}} = \Delta G \tag{1.49}$$

Hence, while the change in Helmholtz energy relates to the total work, the change in Gibbs energy at constant temperature and pressure represents the **maximum non-expansion work** a system can do.

Since $\Delta_r G^\circ$ for the formation of 1 mol of water from hydrogen and oxygen gas at 298 K and 1 bar is -237 kJ mol^{-1}, up to 237 kJ mol^{-1} of 'chemical energy' can be converted into electrical energy in a fuel cell working at these conditions using $H_2(g)$ as fuel. Since the Gibbs energy relates to the energy free for non-expansion work, it has in previous years been called the free energy.

The variation of entropy with temperature

For a reversible change the entropy increment is $dS = dq/T$. The variation of the entropy from T_1 to T_2 is therefore given by

$$S(T_2) = S(T_1) + \int_{T_1}^{T_2} \frac{dq_{\text{rev}}}{T} \tag{1.50}$$

For a process taking place at constant pressure and that does not involve any *non-pV* work

$$dq_{\text{rev}} = dH = C_p dT \tag{1.51}$$

and

$$S(T_2) = S(T_1) + \int_{T_1}^{T_2} \frac{C_p dT}{T} \tag{1.52}$$

The entropy of a particular compound at a specific temperature can be determined through measurements of the heat capacity as a function of temperature, adding entropy increments connected with first-order phase transitions of the compound:

$$S(T) = S(0) + \int_{0}^{T_{\text{trs}}} \frac{C_p(T)}{T} dT + \Delta_{\text{trs}} S_m + \int_{T_{\text{trs}}}^{T} \frac{C_p(T)}{T} dT \tag{1.53}$$

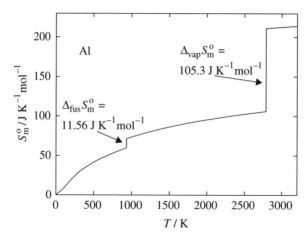

Figure 1.5 Standard entropy of aluminium relative to 0 K. The standard entropy of fusion ($\Delta_{fus} S_m^o$) is significantly smaller than the standard entropy of boiling ($\Delta_{vap} S_m^o$).

The variation of the standard entropy of aluminium from 0 K to the melt at 3000 K is given in Figure 1.5. The standard entropy of fusion and in particular the standard entropy of vaporization contribute significantly to the total entropy increment.

Equation (1.53) applies to each substance in a reaction and a change in the standard entropy of a reaction (p is now $p^o = 1$ bar) going from T_1 to T_2 is given by (neglecting for simplicity first-order phase transitions in reactants and products)

$$\Delta_r S^o(T_2) = \Delta_r S^o(T_1) + \int_{T_1}^{T_2} \frac{\Delta_r C_{p,m}^o(T)}{T} \, dT \tag{1.54}$$

where $\Delta_r C_{p,m}^o(T)$ is given by eq. (1.25).

The third law of thermodynamics

The third law of thermodynamics may be formulated as:

> If the entropy of each element in some perfect crystalline state at $T = 0$ K is taken as zero, then every substance has a finite positive entropy which at $T = 0$ K become zero for all perfect crystalline substances.

In a perfect crystal at 0 K all atoms are ordered in a regular uniform way and the translational symmetry is therefore perfect. The entropy is thus zero. In order to become perfectly crystalline at absolute zero, the system in question must be able to explore its entire phase space: the system must be in internal thermodynamic equilibrium. Thus the third law of thermodynamics does not apply to substances that are not in internal thermodynamic equilibrium, such as glasses and glassy crystals. Such non-ergodic states do have a finite entropy at the absolute zero, called **zero-point entropy** or **residual entropy** at 0 K.

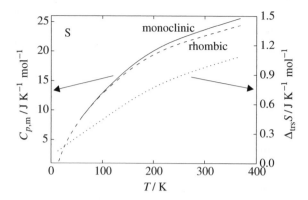

Figure 1.6 Heat capacity of rhombic and monoclinic sulfur [4,5] and the derived entropy of transition between the two polymorphs.

The third law of thermodynamics can be verified experimentally. The stable rhombic low-temperature modification of sulfur transforms to monoclinic sulfur at 368.5 K (p = 1 bar). At that temperature, T_{trs}, the two polymorphs are in equilibrium and the standard molar Gibbs energies of the two modifications are equal. We therefore have

$$\Delta_{trs} G_m^{o} = \Delta_{trs} H_m^{o} - T_{trs} \Delta_{trs} S_m^{o} = 0 \tag{1.55}$$

It follows that the standard molar entropy of the transition can be derived from the measured standard molar enthalpy of transition through the relationship

$$\Delta_{trs} S_m^{o} = \Delta_{trs} H_m^{o} / T_{trs} \tag{1.56}$$

Calorimetric experiments give $\Delta_{trs} H_m^{o}$ = 401.66 J mol^{-1} and thus $\Delta_{trs} S_m^{o}$ = 1.09 J K^{-1} mol^{-1} [4]. The entropies of the two modifications can alternatively be derived through integration of the heat capacities for rhombic and monoclinic sulfur given in Figure 1.6 [4,5]. The entropy difference between the two modifications, also shown in the figure, increases with temperature and at the transition temperature (368.5 K) it is in agreement with the standard entropy of transition derived from the standard enthalpy of melting. The third law of thermodynamics is thereby confirmed. The entropies of both modifications are zero at 0 K.

The Maxwell relations

Maxwell used the mathematical properties of state functions to derive a set of useful relationships. These are often referred to as the Maxwell relations. Recall the first law of thermodynamics, which may be written as

$$dU = dq + dw \tag{1.57}$$

For a reversible change in a closed system and in the absence of any non-expansion work this equation transforms into

$$dU = TdS - pdV \tag{1.58}$$

Since dU is an **exact differential**, its value is independent of the path. The same value of dU is obtained whether the change is reversible or irreversible, and eq. (1.58) applies to any change for a closed system that only does pV work. Equation (1.58) is often called the **fundamental equation.** The equation shows that the internal energy of a closed system changes in a simple way when S and V are changed, and U can be regarded as a function of S and V. We therefore have

$$dU = \left(\frac{\partial U}{\partial S}\right)_V dS + \left(\frac{\partial U}{\partial V}\right)_S dV \tag{1.59}$$

It follows from eqs. (1.58) and (1.59) that

$$\left(\frac{\partial U}{\partial S}\right)_V = T \tag{1.60}$$

and that

$$\left(\frac{\partial U}{\partial V}\right)_S = -p \tag{1.61}$$

Generally, a function $f(x,y)$ for which an infinitesimal change may be expressed as

$$df = gdx + hdy \tag{1.62}$$

is exact if

$$\left(\frac{\partial g}{\partial y}\right)_x = \left(\frac{\partial h}{\partial x}\right)_y \tag{1.63}$$

Thus since the internal energy, U, is a state function, one of the Maxwell relations may be deduced from (eq. 1.58):

$$\left(\frac{\partial T}{\partial V}\right)_S = -\left(\frac{\partial p}{\partial S}\right)_V \tag{1.64}$$

Table 1.6 The Maxwell relations.

Thermodynamic function	Differential	Equilibrium condition	Maxwell's relations
$U\,(S,\,V)$	$dU = TdS - pdV$	$(dU)_{S,V} = 0$	$\left(\dfrac{\partial T}{\partial V}\right)_S = -\left(\dfrac{\partial p}{\partial S}\right)_V$
$H\,(S,\,p)$	$dH = TdS + Vdp$	$(dH)_{S,p} = 0$	$\left(\dfrac{\partial T}{\partial p}\right)_S = \left(\dfrac{\partial V}{\partial S}\right)_p$
$A\,(T,\,V)$	$dA = -SdT - pdV$	$(dA)_{T,V} = 0$	$\left(\dfrac{\partial S}{\partial V}\right)_T = \left(\dfrac{\partial p}{\partial T}\right)_V$
$G\,(T,\,p)$	$dG = -SdT + Vdp$	$(dG)_{T,p} = 0$	$\left(\dfrac{\partial S}{\partial p}\right)_T = -\left(\dfrac{\partial V}{\partial T}\right)_p$

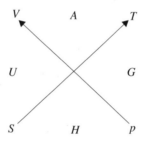

Figure 1.7 The thermodynamic square. Note that the two arrows enable one to get the right sign in the equations given in the second column in Table 1.6.

Using $H = U + pV$, $A = U - TS$ and $G = H - TS$ the remaining three Maxwell relations given in Table 1.6 are easily derived starting with the fundamental equation (eq. 1.58). A convenient method to recall these equations is the thermodynamic square shown in Figure 1.7. On each side of the square appears one of the state functions with the two natural independent variables given next to it. A change in the internal energy dU, for example, is thus described in terms of dS and dV. The arrow from S to T implies that TdS is a positive contribution to dU, while the arrow from p to V implies that pdV is a negative contribution. Hence $dU = TdS - pdV$ follows.

Properties of the Gibbs energy

Thermodynamics applied to real material systems often involves the Gibbs energy, since this is the most convenient choice for systems at constant pressure and temperature. We will thus consider briefly the properties of the Gibbs energy. As the natural variables for the Gibbs energy are T and p, an infinitesimal change, dG, can be expressed in terms of infinitesimal changes in pressure, dp, and temperature, dT.

$$dG = \left(\frac{\partial G}{\partial p}\right)_T dp + \left(\frac{\partial G}{\partial T}\right)_p dT \tag{1.65}$$

The Gibbs energy is related to enthalpy and entropy through $G = H - TS$. For an infinitesimal change in the system

$$dG = dH - TdS - SdT \tag{1.66}$$

Similarly, $H = U + pV$ gives

$$dH = dU + pdV + Vdp \tag{1.67}$$

Thus in the absence of non-expansion work for a closed system, the following important equation

$$dG = Vdp - SdT \tag{1.68}$$

is easily derived using also eq. (1.58). Equations (1.65) and (1.68) implies that the temperature derivative of the Gibbs energy at constant pressure is $-S$:

$$\left(\frac{\partial G}{\partial T} \right)_p = -S \tag{1.69}$$

and thus that

$$G(T_f) = G(T_i) - \int_{T_i}^{T_f} SdT \tag{1.70}$$

where i and f denote the initial and final p and T conditions. Since S is positive for a compound, the Gibbs energy of a compound decreases when temperature is increased at constant pressure. G decreases most rapidly with temperature when S is large and this fact leads to entropy-driven melting and vaporization of compounds when the temperature is raised. The standard molar Gibbs energy of solid, liquid and gaseous aluminium is shown as a function of temperature in Figure 1.8. The corresponding enthalpy and entropy is given in Figures 1.2 and 1.5. The melting (vaporization) temperature is given by the temperature at which the Gibbs energy of the solid (gas) and the liquid crosses, as marked in Figure 1.8.

Equation (1.70) applies to each substance in a reaction and a change in the standard Gibbs energy of a reaction (p is now $p^o = 1$ bar) going from T_i to T_f is given by

$$\Delta_r G^o(T_f) = \Delta_r G^o(T_i) - \int_{T_i}^{T_f} \Delta_r S^o dT \tag{1.71}$$

$\Delta_r S^o$ is not necessarily positive and the Gibbs energy of a reaction may increase with temperature.

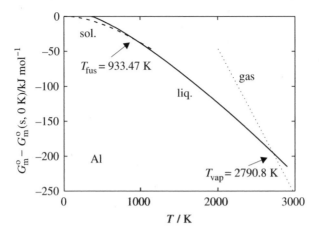

Figure 1.8 Standard Gibbs energy of solid, liquid and gaseous aluminium relative to the standard Gibbs energy of solid aluminium at $T = 0$ K as a function of temperature (at $p = 1$ bar).

The pressure derivative of the Gibbs energy (eq. 1.68) at constant temperature is V:

$$\left(\frac{\partial G}{\partial p} \right)_T = V \tag{1.72}$$

and the pressure variation of the Gibbs energy is given as

$$G(p_f) = G(p_i) + \int_{p_i}^{p_f} V dp \tag{1.73}$$

Since V is positive for a compound, the Gibbs energy of a compound increases when pressure is increased at constant temperature. Thus, while disordered phases are stabilized by temperature, high-density polymorphs (lower molar volumes) are stabilized by pressure. Figure 1.9 show that the Gibbs energy of graphite due to its open structure increases much faster with pressure than that for diamond. Graphite thus transforms to the much denser diamond modification of carbon at 1.5 GPa at 298 K.

Equation (1.73) applies to each substance in a reaction and a change in the Gibbs energy of a reaction going from p_i to p_f is given by

$$\Delta_r G(p_f) = \Delta_r G(p_i) + \int_{p_i}^{p_f} \Delta_r V dp \tag{1.74}$$

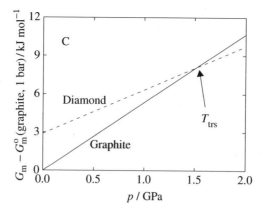

Figure 1.9 Standard Gibbs energy of graphite and diamond at $T = 298$ K relative to the standard Gibbs energy of graphite at 1 bar as a function of pressure.

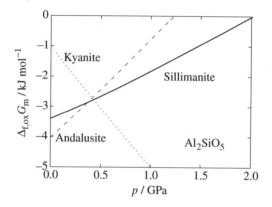

Figure 1.10 The standard Gibbs energy of formation from the binary constitutent oxides of the kyanite, sillimanite and andalusite modifications of Al_2SiO_5 as a function of pressure at 800 K. Data are taken from [3]. All three oxides are treated as incompressible.

$\Delta_r V$ is not necessarily positive, and to compare the relative stability of the different modifications of a ternary compound like Al_2SiO_5 the volume of formation of the ternary oxide from the binary constituent oxides is considered for convenience. The pressure dependence of the Gibbs energies of formation from the binary constituent oxides of kyanite, sillimanite and andalusite polymorphs of Al_2SiO_5 are shown in Figure 1.10. Whereas sillimanite and andalusite have positive volumes of formation and are destabilized by pressure relative to the binary oxides, kyanite has a negative volume of formation and becomes the stable high-pressure phase. The thermodynamic data used in the calculations are given in Table 1.7 [3].[1]

---

1 Note that these three minerals, which are common in the Earth's crust, are not stable at ambient pressure at high temperatures. At ambient pressure, mullite ($3Al_2O_3 \cdot 2SiO_2$), is usually found in refractory materials based on these minerals.

Table 1.7 Thermodynamic properties of the kyanite, sillimanite and andalusite polymorphs of Al_2SiO_5 at 800 K [3].

Compound	$\Delta_f H_m^o$ kJ mol^{-1}	S_m^o J K^{-1} mol^{-1}	V_m^o cm^3 mol^{-1}	$\Delta_{f,ox} H_m^o$ kJ mol^{-1}	$\Delta_{f,ox} S_m^o$ J K^{-1} mol^{-1}	$\Delta_{f,ox} G_m^o$ J mol^{-1}	$\Delta_{f,ox} V_m^o$ cm^3 mol^{-1}
Sillimanite	−2505.57	252.4	50.4	−3.32	0.1	−3400	1.3
Kyanite	−2513.06	240.1	44.8	−10.81	−12.2	−1050	−4.3
Andalusite	−2509.08	248.8	52.2	−6.83	−3.5	−4030	3.1
Al_2O_3	−1622.62	152.2	25.8				
SiO_2	−879.63	100.1	23.3				

1.4 Open systems

Definition of the chemical potential

A homogeneous open system consists of a single phase and allows mass transfer across its boundaries. The thermodynamic functions depend not only on temperature and pressure but also on the variables necessary to describe the size of the system and its composition. The Gibbs energy of the system is therefore a function of T, p and the **number of moles** of the chemical components i, n_i:

$$G = G(T, p, n_i) \tag{1.75}$$

The exact differential of G may be written

$$dG = \left(\frac{\partial G}{\partial T}\right)_{p,n_i} dT + \left(\frac{\partial G}{\partial p}\right)_{T,n_i} dp + \left(\frac{\partial G}{\partial n_i}\right)_{T,p,n_{j \neq i}} dn_i \tag{1.76}$$

The partial derivatives of G with respect to T and p, respectively, we recall are $-S$ and V. The partial derivative of G with respect to n_i is the **chemical potential** of component i, μ_i

$$\mu_i = \left(\frac{\partial G}{\partial n_i}\right)_{T,p,n_{j \neq i}} \tag{1.77}$$

Equation (1.68) can for an open system be expressed as

$$dG = -SdT + Vdp + \sum_i \mu_i dn_i \tag{1.78}$$

The internal energy, enthalpy and Helmholtz energy can be expressed in an analogous manner:

$$dU = TdS - pdV + \sum_i \mu_i dn_i \tag{1.79}$$

$$dH = TdS + Vdp + \sum_i \mu_i dn_i \tag{1.80}$$

$$dA = -SdT - pdV + \sum_i \mu_i dn_i \tag{1.81}$$

The chemical potential is thus defined by any of the following partial derivatives:

$$\mu_i = \left(\frac{\partial G}{\partial n_i}\right)_{T,p,n_{j\neq i}} = \left(\frac{\partial A}{\partial n_i}\right)_{T,V,n_{j\neq i}} = \left(\frac{\partial H}{\partial n_i}\right)_{S,p,n_{j\neq i}} = \left(\frac{\partial U}{\partial n_i}\right)_{S,V,n_{j\neq i}} \tag{1.82}$$

Conditions for equilibrium in a heterogeneous system

Recall that the equilibrium condition for a closed system at constant T and p was given by eq. (1.41). For an open system the corresponding equation is

$$(dG)_{T,p,n_i} = 0 \tag{1.83}$$

For such a system, which allows transfer of both heat and mass, the chemical potential of each species must be the same in all phases present in equilibrium; hence

$$\mu_i^\alpha = \mu_i^\beta = \mu_i^\gamma = ... \tag{1.84}$$

Here α, β and γ denote different phases in the system, whereas i denotes the different components of the system.

Partial molar properties

In open systems consisting of several components the thermodynamic properties of each component depend on the overall composition in addition to T and p. Chemical thermodynamics in such systems relies on the **partial molar properties** of the components. The partial molar Gibbs energy at constant p, T and n_j (eq. 1.77) has been given a special name due to its great importance: the chemical potential. The corresponding partial molar enthalpy, entropy and volume under the same conditions are defined as

$$\overline{H}_i = \left(\frac{\partial H}{\partial n_i}\right)_{T,p,n_{j\neq i}} \tag{1.85}$$

$$\overline{S}_i = \left(\frac{\partial S}{\partial n_i}\right)_{T,p,n_{j\neq i}} \tag{1.86}$$

$$\bar{V}_i = \left(\frac{\partial V}{\partial n_i} \right)_{T,p,n_{j \neq i}} \tag{1.87}$$

Note that the partial molar derivatives may also be taken under conditions other than constant p and T.

The Gibbs–Duhem equation

In the absence of non pV-work, an extensive property such as the Gibbs energy of a system can be shown to be a function of the partial derivatives:

$$G = \sum_i n_i \left(\frac{\partial G}{\partial n_i} \right)_{T,p,n_{j \neq i}} = \sum_i n_i \bar{G}_i = \sum_i n_i \mu_i \tag{1.88}$$

In this context G itself is often referred to as the **integral Gibbs energy**.

For a binary system consisting of the two components A and B the integral Gibbs energy eq. (1.88) is

$$G = n_A \mu_A + n_B \mu_B \tag{1.89}$$

Differentiation of eq. (1.89) gives

$$dG = n_A d\mu_A + dn_A \mu_A + n_B d\mu_B + dn_B \mu_B \tag{1.90}$$

By using eq. (1.78) at constant T and p, G is also given by

$$dG = \mu_A dn_A + \mu_B dn_B \tag{1.91}$$

By combining the two last equations, the Gibbs–Duhem equation for a binary system at constant T and p is obtained:

$$n_A d\mu_A + n_B d\mu_B = 0 \quad \text{i.e.} \quad \sum_i n_i d\mu_i = 0 \tag{1.92}$$

In general, for an arbitrary system with i components, the Gibbs–Duhem equation is obtained by combining eq. (1.78) and eq. (1.90):

$$SdT - Vdp + \sum_i n_i d\mu_i = 0 \tag{1.93}$$

Expressions for the other intensive parameters such as V, S and H can also be derived:

$$\sum_i n_i d\bar{V}_i = 0 \tag{1.94}$$

$$\sum_i n_i d\bar{S}_i = 0 \tag{1.95}$$

$$\sum_i n_i d\bar{H}_i = 0 \tag{1.96}$$

The physical significance of the Gibbs–Duhem equation is that the chemical potential of one component in a solution cannot be varied independently of the chemical potentials of the other components of the solution. This relation will be further discussed and used in Chapter 3.

References

[1] H. D. Beckhaus, T. Ruchardt, M. Kao, F. Diederich and C. S. Foote, *Angew. Chem. Int. Ed.*, 1992, **31**, 63.

[2] H. P. Diogo, M. E. M. da Piedade, A. D. Darwish and T. J. S. Dennis, *J. Phys. Chem. Solids*, 1997, **58**, 1965.

[3] S. K. Saxena, N. Chatterjee, Y. Fei and G. Shen, *Thermodynamic Data on Oxides and Silicates*. Berlin: Springer-Verlag, 1993.

[4] E. D. West, *J. Am. Chem. Soc.*, 1959, **81**, 29.

[5] E. D. Eastman and W. C. McGavock, *J. Am. Chem. Soc.*, 1937, **59**, 145.

Further reading

P. W. Atkins and J. de Paula, *Physical Chemistry*, 7th edn. Oxford: Oxford University Press, 2001.

E. A. Guggenheim, *Thermodynamics: An Advanced Treatment for Chemists and Physicists*, 7th edn. Amsterdam: North-Holland, 1985.

K. S. Pitzer, *Thermodynamics*. New York: McGraw-Hill, 1995. (Based on G. N. Lewis and M. Randall, *Thermodynamics and the free energy of chemical substances*. New York: McGraw-Hill, 1923.)

D. Kondepudi and I. Prigogine, *Modern Thermodynamics: from Heat Engines to Dissipative Structures*. Chichester: John Wiley & Sons, 1998.

F. D. Rossini, *Chemical Thermodynamics*. Chichester: John Wiley & Sons, 1950.

2
Single-component systems

This chapter introduces additional central concepts of thermodynamics and gives an overview of the formal methods that are used to describe single-component systems. The thermodynamic relationships between different phases of a single-component system are described and the basics of phase transitions and phase diagrams are discussed. Formal mathematical descriptions of the properties of ideal and real gases are given in the second part of the chapter, while the last part is devoted to the thermodynamic description of condensed phases.

2.1 Phases, phase transitions and phase diagrams

Phases and phase transitions

In Chapter 1 we introduced the term *phase*. A phase is a state that has a particular composition and also definite, characteristic physical and chemical properties. We may have several different phases that are identical in composition but different in physical properties. A phase can be in the solid, liquid or gas state. In addition, there may exist more than one distinct crystalline phase. This is termed polymorphism, and each crystalline phase represents a distinct polymorph of the substance.

A transition between two phases of the same substance at equilibrium is called a first-order phase transition. At the equilibrium phase transition temperature the equilibrium condition eq. (1.84) yields

$$\mu_i^{\alpha} = \mu_i^{\beta} \tag{2.1}$$

where α and β denote the two coexisting phases. In this chapter we are only considering single component systems ($i = 1$) and for simplicity eq. (2.1) is expressed as

Chemical Thermodynamics of Materials by Svein Stølen and Tor Grande
© 2004 John Wiley & Sons, Ltd ISBN 0 471 492320 2

$$\mu^{\alpha} = \mu^{\beta} \tag{2.2}$$

Thus the molar Gibbs energies of the two phases are the same at equilibrium.

Typical first-order phase transitions are for example melting of ice and vaporization of water at $p = 1$ bar and at $0°$ and $99.999\ °C$, respectively. **First-order phase transitions** are accompanied by discontinuous changes in enthalpy, entropy and volume. H, S and V are thermodynamically given through the first derivatives of the chemical potential with regard to temperature or pressure, and transitions showing discontinuities in these functions are for that reason termed first-order. By using the first derivatives of the Gibbs energy with respect to p and T, defined in eqs. (1.69) and (1.72), the changes in the slopes of the chemical potential at the transition temperature are given as

$$\left(\frac{\partial \mu^{\beta}}{\partial p}\right)_{T} - \left(\frac{\partial \mu^{\alpha}}{\partial p}\right)_{T} = V_{m}^{\beta} - V_{m}^{\alpha} = \Delta_{trs} V_{m} \tag{2.3}$$

$$\left(\frac{\partial \mu^{\beta}}{\partial T}\right)_{p} - \left(\frac{\partial \mu^{\alpha}}{\partial T}\right)_{p} = -S_{m}^{\beta} + S_{m}^{\alpha} = -\Delta_{trs} S_{m} = -\frac{\Delta_{trs} H_{m}}{T_{trs}} \tag{2.4}$$

Here $\Delta_{trs} V_{m}$, $\Delta_{trs} S_{m}$ and $\Delta_{trs} H_{m}$ are the changes in the molar volume, entropy and enthalpy connected with the phase transition. Phases separated by a first-order transition can be present together with a distinct interface, and the phases are thus coexistent under certain conditions. For a single component system like H_2O, ice and water are coexistent at the melting temperature. The same is true at the first-order transition between two crystalline polymorphs of a given compound. The changes in heat capacity at constant pressure, enthalpy, entropy and Gibbs energy at the first-order semi-conductor–metal transition in NiS [1] are shown in Figure 2.1. The heat capacity at constant pressure is the second derivative of the Gibbs energy and is given macroscopically by the temperature increment caused by an enthalpy increment; $C_p = \Delta H/\Delta T$. Since the first-order transition takes place at constant temperature, the heat capacity in theory should be infinite at the transition temperature. This is obviously not observed experimentally, but heat capacities of the order of 10^7–10^8 J K^{-1} mol^{-1} are observed on melting of pure metals [2].

Transformations that involve discontinuous changes in the second derivatives of the Gibbs energy with regard to temperature and pressure are correspondingly termed **second-order transitions**. For these transitions we have discontinuities in the heat capacity, isothermal compressibility and isobaric expansivity:

$$\left(\frac{\partial^2 \mu^{\beta}}{\partial T^2}\right)_{p} - \left(\frac{\partial^2 \mu^{\alpha}}{\partial T^2}\right)_{p} = -\left(\frac{\partial S_{m}^{\beta}}{\partial T}\right)_{p} + \left(\frac{\partial S_{m}^{\alpha}}{\partial T}\right)_{p} \tag{2.5}$$

$$= -(C_{p,m}^{\beta} - C_{p,m}^{\alpha})/T = -\Delta_{trs} C_{p,m}/T$$

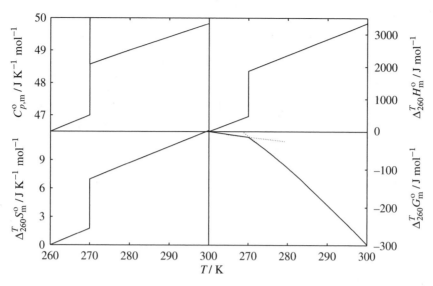

Figure 2.1 The temperature variation of the heat capacity, enthalpy, entropy, and Gibbs energy close to the first-order semiconductor to metal transition in NiS [1].

$$
\left(\frac{\partial^2 \mu^\beta}{\partial p^2}\right)_T - \left(\frac{\partial^2 \mu^\alpha}{\partial p^2}\right)_T = \left(\frac{\partial V_m^\beta}{\partial p}\right)_T - \left(\frac{\partial V_m^\alpha}{\partial p}\right)_T
$$

$$
= -V(\kappa_T^\beta - \kappa_T^\alpha) = -V\Delta_{trs}\kappa_T \tag{2.6}
$$

$$
\left(\frac{\partial^2 \mu^\beta}{\partial T \partial p}\right) - \left(\frac{\partial^2 \mu^\alpha}{\partial T \partial p}\right) = \left(\frac{\partial V_m^\beta}{\partial T}\right)_p - \left(\frac{\partial V_m^\alpha}{\partial T}\right)_p
$$

$$
= V(\alpha^\beta - \alpha^\alpha) = V\Delta_{trs}\alpha \tag{2.7}
$$

where κ_T and α are the isothermal compressibility (eq. 1.19) and isobaric expansivity (eq. 1.18).

Modifications separated by a second-order transition can never be coexistent. One typical second-order transition, the displacive structural transition, is characterized by the distortion of bonds rather than their breaking, and the structural changes that occur are usually small. Typically, there is continuous variation in the positional parameters and the unit cell dimensions as a function of temperature. The structural changes in the system occur gradually as the system moves away from the transition point. As well as a structural similarity, a symmetry relationship

(a)   (b)

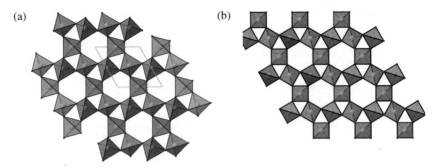

Figure 2.2 Crystal structure of α- (low) and β- (high) quartz (SiO_2).

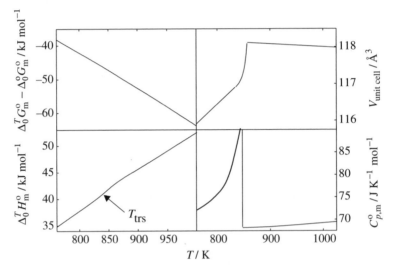

Figure 2.3 The temperature variation of the Gibbs energy [5], unit-cell volume [4] enthalpy and heat capacity [5] at the second-order α- to β-quartz transition of SiO_2. Second-order derivatives of the Gibbs energy like the heat capacity have discontinuities at the transition temperature.

exists between the two modifications.[1] The α- to β-quartz transition may serve as an example, and the two modifications of SiO_2 are illustrated in Figure 2.2. α-quartz is most easily considered as a distorted version of high-temperature β-quartz. When β-quartz is cooled below 573 °C at 1 bar the framework of the structure collapses to the denser α-configuration. The mean Si–O bond distances hardly change, but the Si–O–Si bond angle decreases from 150.9° at 590 °C to 143.61° at room temperature [3]. The variations of the unit cell volume [4], heat capacity, enthalpy, and Gibbs energy with temperature in the transition region [5] are given in Figure 2.3. While the

---

1   Second-order transitions have certain restrictions concerning the symmetry of the space group for each of the two modifications. A second-order transition can only occur between two modifications where the space group of the first is a sub-group of the space group of the second. First-order phase transitions do not have any restrictions concerning the symmetries of the two phases.

transition is barely seen in the Gibbs energy, it gives rise to a change of slope in enthalpy and volume and to a discontinuity in the heat capacity.

It is possible that both the first and second derivatives of the Gibbs energy are continuous, and that the discontinuous changes occur in the third-order derivatives of the Gibbs energy. The corresponding transition would be of third order. In practice, it is difficult to decide experimentally whether or not there is a discontinuity in the heat capacity, thermal expansivity or isothermal compressibility at the transition temperature. Even small jumps in these properties, which are difficult to verify experimentally, will signify a second-order transition. Hence it is common to call all transitions with continuous first-order derivatives second-order transitions. Similarly, it may be difficult to distinguish some first-order transitions from second-order transitions due to kinetics.

Slopes of the phase boundaries

A phase boundary for a single-component system shows the conditions at which two phases coexist in equilibrium. Recall the equilibrium condition for the phase equilibrium (eq. 2.2). Let p and T change infinitesimally but in a way that leaves the two phases α and β in equilibrium. The changes in chemical potential must be identical, and hence

$$d\mu^{\alpha} = d\mu^{\beta} \tag{2.8}$$

An infinitesimal change in the Gibbs energy can be expressed as $dG = Vdp - SdT$ (eq. 1.68) and eq. (2.8) becomes

$$-S_m^{\alpha} dT + V_m^{\alpha} dp = -S_m^{\beta} dT + V_m^{\beta} dp \tag{2.9}$$

Equation (2.9) can be rearranged to the **Clapeyron equation**:

$$\frac{dp}{dT} = \frac{\Delta_{trs} S_m}{\Delta_{trs} V_m} \tag{2.10}$$

At equilibrium $\Delta_{trs} S_m = \Delta_{trs} H_m / T_{trs}$ and the Clapeyron equation may be written

$$\frac{dp}{dT} = \frac{\Delta_{trs} H_m}{T_{trs} \Delta_{trs} V_m} \tag{2.11}$$

The variation of the phase transition temperature with pressure can be calculated from the knowledge of the volume and enthalpy change of the transition. Most often both the entropy and volume changes are positive and the transition temperature increases with pressure. In other cases, notably melting of ice, the density of the liquid phase is larger than of the solid, and the transition temperature decreases

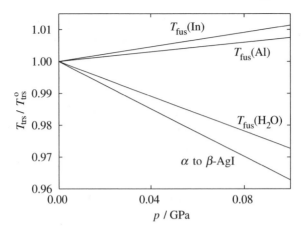

Figure 2.4 The initial dT/dp slope of selected first-order phase transitions relative to the transition temperature at $p = 1$ bar. Data taken from [6,7].

with pressure. The slope of the pT boundary for some first-order transitions is shown in Figure 2.4.

It should be noted that the boiling temperature of all substances varies more rapidly with pressure than their melting temperature since the large volume change during vaporization gives a small dp/dT. For a liquid–vapour or solid–vapour boundary the volume of gas is much larger than the volume of the condensed phase, and $\Delta_{vap}V_m \approx V_m^{gas}$ is a reasonable approximation. For an ideal gas (see eq. 2.23), $V_m^{gas} = RT/p$ and equation (2.11) rearrange to the **Clausius–Clapeyron equation**:

$$\frac{d \ln p}{dT} = \frac{\Delta_{vap}H_m}{RT^2} \tag{2.12}$$

The vapour pressure of Zn as a function of temperature, which implicitly also shows the variation of the boiling temperature with pressure, is shown in Figure 2.5.

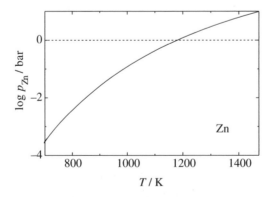

Figure 2.5 The vapour pressure of pure Zn as a function of temperature. The standard boiling (or vaporization) temperature is defined by the temperature at which the pressure of Zn is 1 bar.

Second-order transitions do not involve coexisting phases but are transitions in which the structural properties gradually change within a single phase. The low- and high-temperature modifications are here two modifications of the same phase. Hence, although these transitions often are represented in phase diagrams, they are not heterogeneous phases and do not obey Gibbs' phase rule (see below). There is no discontinuous change in the first derivatives of the Gibbs energy at the transition temperature for a second-order transition, and the volumes of the two phases are thus equal. The change in volume, dV, must be equal for both modifications if the transition is to remain continuous. Taking into account that V is a function of temperature and pressure and by using the definitions of the isobaric expansivity and the isothermal compressibility:

$$dV = \left(\frac{\partial V}{\partial T}\right)_p dT + \left(\frac{\partial V}{\partial p}\right)_T dp = V\alpha\,dT - V\kappa_T\,dp \tag{2.13}$$

Thus the pT slope is for a second-order transition is given as

$$\frac{dp}{dT} = \frac{\Delta_{trs}\alpha}{\Delta_{trs}\kappa_T} \tag{2.14}$$

Some selected examples of the variation with pressure of the transition temperatures of second-order transitions are shown in Figure 2.6.

Phase diagrams and Gibbs phase rule

A **phase diagram** displays the regions of the potential space where the various phases of the system are stable. The **potential space** is given by the variables of the

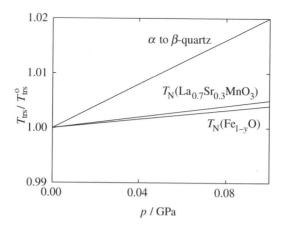

Figure 2.6 The initial dT/dp slope of selected second-order transitions relative to the transition temperature at $p = 1$ bar; α- to β-quartz (SiO_2) [8], the Néel temperature (T_N) of $Fe_{1-y}O$ [9] and the Néel temperature of $La_{0.7}Ca_{0.3}MnO_3$ [10].

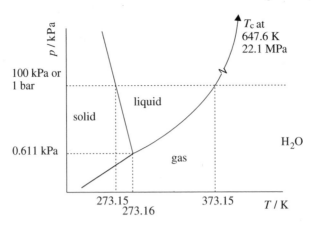

Figure 2.7 The p,T phase diagram of H_2O (the diagram is not drawn to scale).

system: pressure, temperature, composition and, if applicable, other variables such as electric or magnetic field strengths. In this chapter we are considering single component systems only. For a single-component system the phase diagram displays the regions of pressure and temperature where the various phases of this component are stable. The lines separating the regions – the phase boundaries – define the p,T conditions at which two phases of the component coexist in equilibrium.

Let us initially consider a single-component phase diagram involving a solid, a liquid and a gaseous phase. The p,T phase diagram of H_2O is given as an example in Figure 2.7. The transformations between the different phases are of first order. The liquid–vapour phase boundary shows how the vapour pressure of the liquid varies with temperature. Similarly, the solid–vapour phase boundary gives the temperature variation of the sublimation vapour pressure of the solid.

The temperature at which the vapour pressure of a liquid is equal to the external pressure is called the boiling temperature at that pressure. The **standard boiling temperature** is the boiling temperature at 1 bar. Correspondingly, the **standard melting temperature** is the melting temperature at 1 bar. Boiling is not observed when a liquid is heated in a closed vessel. Instead, the vapour pressure increases continuously as temperature is raised. The density of the vapour phase increases while the density of the liquid decreases. At the temperature where the densities of the liquid and the vapour become equal, the interface between the liquid and the gas disappears and we have reached the **critical temperature** of the substance, T_c. This is visualized by using volume (or if preferred, density) as a third variable in a three-dimensional (p,T,V) phase diagram – see Figure 2.8. The vapour pressure at the critical temperature is called the **critical pressure**. A single uniform phase, the **supercritical fluid**, exists above the critical temperature.

For a single-component system p and T can be varied independently when only one phase is present. When two phases are present in equilibrium, pressure and temperature are not independent variables. At a certain pressure there is only one temperature at which the two phases coexist, e.g. the standard melting temperature of water. Hence at a chosen pressure, the temperature is given implicitly. A point

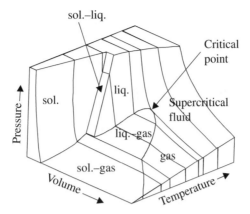

Figure 2.8 Three-dimensional p, T, V representation of a single component phase diagram visualizing the critical point.

where three phases coexist in equilibrium is termed a **triple point**; three phases are in equilibrium at a given temperature and pressure. Ice, water and water vapour are in equilibrium at $T = 273.16$ K and $p = 611$ Pa. None of the intensive parameters can be changed. The observer cannot affect the triple point.

The relationship between the number of degrees of freedom, F, defined as the number of intensive parameters that can be changed without changing the number phases in equilibrium, and the number of phases, Ph, and components, C, in the system is expressed through **Gibbs phase rule**:

$$F = C - Ph + 2 \tag{2.15}$$

In Chapter 4 the determination of the number of components in complex systems will be discussed in some detail. In this chapter we shall only consider single-component systems. For a single-component system, such as pure H_2O, $C = 1$ and $F = 3 - Ph$. Thus, a single phase ($Ph = 1$) is represented by an area in the p, T diagram and the number of degrees of freedom F is 2. A line in the phase diagram represents a heterogeneous equilibrium between two coexisting phases ($Ph = 2$) and $F = 1$, while three phases ($Ph = 3$) in equilibrium are located at a point, $F = 0$.

Field-induced phase transitions

Various types of work in addition to pV work are frequently involved in experimental studies. Research on chemical equilibria for example may involve surfaces or phases at different electric or magnetic potentials [11]. We will here look briefly at field-induced transitions, a topic of considerable interest in materials science. Examples are stress-induced formation of piezoelectric phases, electric polarization-induced formation of dielectrica and field-induced order–disorder transitions, such as for environmentally friendly magnetic refrigeration.

Magnetic contributions to the Gibbs energy due to an internal magnetic field are present in all magnetically ordered materials. An additional energetic contribution arises in a magnetic field with field strength or magnetic flux density B. This contribution is proportional to the magnetic moment, m, of the system and thus is $B \cdot dm$. An important additional complexity of external fields is that the field has a direction; the field can be applied parallel to any of the three principal axes of a single crystal. The magnetic moment and the magnetic field are thus vectors and represented by bold symbols. The fundamental equation for the internal energy for a system involving magnetic polarization is when the pV work is negligible (constant volume):

$$dU = TdS + B \cdot dm \tag{2.16}$$

The corresponding equation for the Helmholtz energy is

$$dA = -SdT + B \cdot dm \tag{2.17}$$

In order to focus on the driving force for phase transitions induced by a magnetic field it is advantageous to use the magnetic flux density as an intensive variable. This can be achieved through what is called a **Legendre transform** [12]. A transformed Helmholtz energy is defined as

$$A' = A - B \cdot m = 0 \tag{2.18}$$

Taking the differential of A' and substituting for dA in eq. (2.17):

$$dA' = dA - B \cdot dm - m \cdot dB = -SdT - m \cdot dB \tag{2.19}$$

Assuming an isotropic system, the following Maxwell relation can be derived from eq. (2.19), since dA' is an exact differential:

$$\left(\frac{\partial S}{\partial B} \right)_{T,V} = \left(\frac{\partial m}{\partial T} \right)_{B,V} \tag{2.20}$$

The entropy of a ferromagnetically ordered phase decreases with increasing magnetic field strength. The decrease is equal to the change in the magnetic moment with temperature and hence is large close to the order–disorder temperature. This implies that a larger change in the magnetic moment with temperature at constant field strength gives a higher entropy change connected with a field change at constant temperature. The effect of a magnetic field on the Helmholtz energy of a magnetic order–disorder transition thus clearly affects phase stability.

The application of n additional thermodynamic potentials (of electric, magnetic or other origin) implies that the Gibbs phase rule must be rewritten to take these new potentials into account:

$$F + Ph = C + 2 + n \tag{2.21}$$

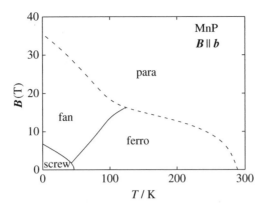

Figure 2.9 The *B*–*T* phase diagram of MnP [13] with the magnetic field along the *b*-axis. Three different magnetically ordered phases – ferro, fan and screw – are separated by first-order phase transitions. The transitions to the disordered paramagnetic state are of second order and given by a dashed line.

The isobaric (1 bar) *T,B*-phase diagram of MnP with magnetic field parallel to the crystallographic *b*-axis [13] is given in Figure 2.9. At isobaric conditions, where one degree of freedom is lost, the number of phases and the number of degrees of freedom are related by $F + Ph = 3$. Thus areas in the *T,B* diagram correspond to a single phase; a line corresponds to two phases in equilibrium; and three phases may exist in equilibrium at an invariant point, the triple point. It should be noted that the fact that a magnetic field can be applied parallel to any of the three principal axes of a single crystal implies that different phase diagrams will result in each case for a non-cubic crystal.

2.2 The gas phase

Ideal gases

The thermodynamic properties of gases are given through **equations of state** (EoS) which in general may be given as

$$p = f(T, V, n) \tag{2.22}$$

For an ideal gas the equation of state is known as the **ideal gas law**:

$$p = \frac{nRT}{V} \tag{2.23}$$

where R is the gas constant and n is the number of moles of gas. The Gibbs energy of a gas at one pressure (p_f) relative to that at another pressure (p_i) is at constant temperature given through

$$G(p_f) = G(p_i) + \int_{p_i}^{p_f} V dp \tag{2.24}$$

Using the ideal gas law the Gibbs energy expression becomes

$$G(p_f) = G(p_i) + nRT \int_{p_i}^{p_f} \frac{dp}{p} = G(p_i) + nRT \ln\left(\frac{p_f}{p_i}\right) \tag{2.25}$$

For any single-component system such as a pure gas the molar Gibbs energy is identical to the chemical potential, and the chemical potential for an ideal gas is thus expressed as

$$\mu(p) = \mu^{\circ}(p^{\circ}) + RT \ln\left(\frac{p}{p^{\circ}}\right) = \mu^{\circ} + RT \ln p \tag{2.26}$$

where the standard chemical potential (μ°) is the standard molar Gibbs energy of the pure ideal gas at the standard pressure 1 bar (p°).

The value of this standard molar Gibbs energy, $\mu^{\circ}(T)$, found in data compilations, is obtained by integration from 0 K of the heat capacity determined by the translational, rotational, vibrational and electronic energy levels of the gas. These are determined experimentally by spectroscopic methods [14]. However, contrary to what we shall see for condensed phases, the effect of pressure often exceeds the effect of temperature. Hence for gases most attention is given to the equations of state.

Real gases and the definition of fugacity

Real gases do not obey the ideal gas law, but the ideal gas law is often a very good approximation. The largest deviation from ideal gas behaviour is observed at high pressures and low temperatures. Figure 2.10 displays schematically the pressure dependence of the chemical potential. For practical reasons, it is advantageous to have an expression for the chemical potential of the real gas, which resembles that used for perfect gases. In order to obtain a simple expression for the chemical potential we replace the ideal pressure in the expression for the chemical potential (eq. 2.26) with the effective pressure, the **fugacity**, f, and we have

$$\mu(p) = \mu^{\circ}(p^{\circ}) + RT \ln\left(\frac{f}{p^{\circ}}\right) = \mu^{\circ} + RT \ln f \tag{2.27}$$

The standard state for a real gas is thus a hypothetical state in which the gas is at a pressure of $p^{\circ} = 1$ bar and behaving ideally.

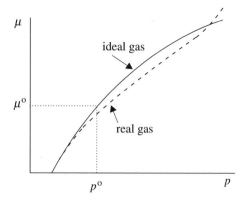

Figure 2.10 Schematic illustration of the pressure dependence of the chemical potential of a real gas showing deviations from ideal gas behaviour at high pressures.

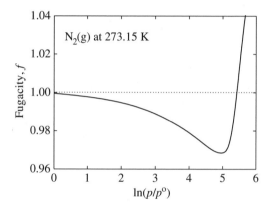

Figure 2.11 Fugacity of N_2(g) at 273.15 K as a function of pressure [15].

Most applications in materials science are carried out under pressures which do not greatly exceed 1 bar and the difference between f and p is small, as can be seen from the fugacity of N_2(g) at 273.15 K [15] given in Figure 2.11. Hence, the fugacity is often set equal to the partial pressure of the gas, i.e. $f \approx p$. More accurate descriptions of the relationship between fugacity and pressure are needed in other cases and here equations of state of real, non-ideal gases are used.

Equations of state of real gases

Purely phenomenological as well as physically based equations of state are used to represent real gases. The deviation from perfect gas behaviour is often small, and the perfect gas law is a natural choice for the first term in a serial expression of the properties of real gases. The most common representation is **the virial equation of state**:

$$pV_m = RT(1 + B'p + C'p^2 + \ldots) \tag{2.28}$$

An alternative formulation is

$$pV_{\mathrm{m}} = RT\left(1 + \frac{B}{V_{\mathrm{m}}} + \frac{C}{V_{\mathrm{m}}^2} + ... \right)$$ (2.29)

The coefficients B and C are the second and third virial coefficients, respectively, the first virial coefficient being 1.

The **compressibility factor** of a gas is defined as

$$Z = \frac{pV_{\mathrm{m}}}{RT}$$ (2.30)

For an ideal gas $Z = 1$. Departures from the value of unity indicate non-ideal behaviour. $Z < 1$ can be related to dominating attractive forces, whereas $Z > 1$ relates to repulsive forces being dominant.

The simplest physically based equation of state for real gases, the **van der Waals equation**, is based on two assumptions. As pressure is increased, the number of atoms per unit volume also increases and the volume available to the molecules in total is reduced, since the molecules themselves take up some space. The volume taken up by the molecules is assumed to be proportional to the number of molecules, n, and the volume occupied per atom, b. The equation of state is accordingly modified initially to

$$p = \frac{nRT}{V - nb}$$ (2.31)

Secondly, since the frequency and force of the collisions with the walls of the container give the pressure, the change in these two factors with concentration must be taken into account. The attractive forces working between the molecules reduce both factors, the reduction being approximately proportional to the molar concentration (n/V). The pressure is hence reduced by a factor proportional to the square of this concentration and is then given as

$$p = \frac{nRT}{V - nb} - a\left(\frac{n}{V}\right)^2$$ (2.32)

The equation can be written in a form resembling the ideal gas law (eq. 2.23):

$$\left[p + a\left(\frac{n}{V}\right)^2\right](V - nb) = nRT$$ (2.33)

For 1 mol of gas, $n = 1$:

$$\left(p + \frac{a}{V^2}\right)(V - b) = RT \tag{2.34}$$

The constants a and b can be related to the pressure (p_c) temperature (T_c) and volume (V_c) at the critical point by noting that at the critical point, by definition (see Section 5.2)

$$\left(\frac{\partial p}{\partial V}\right)_{T_c} = \left(\frac{\partial^2 p}{\partial V^2}\right)_{T_c} = 0 \tag{2.35}$$

The following three equations are obtained:

$$RT_c = \frac{8a}{27b} \tag{2.36}$$

$$p_c = \frac{a}{27b^2} \tag{2.37}$$

$$V_c = 3b \tag{2.38}$$

Hence there must be one relation involving p_c, T_c and V_c which is independent of the parameters a and b. This relation defines the critical compressibility factor Z_c:

$$Z_c = \frac{p_c V_c}{RT_c} = \frac{(a/27b^2)(3b)}{(8a/27b)} = 3/8 \tag{2.39}$$

If p, T and V are measured in units of p_c, T_c and V_c, the van der Waals equation becomes

$$\left(\bar{p} + \frac{3}{\bar{V}^2}\right)(3\bar{V} - 1) = 8\bar{T} \tag{2.40}$$

where $\bar{p} = p/p_c$, $\bar{T} = T/T_c$ and $\bar{V} = V/V_c$. This is a remarkable equation because it does not explicitly contain any free parameter characteristic of the substance and illustrates the **law of corresponding states**. All real gases should, according to this equation, behave in the same manner. The van der Waals equation of state evidently represents an approximation only, and although it works reasonably well for gases composed of spherical molecules it fails in many other cases.

Even though the van der Waals equation is not as accurate for describing the properties of real gases as empirical models such as the virial equation, it has been and still is a fundamental and important model in statistical mechanics and chemical thermodynamics. In this book, the van der Waals equation of state will be used further to discuss the stability of fluid phases in Chapter 5.

2.3 Condensed phases

For condensed phases (liquids and solids) the molar volume is much smaller than for gases and also varies much less with pressure. Consequently the effect of pressure on the chemical potential of a condensed phase is much smaller than for a gas and often negligible. This implies that while for gases more attention is given to the volumetric properties than to the variation of the standard chemical potential with temperature, the opposite is the case for condensed phases.

Variation of the standard chemical potential with temperature

The thermodynamic properties of single-component condensed phases are traditionally given in tabulated form in large data monographs. Separate tables are given for each solid phase as well as for the liquid and for the gas. In recent years analytical representations have been increasingly used to ease the implementation of the data in computations. These polynomial representations typically describe the thermodynamic properties above room temperature (or 200 K) only.

Polynomial expressions are conveniently used to represent a condensed phase which is stable in the whole temperature range of interest and which does not undergo any structural, electronic or magnetic transformations. The Gibbs energy of a compound is in the CALPHAD approach represented relative to the elements in their defined standard state at 298.15 K as a power series in terms of temperature in the form of [16]:

$$G_m^o(T) - H_m^{SER} = a + bT + cT\ln(T) + \sum_{n=2}^{i} d_n T^n \tag{2.41}$$

Here H_m^{SER} is the sum (in the stoichiometric ratio of the compound in question) of $\Delta_0^{298.15} H_m^o$ of the elements in their defined standard state. a, b, c and d_n are coefficients and n integers. This form of expression is useful for storing thermodynamic information in databases. A number of such expressions are often required for a given phase to cover the whole temperature range of interest. From eq. (2.41) all other thermodynamic functions can be derived, e.g.

$$S_m^o(T) = -b - c - c\ln(T) - \sum_n n d_n T^{n-1} \tag{2.42}$$

Table 2.1 Thermodynamic properties of AlN at selected temperatures (data are taken from NIST-JANAF tables [17]). Enthalpy reference temperature = T = 298.15 K; p^o = 1 bar.

$\dfrac{T}{\text{K}}$	$\dfrac{C_{p,\text{m}}}{\text{J K}^{-1}\text{ mol}^{-1}}$	$\dfrac{S_{\text{m}}^o}{\text{J K}^{-1}\text{ mol}^{-1}}$	$\dfrac{\Delta_{298.15}^{T}H_{\text{m}}^o}{\text{kJ mol}^{-1}}$	$\dfrac{\Delta_{\text{f}}H_{\text{m}}^o}{\text{kJ mol}^{-1}}$	$\dfrac{\Delta_{\text{f}}G_{\text{m}}^o}{\text{kJ mol}^{-1}}$	$\log K_{\text{f}}$
0	0.	0.	−3.871	−312.980	−312.980	INFINITE
100	5.678	2.164	−3.711	−314.756	−306.283	159.986
200	19.332	10.267	−2.463	−316.764	−296.990	77.566
298.15	30.097	20.142	0.	−317.984	−286.995	50.280
300	30.254	20.329	0.056	−318.000	−286.803	49.937
400	36.692	29.987	3.428	−318.594	−276.301	36.081
500	40.799	38.647	7.317	−318.808	−265.697	27.757
600	43.538	46.341	11.541	−318.811	−255.072	22.206
700	45.434	53.201	15.994	−318.727	−244.455	18.241
800	46.791	59.361	20.608	−318.648	−233.850	15.269
900	47.792	64.932	25.339	−318.647	−223.252	12.957
1000	48.550	70.008	30.158	−329.363	−211.887	11.068
2000	51.290	104.790	80.490	−328.119	−94.810	2.476

$$H_{\text{m}}^o(T) = a - cT - \sum_{n}(n-1)d_n T^n \tag{2.43}$$

$$C_{p,\text{m}}^o(T) = -c - \sum_{n}n(n-1)d_n T^{n-1} \tag{2.44}$$

These thermodynamic functions implicitly given in analytical representations are given numerically at selected temperatures in monographs, as shown for AlN in Table 2.1 [17]. The analytical approach is exemplified by descriptions of three modifications of aluminium in Table 2.2 [18]. The stable face-centred cubic modification of crystalline aluminium (FCC-Al) melts at 933.473 K. Hexagonal close-packed aluminium is unstable at all temperatures, as evident from the graphical representation of the Gibbs energy relatively to FCC-Al in Figure 2.12. The thermodynamic properties may still be needed to describe alloys with hexagonal closed-packed structure where aluminium is a solute.

Representation of transitions

Thermodynamic representation of transitions often represents a challenge. First-order phase transitions are more easily handled numerically than second-order transitions. The enthalpy and entropy of first-order phase transitions can be calculated at any temperature using the heat capacity of the two phases and the enthalpy and entropy of transition at the equilibrium transition temperature. Small pre-transitional contributions to the heat capacity, often observed experimentally, are most often not included in the polynomial representations since the contribution to the

Table 2.2 CALPHAD-type representation of the thermodynamic properties of face-centred cubic (FCC), liquid and hexagonal close-packed (HCP) aluminium of the form (after Dinsdale [18]):

$$G_m^o(T) - H_m^{SER}(298.15\text{ K}) = a + bT + cT \ln(T) + \sum_n d_n T^n$$

with $H_m^{SER}(298.15\text{ K}) = \Delta_0^{298.15}H_m^o(\text{FCC\_Al}) = 4540 \text{ J mol}^{-1}$.

FCC_Al

$(298.15 < T/\text{K} < 700)$

$-7976.15 + 137.093038\ T - 24.3671976T \ln(T) - 1.884662\text{E}^{-3}\ T^2 - 0.877664\text{E}^{-6}\ T^3 + 74092\ T^{-1}$

$(700 < T/\text{K} < 933.473)$

$-11276.24 + 223.048446\ T - 38.5844296\ T \ln(T) + 18.531982\text{E}^{-3}\ T^2 - 5.764227\text{E}^{-6}\ T^3 + 74092\ T^{-1}$

$(933.473 < T/\text{K} < 2900)$

$-11278.378 + 188.684153\ T - 31.748192\ T \ln(T) - 1.231\text{E}28\ T^{-9}$

Liquid relatively to FCC_Al

$(298.15 < T/\text{K} < 933.473)$

$11005.029 - 11.841867\ T + 7.934\text{E}^{-20}\ T^7$

$(933.473 < T/\text{K} < 2900)$

$10482.382 - 11.253974\ T + 1.231\text{E}28\ T^{-9}$

HCP_Al relative to FCC_Al

$(298.15 < T/\text{K} < 2900)$

$5481 - 1.8\ T$

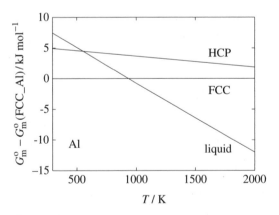

Figure 2.12 $G_m^o - G_m^o(\text{Al\_FCC})$ of hexagonal closed-packed (HCP) aluminium and aluminium melt relative to that of face-centred cubic aluminium [18].

Gibbs energy is small. This contribution is instead incorporated empirically in the enthalpy and entropy of transition.

It is more difficult to describe second-order transitions. Considerable short-range order is in general present far above the transition temperature. Correspondingly,

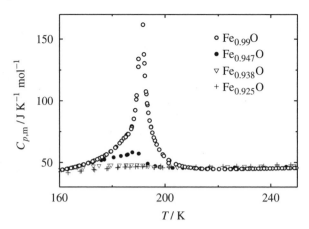

Figure 2.13 Heat capacity of wüstite around the Néel temperature [19]. O: $Fe_{0.99}O$; ●: $Fe_{0.947}O$; ∇: $Fe_{0.938}O$; +: $Fe_{0.925}O$. Reproduced by permission of the Mineralogical Society of America.

considerable disordering has taken place already far below the disordering temperature. The magnetic order–disorder transition in non-stoichiometric wüstite, $Fe_{1-y}O$, may serve as an example. The magnetic transition is largely dependent on the stoichiometry of the compound (see Figure 2.13), and is for the oxygen-rich compositions spread over a considerable temperature range [19]. The disordering is far from abrupt.

The Inden model [20] is frequently used to describe second-order magnetic order–disorder transitions. Inden assumed that the heat capacity varied as a logarithmic function of temperature and used separate expressions above and below the magnetic order–disorder transition temperature (T_{trs}) in order to treat the effects of both long- and short-range order. Thus for $\tau = (T/T_{trs}) < 1$:

$$C_p^{\text{mag}} = K^L R \frac{\ln(1+\tau^3)}{\ln(1-\tau^3)} \tag{2.45}$$

For $\tau > 1$

$$C_p^{\text{mag}} = K^S R \frac{\ln(1+\tau^5)}{\ln(1-\tau^5)} \tag{2.46}$$

The two coefficients K^L and K^S are derived empirically. They are related through the entropy of transition and constrained to reproduce the total enthalpy and entropy increments accompanying the phase transition. Since, the Inden model demands a series expansion in order to calculate the entropy, a simpler related equation by Hillert and Jarl [21] is used in many computer programs.

Second-order structural transitions are less frequently represented in applied thermodynamic calculations. Still, the Landau approach for determination of

Gibbs energy changes connected with second-order transitions has a long tradition. The central concept of the **Landau theory** is the order parameter, Γ, which describes the course of the transition. The order parameter is related to the change in some macroscopic property like strain, average site occupancy or crystallographic distortion through the phase transition. The measured physical property is, however, not necessarily directly proportional to Γ, but most often scales as either Γ or Γ^2. The relationship between the order parameter and the measured physical property is defined by the differences in symmetry between the high- and low-temperature polymorphs [22].

We will consider a phase transition between two crystal structures with different symmetry and where the space group of the low-symmetry structure is a sub-group of the space group of the high-symmetry structure. Hence all symmetry elements of the high-symmetry structure are present in the low-symmetry structure. An order parameter, Γ, is used to describe the thermodynamic state of the low-symmetry phase. The contribution from the phase transition to the total Gibbs energy, here termed the transitional Gibbs energy $\Delta_{trs}G$, is now given as a function of T, p and Γ as

$$\Delta_{trs}G = \Delta_{trs}G(T, p, \Gamma) \tag{2.47}$$

Γ is scaled such that it is assigned the value 0 in the high-temperature modification and 1 in the low-temperature form at 0 K. Thus, $\Delta_{trs}G = 0$ for the high-temperature polymorph. The variation of the order parameter with temperature describes the transition thermodynamically. In general [23]:

$$\Delta_{trs}G = a\Gamma + \frac{1}{2}b\Gamma^2 + \frac{1}{3}c\Gamma^3 + \frac{1}{4}d\Gamma^4 + \ldots \tag{2.48}$$

Here a, b, c, d etc. are coefficients that in general are functions of temperature and pressure. The equilibrium behaviour of Γ through the phase transition is determined by minimizing $\Delta_{trs}G$ with respect to Γ. Furthermore, at equilibrium the $\Delta_{trs}G(\Gamma)$ surface is concave upwards (discussed thoroughly in Section 5.2), hence

$$\frac{\partial \Delta_{trs}G}{\partial \Gamma} = 0 \quad \text{and} \quad \frac{\partial^2 \Delta_{trs}G}{\partial \Gamma^2} > 0 \tag{2.49}$$

These criteria can be used to get information on the coefficients of eq. (2.48). In the high-symmetry phase, stable above the transition temperature, the order parameter $\Gamma = 0$ and the equilibrium conditions imply that the two first constants in the polynomial expansion are restricted to $a = 0$ and $b > 0$. If we assume that $b < 0$, the low-symmetry phase is stable since $\Gamma^2 > 0$ then implies that $\Delta_{trs}G < 0$. The transitional Gibbs energy is thus reduced to

$$\Delta_{trs}G = \frac{1}{2}b\Gamma^2 + \frac{1}{3}c\Gamma^3 + \frac{1}{4}d\Gamma^4 + \dots \tag{2.50}$$

The Landau theory predicts the symmetry conditions necessary for a transition to be thermodynamically of second order. The order parameter must in this case vary continuously from 0 to 1. The presence of odd-order coefficients in the expansion gives rise to two values of the transitional Gibbs energy that satisfy the equilibrium conditions. This is not consistent with a continuous change in Γ and thus corresponds to first-order phase transitions. For this reason all odd-order coefficients must be zero. Furthermore, the sign of b must change from positive to negative at the transition temperature. It is customary to express the temperature dependence of b as a linear function of temperature:

$$b = B(T - T_{trs}) \tag{2.51}$$

Here B is a constant independent of temperature and pressure. The transitional Gibbs energy is thus

$$\Delta_{trs}G = \frac{1}{2}B(T - T_{trs})\Gamma^2 + \frac{1}{4}d\Gamma^4 + \frac{1}{6}f\Gamma^6 + \dots \tag{2.52}$$

where d, f and higher-order coefficients all are assumed to be independent of temperature and pressure. Normally two or three terms of this expression give a satisfactory description of the transitional Gibbs energy using experimentally determined values for the temperature variation of the order parameter.

When $d > 0$ the expansion describes a thermodynamic second-order transition. The equilibrium condition neglecting higher order terms is

$$\frac{\partial \Delta_{trs}G}{\partial \Gamma} = B(T - T_{trs})\Gamma + d\Gamma^3 = 0 \tag{2.53}$$

which gives

$$\Gamma^2 = -\frac{B}{d}(T - T_{trs}) \quad \text{for} \quad T < T_{trs} \tag{2.54}$$

The form of the order parameter is given implicitly since by definition $\Gamma = 1$ at 0 K and hence

$$\frac{B}{d} = \frac{1}{T_{trs}} \tag{2.55}$$

and thus

$$\Gamma = \left[\frac{T_{trs} - T}{T_{trs}} \right]^{1/2} \tag{2.56}$$

The transitional Gibbs energy is for $T \leq T_{trs}$

$$\Delta_{trs} G = -\frac{B^2}{2d}(T - T_{trs})^2 + \frac{B^2}{4d}(T - T_{trs})^2 = -\frac{B^2}{4d}(T - T_{trs})^2 \tag{2.57}$$

The transitional entropy and heat capacity are readily derived by differentiation with respect to temperature. For $T \leq T_{trs}$

$$\Delta_{trs} S = \frac{B^2}{2d}(T - T_{trs}) = -\frac{1}{2}B \cdot \Gamma^2 \tag{2.58}$$

$$C_p^{trs} = \frac{B^2}{2d}T \tag{2.59}$$

The contribution of the transition to the thermodynamic functions can be evaluated once the coefficients B and d have been determined. Experimental determination of the transition temperature and one additional thermodynamic quantity at one specific temperature is sufficient to describe the transition thermodynamically using this model.

It is easily shown that a first-order phase transition is obtained for cases were $d < 0$, whereas behaviour at the borderline between first- and second-order transitions, **tricritical behaviour**, is obtained for $d = 0$. In the latter case the transitional Gibbs energy is

$$\Delta_{trs} G = \frac{1}{2} B(T - T_{trs}) \Gamma^2 + \frac{1}{6} f \Gamma^6 + \dots \tag{2.60}$$

Minimization of the transitional Gibbs energy with respect to Γ gives

$$\Gamma = \left[\frac{T_{trs} - T}{T_{trs}} \right]^{1/4} \tag{2.61}$$

The variation of the order parameter with temperature thus distinguishes second-order transitions from tricritical behaviour. In general the variation of the order parameter with temperature for a continuous transition is described as

$$\Gamma = \left[\frac{T_{trs} - T}{T_{trs}} \right]^{\beta} \tag{2.62}$$

where β is the critical exponent. For our two ideal cases, second-order and tricritical transitions, $\beta = \frac{1}{2}$ (eq. 2.56) and $\frac{1}{4}$ (eq. 2.61), respectively.

The transitional entropy, enthalpy and heat capacity for a tricritical transition is for $T \leq T_{\mathrm{trs}}$:

$$\Delta_{\mathrm{trs}} S = -\frac{1}{2} B \cdot \Gamma^2 \tag{2.63}$$

$$\Delta_{\mathrm{trs}} H = -\frac{1}{2} B \cdot T_{\mathrm{trs}} \cdot \Gamma^2 + \frac{1}{6} f \cdot \Gamma^6 \tag{2.64}$$

$$C_p^{\mathrm{trs}} = \frac{BT}{4\sqrt{T_{\mathrm{trs}}}} (T_{\mathrm{trs}} - T)^{-1/2} \tag{2.65}$$

Orientational disordering of the carbonate groups in $CaCO_3$ above 1260 K may serve as an example of application of Landau theory. Below the transition temperature, alternate layers of planar CO_3 groups point in opposite directions. In the high-temperature modification they are free to rotate and become equivalent. The symmetry reduction on ordering is from space group $R\bar{3}m$ to $R\bar{3}c$ with doubling of the c-axis length. Thus the transition gives rise to superlattice reflections in the diffraction patterns of the low-temperature phase. The intensities of these reflections are according to symmetry considerations proportional to Γ^2. It has been shown by neutron diffraction that the order parameter is proportional to $(T_{\mathrm{trs}} - T)^{1/4}$ and thus that the transition is tricritical [24]. T_{trs} and the excess enthalpy determined by drop calorimetry characterize the transition thermodynamically [25]. The contribution from the transition to the total Gibbs energy and entropy (using $B = 24$ J mol^{-1} K^{-1} and $f = 30$ kJ mol^{-1}) are given in Figure 2.14.

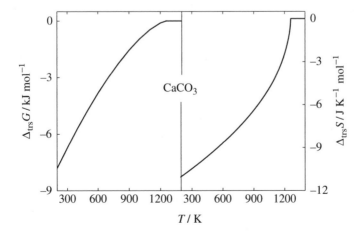

Figure 2.14 The contribution from the order–disorder transition of $CaCO_3(s)$ to the total Gibbs energy and entropy [25].

Equations of state

Equations of state of condensed materials are seldom used in materials science but are frequently used in geophysics and to an increasing degree also in solid state sciences for high-pressure studies of phase transitions. A considerable amount of work on equations of state of minerals has been reported in the geophysical and geochemical literature. In the Earth's mantle the pressure is several orders of magnitude higher than ambient since pressure and temperature increase with increasing depth within the Earth. Thus equation of state data is essential for thermodynamic calculations of phase equilibria in the Earth's interior.

Equations of state for solids are often cast in terms of the bulk modulus, K_T, which is the inverse of the isothermal compressibility, κ_T, and thus defined as

$$K_T = \frac{1}{\kappa_T} = -V \left(\frac{\partial p}{\partial V} \right)_T \tag{2.66}$$

The two most usual equations of state for representation of experimental data at high pressure are the Murnaghan and Birch–Murnaghan equations of state. Both models are based on finite strain theory, the Birch–Murnaghan or Eulerian strain [26]. The main assumption in finite strain theory is the formal relationship between compression and strain [27]:

$$\frac{V}{V_0} = (1 + 2\varepsilon)^{3/2} \tag{2.67}$$

The **Murnaghan equation of state** is given by

$$p = \frac{K_{T,0}}{K'_{T,0}} \left[\left(\frac{V_0}{V} \right)^{K'_{T,0}} - 1 \right] \tag{2.68}$$

while what is termed the third-order **Birch–Murnaghan equation of state** is given by

$$p = \frac{3}{2} K_{T,0} \left[\left(\frac{V_0}{V} \right)^{7/3} - \left(\frac{V_0}{V} \right)^{5/3} \right] \cdot \left[1 - \frac{3}{4}(4 - K'_{T,0}) \cdot \left[\left(\frac{V_0}{V} \right)^{2/3} - 1 \right] + \ldots \right] \tag{2.69}$$

where $K_{T,0}$ and $K'_{T,0}$ are the isothermal bulk modulus and its pressure derivative at $T = 298$ K at zero pressure, respectively. The third-order Birch–Murnaghan EoS reduces to second order when $K'_{T,0} = 4$ and

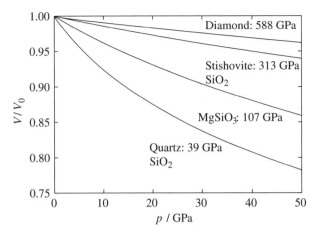

Figure 2.15 Pressure–volume data for diamond, SiO_2-stishovite, $MgSiO_3$ and SiO_2–quartz based on third order Birch–Murnaghan equation of state descriptions. The isothermal bulk modulus at 1 bar and 298 K are given in the figure.

$$p = \frac{3}{2} K_{T,0} \left[\left(\frac{V_0}{V} \right)^{7/3} - \left(\frac{V_0}{V} \right)^{5/3} \right] \tag{2.70}$$

Volumetric data for four different substances represented by the third-order Birch–Murnaghan equation of state are shown in Figure 2.15.

The equations discussed above are reliable for phases where the compressibility does not change too fast with pressure, more specifically for $3.4 < K'_{T,0} < 7$. The equations are thus suitable for a large range of crystalline substances but not for liquids or low-dimensional materials, where $K'_{T,0}$ is often larger than 7. In the latter cases the universal Vinet equation of state seems more appropriate [28].

The effect of temperature on the equation of state is introduced through the isobaric thermal expansivity. It is generally assumed that isobaric expansivity and isobaric compressibility work independently of each order and the volume as a function of T and p is then expressed as

$$V(p,T) = V_{298}^{o} f(p) f(T) \tag{2.71}$$

Finally, it should be noted that the effect of the compressibility on the thermodymanics of solids is small even at relatively high pressures. The molar volume of magnetite, Fe_3O_4, at 1000 K is 46.0 cm^3 mol^{-1} and $V\Delta p$ at $p = 1$ GPa is 46 kJ mol^{-1} if the compressibility of the compound is neglected. Taking compressibility into account reduces this contribution to 45.88 kJ mol^{-1} [29].

References

[1]   F. Grønvold and S. Stølen, *Thermochim. Acta.* 1995, **266**, 213.

[2]   F. Grønvold, *Pure Appl. Chem.* 1993, **65**, 927.

[3]   A. F. Wright and M. S. Lehmann, *J. Solid State Chem.* 1981, **36**, 371.

[4]   M. A. Carpenter, E. K. H. Salje, A. Graeme-Barber, B. Wruck, M. T. Dove and K. S. Knight, *Am. Mineral.* 1998, **83**, 2.

[5]   F. Grønvold, S. Stølen and S. R. Svendsen, *Thermochim. Acta.* 1989, **139**, 225.

[6]   *Supplementary Information for the International Temperature Scale of 1990.* Bureau International des Poids et Mesures: Paris. 1990.

[7]   C. N. R. Rao and J. Gopalakrishnan *New Directions in Solid State Chemistry*, 2nd edn. Cambridge: Cambridge University Press, 1997.

[8]   See e.g. Silica: physical behaviour, geochemistry and materials applications (P. J. Heaney, C. T. Prewitt and G. V. Gibbs, eds), *Reviews in Mineralogy*, vol. 29, Mineralogical Society of America, 1994.

[9]   T. Okamoto, H. Fujii, Y. Hidaka and E. Tatsumoto, *J. Phys. Soc. Japan* 1967, **23**, 1174.

[10] Ulyanov, A. N., Maksimov, I. S., Nyeanchi, E. B., Medvedev, Yu. V., Yu, S. C., Starostyuk, N. Yu. and Sundquist, B. *J. Appl. Phys.* 2002, **91**, 7739.

[11] K. S. Pitzer, *Thermodynamics*. New York: McGraw-Hill, 1995.

[12] R. A. Alberty, *J. Chem. Thermodyn.* 2002, **34**, 1787.

[13] C. C. Becerra, Y. Shapira, N. F. Oliveira Jr. and T. S. Chang, *Phys. Rev. Lett.* 1980, **44**, 1692.

[14] J. E. Meyer and M. G. Meyer, *Statistical Mechanics*, 2nd edn. New York: Wiley, 1977.

[15] J. Otto, A. Michels and H. Wouters, *Physik. Z.* 1934, **35**, 97.

[16] N. Saunders and A. P. Miodownik, *CALPHAD: Calculation of Phase Diagrams: a Comprehensive Guide*. Oxford: Pergamon, Elsevier Science, 1998.

[17] *NIST-JANAF Thermochemical Tables*, 4th edn (M. W. Chase Jr, ed.) *J. Phys. Chem. Ref. Data Monograph* No. 9, 1998.

[18] A. T. Dinsdale, *CALPHAD* 1991, **15**, 317.

[19] S. Stølen, R. Glöckner, F. Grønvold, T. Atake and S. Izumisawa, *Am. Mineral.* 1996, **81**, 973.

[20] G. Inden, *Proceedings of the Fifth conference on Calculation of Phase Diagrams*, CALPHAD V (Dusseldorf) III-(4)-1.

[21] M. Hillert and M. Jarl, *CALPHAD* 1978, **2**, 227.

[22] E. K. H. Salje, *Phase Transitions in Ferroelastic and Co-elastic Crystals*. Cambridge: Cambridge University Press, 1990.

[23] A. Putnis, *Introduction to Mineral Sciences*. Cambridge: Cambridge University Press, 1992.

[24] M. T. Dove and B. M. Powell, *Phys. Chem. Mineral.* 1989, **16**, 503.

[25] S. A. T. Redfern, E. Salje and A. Navrotsky, *Contrib. Mineral. Petrol.* 1989, **101**, 479.

[26] F. Birch, *J. Geophys. Res.* 1952, **57**, 227.

[27] F. D. Stacey, B. J. Brennan and R. D. Irvine, *Geophys. Surv.* 1981, **4**, 189.

[28] P. Vinet, J. H. Rose, J. Ferrante and J. R. Smith, *J. Phys. Condens. Matter* 1989, **1**, 1941.

[29] S. K. Saxena, N. Chatterjee, Y. Fei and G. Shen, *Thermodynamic Data on Oxides and Silicates*. Berlin: Springer-Verlag, 1993.

Further reading

O. L. Anderson, *Equations of State of Solids for Geophysics and Ceramic Science*. Oxford: Oxford University Press, 1995.

P. W. Atkins and J. de Paula, *Physical Chemistry*, 7th edn. Oxford: Oxford University Press, 2001.

J. W. Gibbs, *The Collected Work, Volume I Thermodynamics*. Harvard: Yale University Press, 1948.

L. D. Landau and E. M. Lifshitz, *Statistical Physics*. London: Pergamon Press, 1958.

K. S. Pitzer, *Thermodynamics*. New York: McGraw-Hill, 1995. (Based on G. N. Lewis and M. Randall, *Thermodynamics and the free energy of chemical substances*. New York: McGraw-Hill, 1923.

E. K. H. Salje *Phase Transitions in Ferroelastic and Co-elastic Crystals*. Cambridge: Cambridge University Press, 1990.

3

Solution thermodynamics

So far we have discussed the thermodynamic properties of materials, which have been considered as pure and to consist of only a single component. We will now continue with systems containing two or more components and thereby solutions. Solutions are thermodynamic phases with variable composition, and are common in chemical processes, in materials and in daily life. Alloys – solutions of metallic elements – have played a key role in the development of human civilisation from the Bronze Age until today. Many new advanced materials are also solutions. Examples are tetragonal or cubic ZrO_2, stabilized by CaO or Y_2O_3, with high toughness or high ionic conductivity, and piezoelectric and dielectric materials based on $BaTiO_3$ or $PbZrO_3$. In all these cases the mechanical or functional properties are tailored by controlling the chemical composition of the solid solution. The chemical and thermal stability of these complex materials can only be understood if we know their thermodynamic properties.

The understanding of how the chemical potential of a component is changed by mixing with other components in a solution is an old and fascinating problem. The aim of this chapter is to introduce the formalism of solution thermodynamics. Models in which the solution is described in terms of the end members of the solution, **solution models**, are given special attention. While the properties of the end members must be described following the methods outlined in the previous chapter, the present chapter is devoted to the changes that occur on formation of the solutions. In principle one could describe the Gibbs energy of a mixture without knowing the properties of the end members, but since it is often of interest to apply a solution model in thermodynamic calculations involving other phases, the solution model often is combined with descriptions of the Gibbs energies of the end members to give a complete thermodynamic description of the system.

Chemical Thermodynamics of Materials by Svein Stølen and Tor Grande
© 2004 John Wiley & Sons, Ltd ISBN 0 471 492320 2

3.1 Fundamental definitions

Measures of composition

The most important characteristic of a solution is its composition, i.e. the concentration of the different components of the phase. The composition of a solution is best expressed by the ratio of the number of moles of each component to the total number of moles. This measure of the composition is the **mole fraction** of a component. In the case of a binary solution consisting of the components A and B, the mole fractions of the two components are defined as

$$x_A = \frac{n_A}{n_A + n_B} \quad \text{and} \quad x_B = \frac{n_B}{n_A + n_B} \tag{3.1}$$

and it is evident that

$$x_A + x_B = 1 \tag{3.2}$$

For an infinitesimal change in composition of a binary solution the differentials of the two mole fractions are related as

$$dx_A = -dx_B \tag{3.3}$$

In dealing with dilute solutions it is convenient to speak of the component present in the largest amount as the **solvent**, while the diluted component is called the **solute**.

While the mole fraction is a natural measure of composition for solutions of metallic elements or alloys, the mole fraction of each molecule is chosen as the measure of composition in the case of solid or liquid mixtures of molecules.[1] In ionic solutions cations and anions are not randomly mixed but occupy different sub-lattices. The mole fractions of the atoms are thus an inconvenient measure of composition for ionic substances. Since cations are mixed with cations and anions are mixed with anions, it is convenient for such materials to define composition in terms of **ionic fractions** rather than mole fractions. In a mixture of the salts AB and AC, where A is a cation and B and C are anions, the ionic fractions of B and C are defined through

$$X_B = \frac{n_B}{n_B + n_C} = 1 - X_C \tag{3.4}$$

---

1   Note that **volume fraction** rather than mole fraction is recommended in mixtures of molecules with significant different molecular mass. This will be discussed in Chapter 9.

In a binary solution AB–AC, the ionic fractions of B and C are identical to the mole fractions of AB and AC. It may therefore seem unnecessary to use the ionic fractions. However, in the case of multi-component systems the advantage of ionic fractions is evident, as will be shown in Chapter 9.

Mixtures of gases

The simplest solution one can imagine is a mixture of ideal gases. Let us simplify the case by assuming only two types of ideal gas molecules, A and B, in the mixture. The total pressure in this case is the sum of the partial pressures of the two components (this is termed Dalton's law). Thus,

$$p_{tot} = p_A + p_B \tag{3.5}$$

where p_A and p_B are the partial pressures of the two gases and p_{tot} is the total pressure. By applying the ideal gas law (eq. 2.23), the volume of the gas mixture is

$$V_{tot} = n_A V_{m,A} + n_B V_{m,B} \tag{3.6}$$

where n_A and n_B are the number of moles of A and B in the mixture and $V_{m,A}$ and $V_{m,B}$ are the molar volumes of pure A(g) and B(g). In this case, where both A and B are ideal gases, $V_{m,A} = V_{m,B}$. It follows that, for a mixture of ideal gases

$$p_A = x_A p_{tot} \tag{3.7}$$

The chemical potential of an ideal gas A is given by eq. (2.26) as

$$\mu_A(p_A) = \mu_A^o + RT \ln\left(\frac{p_A}{p_A^o}\right) = \mu_A^o + RT \ln p_A \tag{3.8}$$

where μ_A^o is the standard chemical potential of the pure ideal gas A at $p_A^o = 1$ bar at a given temperature T. For a mixture of the ideal gases A and B at constant pressure ($p_{tot} = p_A^o = 1$ bar) the chemical potential of A for a given composition of the solution, x_A, is, by using eq. (3.7)

$$\mu_A(x_A) = \mu_A^o + RT \ln\left(\frac{x_A p_{tot}}{p_A^o}\right) = \mu_A^o + RT \ln x_A \tag{3.9}$$

The difference between the chemical potential of a pure and diluted ideal gas is simply given in terms of the logarithm of the mole fraction of the gas component. As we will see in the following sections this relationship between the chemical potential and composition is also valid for ideal solid and liquid solutions.

In mixtures of real gases the ideal gas law does not hold. The chemical potential of A of a mixture of real gases is defined in terms of the fugacity of the gas, f_A. The fugacity is, as discussed in Chapter 2, the thermodynamic term used to relate the chemical potential of the real gas to that of the (hypothetical) standard state of the gas at 1 bar where the gas is ideal:

$$\mu_A(x_A) = \mu_A^o + RT \ln\left(\frac{f_A}{p_A^o}\right) = \mu_A^o + RT \ln(f_A) \qquad (3.10)$$

Solid and liquid solutions – the definition of activity

In the solid or liquid state the **activity**, a, is introduced to express the chemical potential of the components of a solution. It is defined by

$$\mu_A = \mu_A^* + RT \ln a_A \qquad (3.11)$$

where μ_A^* is the chemical potential of A in the reference state. For $p = 1$ bar $\mu_A^* = \mu_A^o$. One of the most important tasks of solution thermodynamics is the choice of an appropriate reference state, and this is the topic of one of the following sections.

3.2 Thermodynamics of solutions

Definition of mixing properties

The volume of an ideal gas mixture is given by eq. (3.6). Let us now consider only solid or liquid mixtures. Our starting point is an arbitrary mixture of n_A mole of pure A and n_B mole of pure B. The mixing process is illustrated in Figure 3.1. We

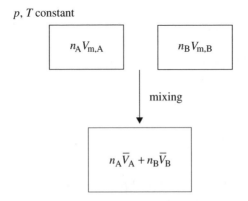

Figure 3.1 Mixing of n_A moles of A and n_B moles of B at constant p and T. The molar volumes of pure A and B are V_A and V_B. The partial molar volumes of A and B in the solution are $\overline{V}_A$ and $\overline{V}_B$, respectively.

will first derive the expressions for the volume of the system before and after the mixing. The volume before mixing is

$$V(\text{before}) = n_A V_{m,A} + n_B V_{m,B} \qquad (3.12)$$

where $V_{m,A}$ and $V_{m,B}$ are the molar volumes of pure A and B. We now mix A and B at constant pressure p and temperature T and form the solution as illustrated in Figure 3.1. The expression for the volume of the solution is then

$$V(\text{after}) = n_A \overline{V}_A + n_B \overline{V}_B \qquad (3.13)$$

where $\overline{V}_A$ and $\overline{V}_B$ represent the partial molar volumes of A and B (defined by eq. 1.87) in the solution. These partial molar volumes may be seen as apparent volumes that when weighted with the number of A and B atoms give the observed total volume of the solution. The difference in the volume of the solution before and after mixing, the **volume of mixing**, is designated $\Delta_{mix}V$:

$$\Delta_{mix}V = V(\text{after}) - V(\text{before}) = n_A (\overline{V}_A - V_A) + n_B(\overline{V}_B - V_B) \qquad (3.14)$$

The volume of mixing for one mole of solution is termed the **molar volume of mixing**, $\Delta_{mix}V_m$, and is derived by dividing eq. (3.14) by the total number of moles $(n_A + n_B)$ in the system

$$\Delta_{mix}V_m = \frac{\Delta_{mix}V}{(n_A + n_B)} = x_A \Delta_{mix}\overline{V}_A + x_B \Delta_{mix}\overline{V}_B \qquad (3.15)$$

The molar volume of mixing of two binary systems is shown in Figure 3.2. Pb–Sn shows positive deviation from the ideal behaviour at 1040 K [1] while the volume of mixing of Pb–Sb at 907 K is negative, with a minimum at $x_{Pb} \neq 0.5$ and asymmetric with respect to the composition [2].

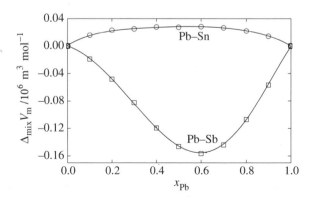

Figure 3.2 Molar volume of mixing of molten Pb–Sn at 1040 K [1] and Pb–Sb at 907 K [2] as a function of composition.

The phenomenology described above can be applied to any thermodynamic extensive function, $\overline{Y}_i$, for a solution. The integral molar enthalpy, entropy and Gibbs energy of mixing are thus

$$\Delta_{mix} H_m = x_A \Delta_{mix} \overline{H}_A + x_B \Delta_{mix} \overline{H}_B \tag{3.16}$$

$$\Delta_{mix} S_m = x_A \Delta_{mix} \overline{S}_A + x_B \Delta_{mix} \overline{S}_B \tag{3.17}$$

$$\Delta_{mix} G_m = x_A \Delta_{mix} \overline{G}_A + x_B \Delta_{mix} \overline{G}_B \tag{3.18}$$

The three functions are interrelated by

$$\Delta_{mix} G_m = \Delta_{mix} H_m - T \Delta_{mix} S_m \tag{3.19}$$

Since $\Delta_{mix} \overline{G}_A = \mu_A - \mu_A^o$, the integral molar Gibbs energy of mixing can alternatively be expressed in terms of the chemical potentials as

$$\Delta_{mix} G_m = x_A (\mu_A - \mu_A^o) + x_B (\mu_B - \mu_B^o) = RT(x_A \ln a_A + x_B \ln a_B) \tag{3.20}$$

where μ_A^o and μ_B^o are the chemical potentials of pure A and B, whereas μ_A and μ_B are the chemical potentials of A and B in the given solution. Using $G = H - TS$, the partial molar Gibbs energy of mixing is given as

$$\Delta_{mix} \overline{G}_A = \Delta_{mix} \overline{H}_A - T \Delta_{mix} \overline{S}_A = \mu_A - \mu_A^o = RT \ln a_A \tag{3.21}$$

The partial molar entropy, enthalpy or volume of mixing can be derived from eq. (3.21) and are given by the relations

$$\Delta_{mix} \overline{S}_A = -\left(\frac{\partial \Delta_{mix} \overline{G}_A}{\partial T} \right)_p = -R \ln a_A - RT \left(\frac{\partial \ln a_A}{\partial T} \right)_p \tag{3.22}$$

$$\Delta_{mix} \overline{H}_A = R \left(\frac{\partial \ln a_A}{\partial (1/T)} \right)_p \tag{3.23}$$

$$\Delta_{mix} \overline{V}_A = \left(\frac{\partial \Delta_{mix} \overline{G}_A}{\partial p} \right)_T = RT \left(\frac{\partial \ln a_A}{\partial p} \right)_T \tag{3.24}$$

Corresponding equations can be derived for the partial molar properties of B.

Ideal solutions

In Section 3.1 we showed that the chemical potential of an ideal gas in a mixture with other ideal gases is simply given in terms of a logarithmic function of the mole fraction. By comparing eqs. (3.9) and (3.10) we see that the fugacity/activity of the ideal gas is equal to the mole fraction. A solution (gas, liquid or solid) is in general called ideal if there are no extra interactions between the different species in addition to those present in the pure components. Thermodynamically this implies that the chemical activity is equal to the mole fraction, $a_i = x_i$, over the entire composition range. The molar Gibbs energy of mixing for an ideal solution then becomes

$$\Delta_{mix}^{id} G_m = RT(x_A \ln x_A + x_B \ln x_B) \tag{3.25}$$

The Gibbs energy of mixing of an ideal solution is negative due to the positive entropy of mixing obtained by differentiation of $\Delta_{mix}^{id} G_m$ with respect to temperature:

$$\Delta_{mix}^{id} S_m = -\left(\frac{\partial \Delta_{mix}^{id} G_m}{\partial T} \right)_p = -R(x_A \ln x_A + x_B \ln x_B) \tag{3.26}$$

In the absence of additional chemical interactions between the different species that are mixed the solution is stabilized entropically; the solution is more disordered than a mechanical mixture of the components. The origin of the entropy contribution is most easily understood by considering the distribution of two species on a crystalline lattice where the number of lattice sites is equal to the sum of the number of the two species A and B. For an ideal solution, a specific number of A and B atoms can be distributed randomly at the available sites, i.e. in a large number of different ways. This gives rise to a large number of different structural configurations with the same enthalpy and thus to the **configurational entropy** given by eq. (3.26). This will be discussed further in Chapter 9.

Two other characteristic properties of ideal solutions are

$$\Delta_{mix}^{id} H_m = \Delta_{mix}^{id} G_m + T\Delta_{mix}^{id} S_m = 0 \tag{3.27}$$

$$\Delta_{mix}^{id} V_m = \left(\frac{\partial \Delta_{mix}^{id} G_m}{\partial p} \right)_T = 0 \tag{3.28}$$

Or in words: in the absence of additional chemical interactions between the two types of atom, the enthalpy and volume of mixing are both zero.

The partial molar properties of a component i of an ideal solution are readily obtained:

$$\Delta^{id}_{mix} \overline{G}_i = \left(\frac{\partial \Delta^{id}_{mix} G_m}{\partial n_i} \right)_{p,T,n_{j \neq i}} = RT \ln x_i \tag{3.29}$$

$$\Delta^{id}_{mix} \overline{S}_i = \left(\frac{\partial \Delta^{id}_{mix} S_m}{\partial n_i} \right)_{p,T,n_{j \neq i}} = -\left(\frac{\partial \Delta^{id}_{mix} \overline{G}_i}{\partial T} \right)_{p,n_{j \neq i}} = -R \ln x_i \tag{3.30}$$

The thermodynamic properties of an ideal binary solution at 1000 K are shown in Figure 3.3. The integral enthalpy, entropy and Gibbs energy are given in Figure 3.3(a), while the integral entropy of mixing and the partial entropy of mixing of component A are given in Figure 3.3(b). Corresponding Gibbs energies are given in Figure 3.3(c). The largest entropic stabilization corresponds to the minimum Gibbs energy of mixing, which for an ideal solution is $RT \ln(\frac{1}{2})$ or $-RT \ln 2$, or about 0.7 times the thermal energy (RT) at 1000 K.

Excess functions and deviation from ideality

Most real solutions cannot be described in the ideal solution approximation and it is convenient to describe the behaviour of real systems in terms of deviations from the ideal behaviour. Molar excess functions are defined as

$$\Delta^{exc}_{mix} Y_m = \Delta_{mix} Y_m - \Delta^{id}_{mix} Y_m \tag{3.31}$$

The excess molar Gibbs energy of mixing is thus

$$\begin{aligned} \Delta^{exc}_{mix} G_m &= \Delta_{mix} G_m - RT(x_A \ln x_A + x_B \ln x_B) \\ &= RT(x_A \ln a_A + x_B \ln a_B) - RT(x_A \ln x_A + x_B \ln x_B) \end{aligned} \tag{3.32}$$

The **activity coefficient** of component i, γ_i, is now defined as a measure of the deviation from the ideal solution behaviour as the ratio between the chemical activity and the mole fraction of i in a solution.

$$\gamma_i = \frac{a_i}{x_i} \quad \text{or} \quad a_i = \gamma_i x_i \tag{3.33}$$

For an ideal solution $\gamma_i = 1$.

The partial molar Gibbs energy of mixing of a component i in a non-ideal mixture can in general be expressed in terms of activity coefficients as

$$\Delta_{mix} \overline{G}_i = RT \ln a_i = RT \ln x_i + RT \ln \gamma_i \tag{3.34}$$

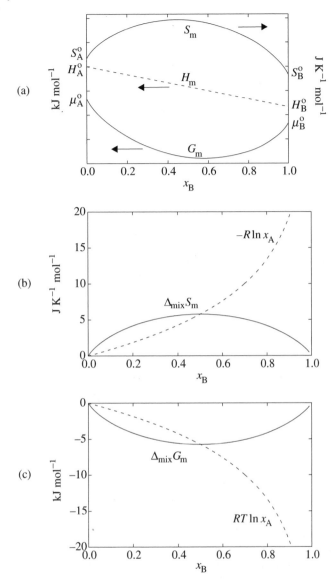

Figure 3.3 Thermodynamic properties of an arbitrary ideal solution A–B at 1000 K. (a) The Gibbs energy, enthalpy and entropy. (b) The entropy of mixing and the partial entropy of mixing of component A. (c) The Gibbs energy of mixing and the partial Gibbs energy of mixing of component A.

Using eq. (3.34) the excess Gibbs energy of mixing is given in terms of the mole fractions and the activity coefficients as

$$\Delta_{mix}G_m = RT(x_A \ln a_A + x_B \ln a_B) = \Delta_{mix}^{id}G_m + \Delta_{mix}^{exc}G_m \qquad (3.35)$$
$$= RT(x_A \ln x_A + x_B \ln x_B) + RT(x_A \ln \gamma_A + x_B \ln \gamma_B)$$

Implicitly:

$$\Delta_{mix}^{exc} G_m = RT(x_A \ln \gamma_A + x_B \ln \gamma_B) \tag{3.36}$$

Since $\Delta_{mix}^{id} H_m = \Delta_{mix}^{id} V_m = 0$ (eqs. 3.27 and 3.28), the excess molar enthalpy and volume of mixing are simply

$$\Delta_{mix}^{exc} V_m = \Delta_{mix} V_m \tag{3.37}$$

$$\Delta_{mix}^{exc} H_m = \Delta_{mix} H_m \tag{3.38}$$

The excess molar entropy of mixing is the real entropy of mixing minus the ideal entropy of mixing. Using a binary A–B solution as an example, $\Delta_{mix}^{exc} S_m$ is

$$\Delta_{mix}^{exc} S_m = \Delta_{mix} S_m - R(x_A \ln x_A + x_B \ln x_B) \tag{3.39}$$

For a large number of the more commonly used microscopic solution models it is assumed, as we will see in Chapter 9, that the entropy of mixing is ideal. The different atoms are assumed to be randomly distributed in the solution. This means that the excess Gibbs energy is most often assumed to be purely enthalpic in nature. However, in systems with large interactions, the excess entropy may be large and negative.

As shown above, the activity coefficients express the departure from ideality and thus define the excess Gibbs energy of the solution. Deviation from ideality is said to be positive when $\gamma > 1$ (ln γ is positive) and negative when $\gamma < 1$ (ln γ is negative). A negative deviation implies a negative contribution to the Gibbs energy relatively to an ideal solution and hence a stabilization of the solution relative to ideal solution behaviour. Similar arguments imply that positive deviations from ideality result in destabilization relative to ideal solution behaviour.

The activities of Fe and Ni in the binary system Fe–Ni [3] and the corresponding Gibbs energy and excess Gibbs energy of mixing are shown in Figures 3.4 and 3.5, respectively. The Fe–Ni system shows negative deviation from ideality and is thus stabilized relative to an ideal solution. This is reflected in the negative excess Gibbs energy of mixing. The activity coefficients γ_i, defined by eq. (3.33) as a_i/x_i, are readily determined from Figure 3.4. γ_{Ni} for the selected composition $x_{Ni} = 0.4$ is given by the ratio MQ/PQ. At the point of infinite dilution, $x_i = 0$, the activity coefficient takes the value γ_i^∞. γ_i^∞ is termed the **activity coefficient at infinite dilution** and is, as will be discussed in Chapter 4, an important thermodynamic characteristic of a solution. The activity coefficient of a given solute at infinite dilution will generally depend on the nature of the solvent, since the solute atoms at infinite dilution are surrounded on average by solvent atoms only. This determines the properties of the solute in the solution and thus γ_i^∞.

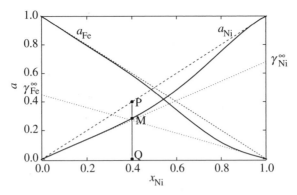

Figure 3.4 The activity of Fe and Ni of molten Fe–Ni at 1850 K [3]. At $x_{Ni} = 0.4$ the activity coefficient of Ni is given by MQ/PQ.

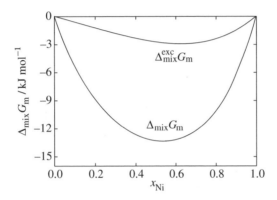

Figure 3.5 The molar Gibbs energy of mixing and the molar excess Gibbs energy of mixing of molten Fe–Ni at 1850 K. Data are taken from reference [3].

The formalism shown above is in general easily extended to multi-component systems. All thermodynamic mixing properties may be derived from the integral Gibbs energy of mixing, which in general is expressed as

$$\Delta_{mix} G_m = \Delta_{mix}^{id} G_m + \Delta_{mix}^{exc} G_m = RT \sum_i x_i \ln a_i$$
$$= RT \sum_i x_i \ln x_i + RT \sum_i x_i \ln \gamma_i$$

(3.40)

3.3 Standard states

In solution thermodynamics the standard or reference states of the components of the solution are important. Although the standard state in principle can be chosen freely, the standard state is in practice not taken by chance, but does in most cases reflect the type of model one wants to fit to experimental data. The choice of standard state is naturally influenced by the data available. In some cases the vapour pressure of one of the components is known in the whole compositional interval. In other cases the activity of the solute is known for dilute solutions only.

In the following, the Raoultian and Henrian standard states will be presented. These two are the far most frequent standard states applied in solution thermodynamics. Before discussing these standard states we need to consider Raoult's and Henry's laws, on which the Raoultian and Henrian standard states are based, in some detail.

Henry's and Raoult's laws

In the development of physical chemistry, investigations of dilute solutions have been very important. A dilute solution consists of the main constituent, the solvent, and one or more solutes, which are the diluted species. As early as in 1803 William Henry showed empirically that the vapour pressure of a solute i is proportional to the concentration of solute i:

$$p_i = x_i k_{H,i} \qquad (3.41)$$

where x_i is the mole fraction solute and $k_{H,i}$ is known as the **Henry's law constant**. Here we have used mole fraction as the measure of the concentration (alternatively the mass fraction or other measures may be used).

More than 80 years later François Raoult demonstrated that at low concentrations of a solute, the vapour pressure of the solvent is simply

$$p_i = x_i p_i^* \qquad (3.42)$$

where x_i is the mole fraction solvent and p_i^* is the vapour pressure of the pure solvent.

Raoult's and Henry's laws are often termed 'limiting laws'. This use reflects that real solutions often follow these laws at infinite dilution only. The vapour pressure above molten Ge–Si at 1723 K [4] is shown in Figure 3.6 as an example. It is evident that at dilute solution of Ge or Si, the vapour pressure of the dominant component follows Raoult's law. Raoult's law is expressing that a real non-ideal solution approaches an ideal solution when the concentration of the solvent approaches unity. In the corresponding concentration region Henry's law is valid for the solute. The Ge–Si system shows positive deviation from ideality and the activity coefficients of the two components, given as a function of x_i in Figure 3.6(b), are thus positive for all compositions (using Si and Ge as standard states).

Raoult's law is obeyed for a solvent at infinite dilution of a solute. Mathematically this implies

$$(da_A/dx_A)_{x_A \to 1} = 1 \qquad (3.43)$$

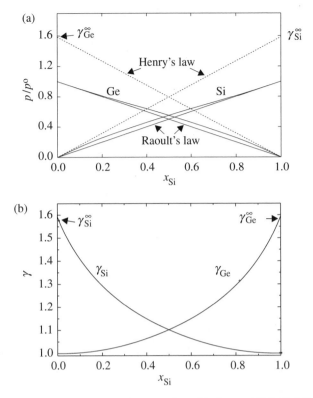

Figure 3.6 (a) The vapour pressure above molten Si–Ge at 1723 K [4]. (b) The corresponding activity coefficients of the two components.

In terms of activity coefficients eq. (3.43) can be transformed to

$$\left(\frac{d(x_A \gamma_A)}{dx_A}\right)_{x_A \to 1} = \left(\gamma_A + x_A \frac{d\gamma_A}{dx_A}\right)_{x_A \to 1} = 1 \tag{3.44}$$

Since $\gamma_A \to 1$ when $x_A \to 1$ the expression for Raoult's law becomes

$$\left(\frac{d\gamma_A}{dx_A}\right)_{x_A \to 1} = 0 \tag{3.45}$$

This is a necessary and sufficient condition for Raoult's law.

A solute B obeys Henry's law at infinite dilution if the slope of the activity curve a_B versus x_B has a nonzero finite value when $x_B \to 0$:

$$(da_B/dx_B)_{x_B \to 0} = \gamma_B^{\infty} \tag{3.46}$$

The finite value of the slope when $x_B \to 0$, γ_B^∞, is the activity coefficient at infinite dilution defined earlier. In terms of activity coefficients eq. (3.46) becomes

$$\left(\frac{d(x_B \gamma_B)}{dx_B} \right)_{x_B \to 0} = \left(\gamma_B + x_B \frac{d\gamma_B}{dx_B} \right)_{x_B \to 0} = \gamma_B^\infty \qquad (3.47)$$

It follows that if Henry's law behaviour is obeyed at infinite dilution:

$$\left(x_B \frac{d\gamma_B}{dx_B} \right)_{x_B \to 0} = 0 \qquad (3.48)$$

Equation (3.48) is a necessary consequence of Henry's law, but it is not a sufficient condition. It can be shown that Raoult's law behaviour of the solvent follows as a consequence of Henry's law behaviour for the solute, while the reverse does not follow.

Raoultian and Henrian standard states

The Raoultian standard state is the most frequently used standard state for a component in a solution. The Raoultian standard state implies that all thermodynamic properties are described relative to those of the pure component with the same structure as the solution. For liquids the specification of the structure seems artificial, but for solid solutions, which may have different crystal structures, this is of great importance. The activity of Ni in molten Fe–Ni at 1850 K using the Raoultian standard state is given in Figure 3.7 (ordinate given on the left-hand y-axis). The activity of pure Ni is set as standard state and is thus unity. While the Raoultian standard state represents a real physical reachable state, the Henrian standard state is a hypothetical one. The Henrian standard state for Ni in the Fe–Ni solution is found by extrapolation of the Henrian law behaviour at $x_{Ni} \to 0$ to $x_{Ni} = 1$; see Figure

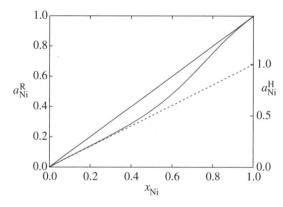

Figure 3.7 The activity of Ni of molten Fe–Ni at 1850 K using both a Raoultian and a Henrian standard state. Data are taken from reference [3].

3.7. The activity of Ni in molten Fe–Ni at 1850 K using the Henrian standard state is also given in the figure (ordinate given on the right-hand y-axis).

If an arbitrary standard state is marked with *, a formal definition of a Raoultian standard state for component A of a solution is

$$\mu_A^* = \mu_A^R \tag{3.49}$$

It follows that the activity coefficient with this standard state:

$$\gamma_A^R = \frac{a_A^R}{x_A} \tag{3.50}$$

approaches 1 when the mole fraction x_A approaches 1 or

$$(\gamma_A^R)_{x_A \to 1} = 1 \tag{3.51}$$

Correspondingly, a formal definition of a Henrian standard state for component B of a solution is

$$\mu_B^* = \mu_B^H \tag{3.52}$$

The activity coefficient with this standard state:

$$\gamma_B^H = \frac{a_B^H}{x_B} \tag{3.53}$$

approaches 1 when x_B approaches 0 or

$$(\gamma_B^H)_{x_B \to 0} = 1 \tag{3.54}$$

The activities on the two standard states are related since

$$\mu_i = \mu_i^R + RT \ln a_i^R = \mu_i^H + RT \ln a_i^H \tag{3.55}$$

which gives

$$\frac{a_i^R}{a_i^H} = \exp\left[-\frac{(\mu_i^R - \mu_i^H)}{RT}\right] \tag{3.56}$$

The ratio of two activities defined on the basis of two different standard states is constant and does not vary with the composition of the solution:

$$\frac{a_B^R}{a_B^H} = \frac{\gamma_B^R}{\gamma_B^H} \frac{x_B}{x_B} = \frac{\gamma_B^R}{\gamma_B^H} \tag{3.57}$$

For the present case this constant can be deduced by using the conditions at infinite dilution as a constraint, thus:

$$\frac{\gamma_B^R}{\gamma_B^H} = \left(\frac{\gamma_B^R}{\gamma_B^H} \right)_{x_B \to 0} = \frac{\gamma_B^{R,\infty}}{1} \tag{3.58}$$

$$\gamma_B^H = \frac{\gamma_B^R}{\gamma_B^{R,\infty}} \tag{3.59}$$

Whereas the total Gibbs energy of the solution is independent of the choice of the standard state, the standard state must be explicitly given when it comes to the mixing properties of a solution. The molar Gibbs energy of mixing of the Fe–Ni system for which the activity of Ni is shown in Figure 3.7 is given in Figure 3.8. The solid and dashed lines represent Gibbs energies of mixing based on the Raoultian and Henrian standard states for Ni, respectively.

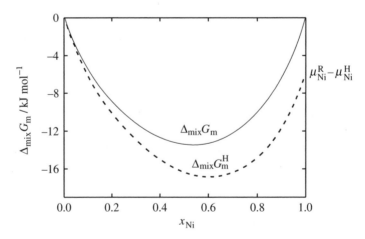

Figure 3.8 The molar Gibbs energy of mixing of molten Fe–Ni at 1850 K using both the Raoultian (solid line) and Henrian (dashed line) standard states for Ni as defined in Figure 3.7. The Raoultian standard state is used for Fe. Data are taken from reference [3].

3.4 Analytical solution models

Dilute solutions

Binary solutions have been extensively studied in the last century and a whole range of different analytical models for the molar Gibbs energy of mixing have evolved in the literature. Some of these expressions are based on statistical mechanics, as we will show in Chapter 9. However, in situations where the intention is to find mathematical expressions that are easy to handle, that reproduce experimental data and that are easily incorporated in computations, polynomial expressions obviously have an advantage.

Simple polynomial expressions constitute the most common analytical model for partial or integral thermodynamic properties of solutions:

$$Y(x_B) = Q_0 + Q_1 x_B + Q_2 x_B^2 + \ldots + Q_n x_B^n = \sum_{i=0}^{n} Q_i x_B^i \tag{3.60}$$

or

$$Y(x_B) = x_A x_B \sum_{i=0}^{n} R_i (x_A - x_B)^i = x_B(1 - x_B) \sum_{i=0}^{n} R_i (1 - 2x_B)^i \tag{3.61}$$

The variable x is usually the mole fraction of the components. The last expression was first introduced by Guggenheim [5]. Equation (3.60) is a particular case of the considerably more general Taylor series representation of Y as shown by Lupis [6]. Let us apply a Taylor series to the activity coefficient of a solute in a dilute binary solution:

$$\ln \gamma_B = \ln \gamma_B^\infty + \left(\frac{\partial \ln \gamma_B}{\partial x_B} \right)_{x_B \to 0} x_B + \frac{1}{2} \left(\frac{\partial^2 \ln \gamma_B}{\partial x_B^2} \right)_{x_B \to 0} x_B^2 + \ldots$$

$$+ \frac{1}{i!} \left(\frac{\partial^i \ln \gamma_B}{\partial x_B^i} \right)_{x_B \to 0} x_B^i \tag{3.62}$$

The derivatives of the Taylor series are all finite. It is not necessary to expand the series at $x_B = 0$, but it is most common and convenient for dilute solutions. The Taylor series expansion of $\ln \gamma_B$ may be expressed in a different notation as

$$\ln \gamma_B = \sum_{i=0}^{n} J_i^B x_B^i \tag{3.63}$$

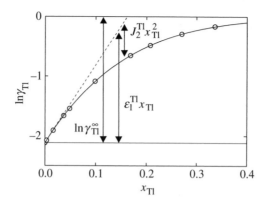

Figure 3.9 An illustration of low-order terms in the Taylor series expansion of $\ln \gamma_i$ for dilute solutions using $\ln \gamma_{Tl}$ for the binary system Tl–Hg at 293 K as example. Here $\ln \gamma_{Tl}^\infty = -2.069$, $\varepsilon_1^{Tl} = 10.683$ and $J_2^{Tl} = -14.4$. Data are taken from reference [8].

where

$$J_i^B = \frac{1}{i!}\left(\frac{\partial^i \ln \gamma_B}{\partial x_B^i}\right)_{x_B \to 0} \tag{3.64}$$

The coefficients J_i^B are called interaction coefficients of order i. The coefficient of zeroth order is just the value of $\ln \gamma_B$ at infinite solution. The first-order coefficient is the most used and is often designated by ε_1^B [7]. This coefficient is a measure of how an increase in the concentration of B changes $\ln \gamma_B$, which explains why it is called the **self-interaction coefficient**. The expression for $\ln \gamma_B$ with only three coefficients is

$$\ln \gamma_B = \ln \gamma_B^\infty + \varepsilon_1^B x_B + J_2^B x_B^2 \tag{3.65}$$

The orders of magnitude of the coefficients depend very much on the system studied. Generally stronger atomic interactions give larger interaction coefficients. An illustration of low order terms in the Taylor series expansion of $\ln \gamma_{Tl}$ in the binary system Tl–Hg is given in Figure 3.9 [8].

The same type of polynomial formalism may also be applied to the partial molar enthalpy and entropy of the solute and converted into integral thermodynamic properties through use of the Gibbs–Duhem equation; see Section 3.5.

Solution models

The simplest model beyond the ideal solution model is the **regular solution** model, first introduced by Hildebrant [9]. Here $\Delta_{mix} S_m$ is assumed to be ideal, while $\Delta_{mix} H_m$ is not. The molar excess Gibbs energy of mixing, which contains only a single free parameter, is then

$$\Delta_{mix}^{exc} G_m = \Omega x_A x_B \tag{3.66}$$

where Ω is named the **regular solution constant** or the **interaction coefficient**. The molar Gibbs energy is in this approximation

$$G_m = x_A \mu_A^o + x_B \mu_B^o + RT(x_A \ln x_A + x_B \ln x_B) + \Omega x_A x_B \tag{3.67}$$

The molar Gibbs energy of mixing

$$\Delta_{mix} G_m = RT(x_A \ln x_A + x_B \ln x_B) + \Omega x_A x_B \tag{3.68}$$

thus consists of one entropic and one enthalpic contribution:

$$\Delta_{mix} S_m = -R(x_A \ln x_A + x_B \ln x_B) \tag{3.69}$$

$$\Delta_{mix} H_m = \Omega x_A x_B \tag{3.70}$$

For ideal solutions Ω is zero and there are no extra interactions between the species that constitute the solution. In terms of nearest neighbour interactions only, the energy of an A–B interaction, u_{AB}, equals the average of the A–A, u_{AA}, and B–B, u_{BB}, interactions or

$$\Omega = zL \left[u_{AB} - \frac{1}{2}(u_{AA} + u_{BB}) \right] \tag{3.71}$$

where z is the coordination number and L is Avogadro's number. For the general case of a non-ideal solution $\Omega < 0$ gives an increased stability of the solution relative to an ideal solution, while $\Omega > 0$ destabilizes the solution. It follows that $\Omega < 0$ and $\Omega > 0$ are usually interpreted as attraction and repulsion, respectively, between the A and B atoms. Repulsion between the different atoms of the solution will imply that the atoms do not mix at absolute zero, where the entropic contribution is zero. Complete solubility will be obtained when the temperature is raised sufficiently so that the entropy gain due to randomization of the atoms is larger than the positive enthalpic contribution to the Gibbs energy. The integral Gibbs energies of systems with Ω/RT larger and smaller than zero are shown in Figure 3.10.

The regular solution model can be extended to multi-component systems, in which case the excess Gibbs energy of mixing is expressed as

$$\Delta_{mix}^{exc} G_m = \sum_{i=1}^{m-1} \sum_{j>1}^{m} x_i x_j \Omega_{ij} \tag{3.72}$$

Thus for a ternary system

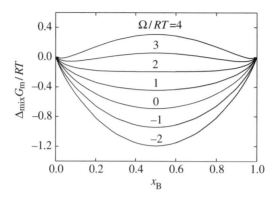

Figure 3.10 The molar Gibbs energy of mixing of a regular solution A–B for different values of Ω/RT.

$$\Delta_{\text{mix}}^{\text{exc}} G_{\text{m}} = x_1 x_2 \Omega_{12} + x_1 x_3 \Omega_{13} + x_2 x_3 \Omega_{23} \qquad (3.73)$$

An additional ternary interaction term, Ω_{123}, may be incorporated.

The regular solution model (eq. 3.68) is symmetrical about $x_A = x_B = 0.5$. In cases where the deviation from ideality is not symmetrical, the regular solution model is unable to reproduce the properties of the solutions and it is then necessary to introduce models with more than one free parameter. The most convenient polynomial expression with two parameters is termed the **sub-regular solution** model.

$$\Delta_{\text{mix}}^{\text{exc}} G_{\text{m}} = x_A x_B (A_{21} x_A + A_{12} x_B) \qquad (3.74)$$

If more than two parameters are necessary a general polynomial expression may be applied:

$$\Delta_{\text{mix}}^{\text{exc}} G_{\text{m}} = \sum_{i=1}^{m} \sum_{j=1}^{n} x_A^i x_B^j A_{ij} \qquad (3.75)$$

The **Redlich–Kister expression**

$$\Delta_{\text{mix}}^{\text{exc}} G_{\text{m}} = x_A x_B [\Omega + A_1 (x_A - x_B) + A_2 (x_A - x_B)^2$$
$$+ A_3 (x_A - x_B)^3 + \ldots] \qquad (3.76)$$

is a frequently used special case of this general polynomial approach. While the first term is symmetrical about $x = 0.5$, the second term changes sign for $x = 0.5$. The compositional variation of the third and fourth terms is given in Figure 3.11. In all these models the entropy of mixing is assumed to be ideal and the excess Gibbs energy is an analytical model for the enthalpy of mixing.

The entropy of mixing of many real solutions will deviate considerably from the ideal entropy of mixing. However, accurate data are available only in a few cases. The simplest model to account for a non-ideal entropy of mixing is the **quasi-regular model**, where the excess Gibbs energy of mixing is expressed as

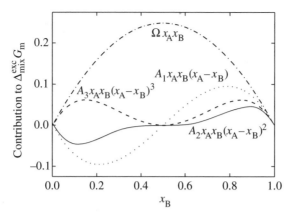

Figure 3.11 Contributions to the molar excess Gibbs energy of mixing from the four first terms of the Redlich–Kister expression (eq. 3.76). For convenience $\Omega = A_1 = A_2 = A_3 = 1$.

$$\Delta_{\text{mix}}^{\text{exc}} G_{\text{m}} = x_A x_B \Omega \left(1 - \frac{T}{\tau} \right) \tag{3.77}$$

Thus

$$\Delta_{\text{mix}}^{\text{exc}} S_{\text{m}} = -\frac{\partial (\Delta_{\text{mix}}^{\text{exc}} G_{\text{m}})}{\partial T} = x_A x_B \left(\frac{\Omega}{\tau} \right) \tag{3.78}$$

The sign of the excess entropy is given by the sign of τ.

Derivation of partial molar properties

The partial molar properties of binary solutions may be determined by both analytical and graphical methods. In cases where analytical expressions for integral extensive thermodynamic quantities are available, the partial molar quantities are obtained by differentiation, but graphical determination of partial molar properties also has a long history in thermodynamics. The molar Gibbs energy of mixing of molten Si–Ge at 1500 K is given as a function of the mole fraction of Ge in Figure 3.12. Pure solid Si and pure liquid Ge are chosen as standard states. If we draw a tangent to the curve at any composition, the intercept of this tangent upon the ordinate $x_{\text{Si}} = 1$ equals μ_{Si} and the intercept for $x_{\text{Ge}} = 1$ equals μ_{Ge}.

In mathematical terms the partial molar properties of a binary system will in general be given through

$$\overline{Y}_A = Y_{\text{m}} - x_B \frac{dY_{\text{m}}}{dx_B} \tag{3.79}$$

and

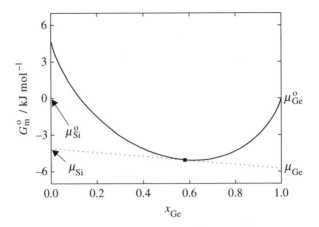

Figure 3.12 The integral molar Gibbs energy of liquid Ge–Si at 1500 K with pure liquid Ge and solid Si as standard states. Data are taken from reference [4].

$$\overline{Y}_B = Y_m + (1 - x_B)\frac{dY_m}{dx_B} \tag{3.80}$$

Application to the Gibbs energy of the two components of a binary solution therefore gives

$$\mu_A = G_m - x_B \frac{dG_m}{dx_B} \tag{3.81}$$

$$\mu_B = G_m + (1 - x_B)\frac{dG_m}{dx_B} \tag{3.82}$$

Taking the excess Gibbs energy of a regular solution as an example:

$$\Delta_{mix}^{exc} G_m = \Omega x_A x_B \tag{3.83}$$

the partial excess Gibbs energies of the two components are

$$\frac{\Delta_{mix}^{exc} \overline{G}_A}{RT} = \ln \gamma_A = \frac{\Omega}{RT} x_B^2 \tag{3.84}$$

$$\frac{\Delta_{mix}^{exc} \overline{G}_B}{RT} = \ln \gamma_B = \frac{\Omega}{RT} x_A^2 \tag{3.85}$$

In general, the chemical potential of species i for a multi-component system is given as

$$\mu_i = G_m + \sum_{j=2}^{r} (\delta_{ij} - x_j) \frac{dG_m}{dx_j} \tag{3.86}$$

where $\delta_{ij} = 0$ for $i \neq j$ and $\delta_{ij} = 1$ for $i = j$.

3.5 Integration of the Gibbs–Duhem equation

In experimental investigations of thermodynamic properties of solutions, it is common that one obtains the activity of only one of the components. This is in particular the case when one of the components constitutes nearly the complete vapour above a solid or liquid solution. A second example is when the activity of one of the components is measured by an electrochemical method. In these cases we can use the Gibbs–Duhem equation to find the activity of the second component.

We have already derived the Gibbs–Duhem equation in Chapter 1.4. At constant p and T:

$$n_A d\mu_A + n_B d\mu_B = 0 \tag{3.87}$$

In terms of activity and mole fractions this yields

$$x_A d \ln a_A + x_B d \ln a_B = 0 \tag{3.88}$$

or

$$x_A d \ln x_A + x_A d \ln \gamma_A + x_B d \ln x_B + x_B d \ln \gamma_B = 0 \tag{3.89}$$

Since

$$x_A d \ln x_A + x_B d \ln x_B = x_A \frac{dx_A}{x_A} + x_B \frac{dx_B}{x_B} = dx_A + dx_B = 0 \tag{3.90}$$

eq. (3.89) may be rewritten

$$x_A d \ln \gamma_A + x_B d \ln \gamma_B = 0 \tag{3.91}$$

or by integration

$$\ln \gamma_B - \ln \gamma_B(x_B = 1) = - \int_{x_B=1}^{x_B} \frac{x_A}{x_B} d \ln \gamma_A \tag{3.92}$$

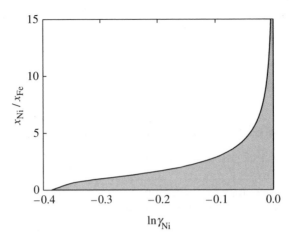

Figure 3.13 x_{Ni}/x_{Fe} versus $\ln \gamma_{Ni}$ of molten Fe–Ni at 1850 K. Data are taken from reference [3].

If a Raoultian reference state is chosen for both A and B, $\ln \gamma_B = 0$ when $x_B = 1$. Now by plotting x_A/x_B against $\ln \gamma_A$, as done for the activity coefficient of Fe of molten Fe–Ni at 1850 K in Figure 3.13, the Gibbs–Duhem equation may be integrated graphically by determining the area between the limits. The challenge in our case is that when $x_{Fe} \rightarrow 0$, $x_{Ni}/x_{Fe} \rightarrow \infty$ and $\ln \gamma_{Ni} \rightarrow 0$. It may therefore be difficult to evaluate the integral accurately since this demands a large amount of experimental data for $x_{Fe} \rightarrow 0$.

We may also integrate the Gibbs–Duhem equation using an Henrian reference state for B:

$$\ln \gamma_B - \ln \gamma_B(x_B = 0) = - \int_{x_B = 0}^{x_B} \frac{x_A}{x_B} d \ln \gamma_A \qquad (3.93)$$

Henry's law for B leads to $\ln \gamma_B = 0$ when $x_B = 0$.

An alternative method of integrating the Gibbs–Duhem equation was developed by Darken and Gurry [10]. In order to calculate the integral more accurately, a new function, α, defined as

$$\alpha_i = \frac{\ln \gamma_i}{(1 - x_i)^2} \quad (i = A \text{ or } B) \qquad (3.94)$$

was introduced for binary solutions, since this gave a convenient expression for the much used regular solution model. An expression for $d \ln \gamma_A$ is obtained by differentiation:

$$d \ln \gamma_A = d(\alpha_A x_B^2) = 2\alpha_A x_B dx_B + x_B^2 d\alpha_A \qquad (3.95)$$

By substituting eq. (3.95) into the Gibbs–Duhem equation (eq. 3.92) we obtain

$$\ln \gamma_B = - \int_{x_B=1}^{x_B} 2\alpha_A \, x_A \, dx_B - \int_{x_B=1}^{x_B} x_A \, x_B d\alpha_A \tag{3.96}$$

Integrating by parts the second integral, we obtain

$$\int_{x_B=1}^{x_B} x_A \, x_B d\alpha_A = [\alpha_A \, x_A \, x_B]_{x_B=1}^{x_B} - \int_{x_B=1}^{x_B} \alpha_A \, x_A \, dx_B + \int_{x_B=1}^{x_B} \alpha_A \, x_B dx_B \tag{3.97}$$

which gives the following expression for $\ln \gamma_B$ (eq. 3.96):

$$\ln \gamma_B = -\alpha_A \, x_A \, x_B - \int_{x_B=1}^{x_B} \alpha_A (x_A + x_B) dx_B = -\alpha_A \, x_A \, x_B - \int_{x_B=1}^{x_B} \alpha_A \, dx_B \tag{3.98}$$

Integration of the Gibbs–Duhem equation applying the method by Darken and Gurry is illustrated by using the Fe–Ni system as an example: see Figure 3.14. α_{Ni} plotted against x_{Fe} gives a curve that is more easily integrated.

A graphical integration of the Gibbs–Duhem equation is not necessary if an analytical expression for the partial properties of mixing is known. Let us assume that we have a dilute solution that can be described using the activity coefficient at infinite dilution and the self-interaction coefficients introduced in eq. (3.64).

$$\ln \gamma_B = \ln \gamma_B^\infty + \varepsilon_1^B x_B + J_2^B x_B^2 \tag{3.99}$$

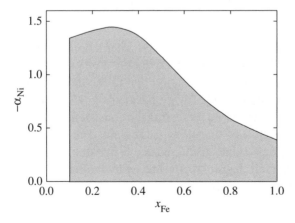

Figure 3.14 α_{Ni} plotted versus x_{Fe} for molten Fe–Ni at 1850 K. Data are taken from reference [3]. The area under the curve represents integration from $x_{Fe} = 1.0$ to 0.1.

The Gibbs–Duhem equation can be modified to

$$(1-x_B)\frac{\partial \ln \gamma_A}{\partial x_B} + x_B \frac{\partial \ln \gamma_B}{\partial x_B} = 0 \tag{3.100}$$

and thus if the first- and second-order terms in the Taylor series for the solvent are termed J_1^A and J_2^A (i.e. $\ln \gamma_A = J_1^A x_B + J_2^A x_B^2$):

$$(1-x_B)(J_1^A + 2J_2^A x_B) + x_B(\varepsilon_1^B + 2J_2^B x_B) = 0 \tag{3.101}$$

Hence this implies that

$$J_1^A = 0 \tag{3.102}$$

and

$$J_2^A = -\frac{1}{2}\varepsilon_1^B \tag{3.103}$$

We are thus able to express the activity coefficient of the second component, A, in terms of ε_2^B:

$$\ln \gamma_A = -\frac{1}{2}\varepsilon_1^B x_B^2 \tag{3.104}$$

All other properties follow. For example, the excess Gibbs energy of mixing is

$$\Delta_{mix}^{exc} G_m = RT\left(x_B \ln \gamma_B^\infty + \frac{1}{2}\varepsilon_1^B x_B^2 + \text{higher order terms}\right) \tag{3.105}$$

The relationship between the different self-interaction coefficients of component A and B, J_i^A and J_i^B, may in general be obtained in a similar way.

Although the Gibbs–Duhem equation due to the development of versatile and user-friendly thermodynamic software packages is less central than before, it is still of great value, for example for testing the consistency of experimental data and also for systematization of thermodynamic data. The order of magnitude of the major interaction coefficients discussed above may for alloy systems, for instance, be estimated with a fair degree of confidence by looking at trends and by comparison with data on similar systems.

References

[1]   H. R. Thresh, A. F. Crawley and D. W. G. White, *Trans. Min. Soc. AIME*, 1968, **242**, 819.

[2]   O. Sato, *Bull. Res. Inst. Mineral. Dress. Metall.* 1955, **11**, 183.

[3]   G. R. Zellars, S. L. Payne, J. P. Morris and R. L. Kipp, *Trans. Met. Soc. AIME* 1959, **215**, 181.

[4]   C. Bergman, R. Chastel and R. Castanet, *J. Phase Equil.* 1992, **13**, 113.

[5]   E. A. Guggenheim, *Proc. Roy. Soc. London, Ser. A* 1935, **148**, 304.

[6]   C. H. P. Lupis, *Acta Metall.*, 1968, **16**, 1365.

[7]   C. Wagner, *Thermodynamics of Alloys*. Reading, MA, Addison-Wesley, 1962.

[8]   T. W. Richards and F. Daniels, *J. Am. Chem. Soc.* 1919, **41**, 1732.

[9]   J. H. Hildebrand, *J. Am. Chem. Soc.* 1929, **51**, 66.

[10] L. S. Darken and R. W. Gurry, *Physical Chemistry of Metals.* New York: McGraw-Hill, 1953, p. 264.

Further reading

P. W. Atkins and J. de Paula, *Physical Chemistry*, 7th edn. Oxford: Oxford University Press, 2001.

E. A. Guggenheim, *Thermodynamics: An Advanced Treatment for Chemists and Physicists*, 7th edn. Amsterdam: North-Holland, 1985.

C. H. P. Lupis, *Chemical Thermodynamics of Materials*. New York: North-Holland, 1983.

K. S. Pitzer, *Thermodynamics*. New York: McGraw-Hill, 1995. (Based on G. N. Lewis and M. Randall, *Thermodynamics and the free energy of chemical substances*. New York: McGraw-Hill, 1923.

4

Phase diagrams

Thermodynamics in materials science has often been used indirectly through phase diagrams. Knowledge of the equilibrium state of a chemical system for a given set of conditions is a very useful starting point for the synthesis of any material, for processing of materials and in general for considerations related to material stability. A phase diagram is a graphical representation of coexisting phases and thus of stability regions when equilibrium is established among the phases of a given system. A material scientist will typically associate a 'phase diagram' with a plot with temperature and composition as variables. Other variables, such as the partial pressure of a component in the system, may be given explicitly in the phase diagram; for example, as a line indicating a constant partial pressure of a volatile component. In other cases the partial pressure may be used as a variable. The stability fields of the condensed phases may then be represented in terms of the chemical potential of one or more of the components.

The Gibbs phase rule introduced in Section 2.1 is an important guideline for the construction and understanding of phase diagrams, and the phase rule is therefore referred to frequently in the present chapter. The main objective of the chapter is to introduce the quantitative link between phase diagrams and chemical thermodynamics. With the use of computer programs the calculation of phase diagrams from thermodynamic data has become a relatively easy task. The present chapter focuses on the theoretical basis for the calculation of heterogeneous phase equilibria with particular emphasis on binary phase diagrams.

4.1 Binary phase diagrams from thermodynamics

Gibbs phase rule

In chemical thermodynamics the system is analyzed in terms of the potentials defining the system. In the present chapter the potentials of interest are T (thermal

Chemical Thermodynamics of Materials by Svein Stølen and Tor Grande
© 2004 John Wiley & Sons, Ltd ISBN 0 471 492320 2

potential), p (mechanical potential) and the chemical potential of the components $\mu_1, \mu_2, \ldots, \mu_N$. We do not consider other potentials, e.g. the electrical and magnetic potentials treated briefly in Section 2.1. In a system with C components there are therefore $C + 2$ potentials. The potentials of a system are related through the Gibbs–Duhem equation (eq. 1.93):

$$S\mathrm{d}T - V\mathrm{d}p + \sum_i n_i \mathrm{d}\mu_i = 0 \tag{4.1}$$

and also through the Gibbs phase rule (eq. 2.15):

$$F + Ph = C + 2 \tag{4.2}$$

The latter is used as a guideline to determine the relationship between the number of potentials that can be varied independently (the number of degrees of freedom, F) and the number of phases in equilibrium, Ph. Varied independently in this context means varied without changing the number of phases in equilibrium.

For a single-component system, the Gibbs phase rule reads $F + Ph = C + 2 = 3$, and we can easily construct a p,T-phase diagram in two dimensions (see Figure 2.7, for example). To apply the Gibbs phase rule to a system containing two or more components ($C > 1$) it is necessary to take into consideration the nature of the different variables (potentials), the number of components, chemical reactions and compositional constraints. Initially we will apply the Gibbs phase rule to a binary system ($C = 2$). The Gibbs phase rule is then $F + Ph = C + 2 = 4$, and since at least one phase must be present, F is at most 3. Three dimensions are needed to show the phase relations as a function of T, p and a compositional variable (or a chemical potential). Here, we will use the mole fraction as a measure of composition although in some cases the weight fraction and other compositional variables are more practical. When a single phase is present ($F = 3$), T, p and the composition may be varied independently. With two phases present ($F = 2$) a set of two intensive variables can be chosen as independent; for example temperature and a composition term, or pressure and a chemical potential. With three phases present only a single variable is independent ($F = 1$); the others are given implicitly. Finally, with four phases present at equilibrium none of the intensive variables can be changed. The observer cannot affect the chemical equilibrium between these four phases.

It is sometimes convenient to fix the pressure and decrease the degrees of freedom by one in dealing with condensed phases such as substances with low vapour pressures. Gibbs phase rule then becomes

$$F = C - Ph + 1 \tag{4.3}$$

often called the reduced or condensed phase rule in metallurgical literature.

For a binary system at constant pressure the phase rule gives $F = 3 - Ph$ and we need only two independent variables to express the stability fields of the phases. It is most often convenient and common to choose the temperature and composition, given for example as the mole fraction. An example is the phase diagram of the system Ag–Cu shown in Figure 4.1 [1]. There are only three phases in the system: the solid solutions Cu(ss) and Ag(ss) and the Ag–Cu liquid solution. Cu(ss) and Ag(ss) denote solid solutions with Cu and Ag as solvents and Ag and Cu as solutes, respectively. When a single phase is present, for example the liquid, $F = 2$ and both composition and temperature may be varied independently. The stability fields for the liquid and the two solid solutions are therefore two-dimensional regions in the phase diagram. With two phases in equilibrium, the temperature and composition are no longer independent of each other. It follows that the compositions of two phases in equilibrium at a given temperature are fixed. In the case of a solid–liquid equilibrium the composition of the coexisting phases are defined by the **solidus** and **liquidus lines**, respectively. This is illustrated in Figure 4.1 where the composition of Cu(ss) in equilibrium with the liquid (also having a distinct composition) for a given temperature, T_2, is indicated by open circles. Since $F = 1$ this is called a **univariant equilibrium**. Finally, when three phases are present at equilibrium, $F = 0$ and the compositions of all three phases and the temperature are fixed. In this situation there are no degrees of freedom and the three phases are therefore present in an **invariant equilibrium**. In the present example, the system Ag–Cu, the two solid and the liquid phases coexist in an invariant eutectic equilibrium at 1040 K. The **eutectic reaction** taking place is defined in general for a two-component system as one in which a liquid on cooling solidifies under the formation of two solid phases. Hence for the present example the eutectic reaction is

$$\text{liquid} \rightarrow \text{Cu(ss)} + \text{Ag(ss)} \tag{4.4}$$

The temperature of the eutectic equilibrium is called the **eutectic temperature** and is shown as a horizontal line in Figure 4.1.

It should be noted that we have here considered the system at constant pressure. If we are not considering the system at isobaric conditions, the invariant equilibrium becomes univariant, and a univariant equilibrium becomes divariant, etc. A

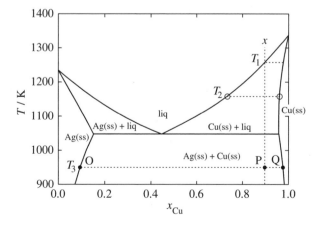

Figure 4.1 Phase diagram of the system Ag–Cu at 1 bar [1].

consequence is that the eutectic temperature in the Ag–Cu system will vary with pressure. However, as discussed in Section 2.3, small variations in pressure give only minor variations in the Gibbs energy of condensed phases. Therefore minor variations in pressure (of the order of 1–10 bar) are not expected to have a large influence on the eutectic temperature of a binary system.

One of the useful aspects of phase diagrams is that they define the equilibrium behaviour of a sample on cooling from the liquid state. Assume that we start at high temperatures with a liquid with composition x in the diagram shown in Figure 4.1. On cooling, the liquidus line is reached at T_1. At this temperature the first crystallites of the solid solution Cu(ss) are formed at equilibrium. The composition of both the liquid and Cu(ss) changes continuously with temperature. Further cooling produces more Cu(ss) at the expense of the liquid. If equilibrium is maintained the last liquid disappears at the eutectic temperature. The liquid with eutectic composition will at this particular temperature precipitate Cu(ss) and Ag(ss) simultaneously. The system is invariant until all the liquid has solidified. Below the eutectic temperature the two solid solutions Cu(ss) and Ag(ss) are in equilibrium, and for any temperature the composition of both solid solutions can be read from the phase diagram, as shown for the temperature T_3.

The relative amount of two phases present at equilibrium for a specific sample is given by the **lever rule**. Using our example in Figure 4.1, the relative amount of Cu(ss) and Ag(ss) at T_3 when the overall composition is x_{Cu}, is given by the ratio

$$\frac{OP}{OQ} = \frac{x_{Cu} - x_{Cu}^{Ag(ss)}}{x_{Cu}^{Cu(ss)} - x_{Cu}^{Ag(ss)}} \tag{4.5}$$

where $x_{Cu}^{Cu(ss)}$ and $x_{Cu}^{Ag(ss)}$ denote the mole fractions of Cu in the two coexisting solid solutions. The lines OP and OQ are shown in the figure. An isothermal line in a two-phase field, like the line OQ, is called a **tieline** or **conode**. As the overall composition is varied at constant temperature between the points O and Q, the compositions of the two solid phases remain fixed at O and Q; only the relative amount of the two phases changes.

Conditions for equilibrium

Phase diagrams show coexistent phases in equilibrium. We have seen in Chapter 1 that the conditions for equilibrium in a heterogeneous closed system at constant pressure and temperature can be expressed in terms of the chemical potential of the components of the phases in equilibrium:

$$\mu_i^\alpha = \mu_i^\beta = \mu_i^\gamma = \dots \quad \text{for } i = 1, 2, \dots, C \tag{4.6}$$

Here α, β and γ denote the different phases, whereas i denotes the different components of the system and C the total number of components. The conditions for equilibrium between two phases α and β in a binary system A–B (at a given temperature and pressure) are thus

$$\mu_A^\alpha(x_A^\alpha) = \mu_A^\beta(x_A^\beta) \tag{4.7}$$

and

$$\mu_B^\alpha(x_A^\alpha) = \mu_B^\beta(x_A^\beta) \tag{4.8}$$

where x_A^α and $x_{A_i}^\beta$ are the mole fractions of A in the phases α and β at equilibrium (remember that $x_A^i + x_B^i = 1$).

At a given temperature and pressure eqs. (4.7) and (4.8) must be solved simultaneously to determine the compositions of the two phases α and β that correspond to coexistence. At isobaric conditions, a plot of the composition of the two phases in equilibrium versus temperature yields a part of the equilibrium T, x-phase diagram.

Equations (4.7) and (4.8) may be solved numerically or graphically. The latter approach is illustrated in Figure 4.2 by using the Gibbs energy curves for the liquid and solid solutions of the binary system Si–Ge as an example. The chemical potentials of the two components of the solutions are given by eqs. (3.79) and (3.80) as

$$\mu_{Ge} = G_m + (1 - x_{Ge}) \frac{dG_m}{dx_{Ge}} \tag{4.9}$$

$$\mu_{Si} = G_m - x_{Ge} \frac{dG_m}{dx_{Ge}} \tag{4.10}$$

Here G_m is the Gibbs energy of the given solution at a particular composition x_{Ge}. The equilibrium conditions can now be derived graphically from Gibbs energy versus composition curves by finding the compositions on each curve linked by a common tangent (the **common tangent construction**). In the case shown in Figure 4.2(a) the solid and liquid solutions are in equilibrium; they are not in the case shown in Figure 4.2(b). The compositions of the coexisting solid and liquid solutions are marked by arrows in Figure 4.2(a).

The relationship between the Gibbs energy of the phases present in a given system and the phase diagram may be further illustrated by considering the variation of the Gibbs energy of the phases in the system Si–Ge with temperature. Similar common tangent constructions can then be made at other temperatures as well using thermodynamic data by Bergman *et al.* [2]. The phase diagram of the system is given in Figure 4.3(a). A sequence of Gibbs energy–composition curves for the liquid and solid solutions are shown as a function of decreasing temperature in Figures 4.3(b)–(f). The two Gibbs energy curves are broad and have shallow minima and the excess Gibbs energies of mixing are small since Ge and Si are chemically closely

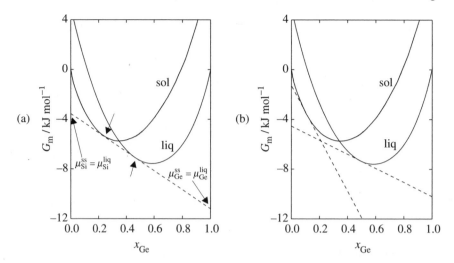

Figure 4.2 Gibbs energy curves for the liquid and solid solution in the binary system Si–Ge at 1500 K. (a) A common tangent construction showing the compositions of the two phases in equilibrium. (b) Tangents at compositions that do not give two phases in equilibrium. Thermodynamic data are taken from reference [2].

related. This is often termed near-ideal behaviour. At high temperatures, e.g. at T_1, where the liquid solution is stable over the whole composition region, the Gibbs energy of the liquid is more negative than that of the solid solution for all compositions (Figure 4.3(b)). On cooling, the Gibbs energy of the solid solution, having lower entropy than the liquid solution, increases more slowly than that of the liquid solution. At T_2, the Gibbs energies of pure liquid Si and pure solid Si are equal, and the melting temperature of pure Si is reached (Figure 4.3(c)). For $x_{Si} < 1$ the liquid solution is more stable than the crystalline phase. Further cooling gives situations corresponding to T_3 or T_4, where the solid solution is stable for the Si-rich compositions and the liquid solution for the Ge-rich compositions. The Gibbs energy curves at these two temperatures are shown in Figures 4.3(d) and (e). The compositions of the two phases in equilibrium at these temperatures are given by the common tangent construction, as illustrated in Figure 4.3(d). At T_5 the liquid has been cooled down to the melting temperature of pure Ge (see Figure 4.3(f)). Below this temperature the solid solution is stable for all compositions. Since Ge and Si are chemically closely related, Si–Ge forms a complete solid solution at low temperatures. The resulting equilibrium phase diagram, shown in Figure 4.3(a), is a plot of the locus of the common tangent constructions and defines the compositions of the coexisting phases as a function of temperature. The solidus and liquidus curves here define the stability regions of the solid and liquid solutions, respectively.

Ideal and nearly ideal binary systems

Let us consider a binary system for which both the liquid and solid solutions are assumed to be ideal or near ideal in a more formal way. It follows from their near-

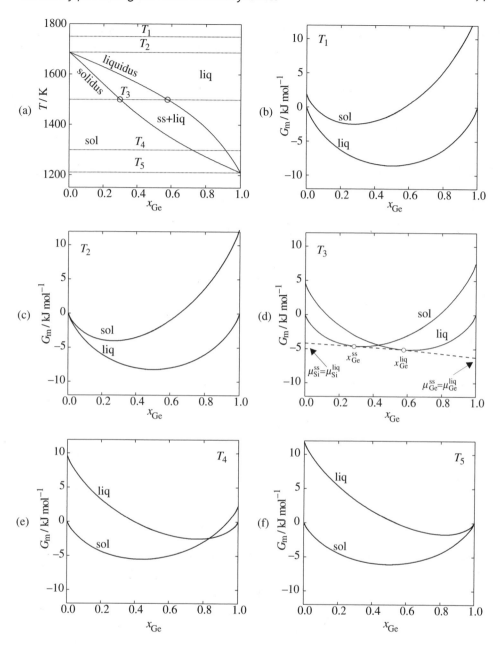

Figure 4.3 (a) Phase diagram for the system Si–Ge at 1 bar defining the five temperatures for the Gibbs energy curves shown in (b) T_1; (c) T_2; (d) T_3; (e) T_4; (f) T_5. Thermodynamic data are taken from reference [2].

ideal behaviour that the two components must have similar physical and chemical properties in both the solid and liquid states. Two systems which show this type of behaviour are the Si–Ge system discussed above and the binary system

FeO–MnO[1,2]. However, we will initially look at a general A–B system. The chemical potentials of component A in the liquid and solid states are given as

$$\mu_A^{ss} = \mu_A^{s,o} + RT \ln a_A^{ss} \tag{4.11}$$

$$\mu_A^{liq} = \mu_A^{l,o} + RT \ln a_A^{liq} \tag{4.12}$$

Similar expressions are valid for the chemical potential of component B of the two phases. According to the equilibrium conditions given by eqs. (4.7) and (4.8), the solid and liquid solutions are in equilibrium when $\mu_A^{ss} = \mu_A^{liq}$ and $\mu_B^{ss} = \mu_B^{liq}$, giving the two expressions

$$\mu_A^{s,o} + RT \ln a_A^{ss} = \mu_A^{l,o} + RT \ln a_A^{liq} \tag{4.13}$$

$$\mu_B^{s,o} + RT \ln a_B^{ss} = \mu_B^{l,o} + RT \ln a_B^{liq} \tag{4.14}$$

which may be rearranged as

$$\ln\left(\frac{a_A^{liq}}{a_A^{ss}}\right) = -\frac{\Delta\mu_A^{o(s\to l)}}{RT} \tag{4.15}$$

and

$$\ln\left(\frac{a_B^{liq}}{a_B^{ss}}\right) = -\frac{\Delta\mu_B^{o(s\to l)}}{RT} \tag{4.16}$$

Here $\Delta\mu_i^{o(s\to l)}$ is the change in chemical potential or Gibbs energy on fusion of pure i. By using $G = H - TS$ we have

$$\Delta\mu_i^{o(s\to l)} = \mu_i^{l,o} - \mu_i^{s,o} = \Delta_{fus}G_i^o = \Delta_{fus}H_i^o - T\Delta_{fus}S_i^o \tag{4.17}$$

At the melting temperature we have $\Delta_{fus}G_i^o = 0$, which implies that $\Delta_{fus}S_i^o = \Delta_{fus}H_i^o/T_{fus,i}$. If the heat capacity of the solid and the liquid are assumed to be equal, the enthalpy of fusion is independent of temperature and eq. (4.17) becomes

---

1 The FeO–MnO system is in principle a three-component system, but can be treated as a two-component system. This requires that the chemical potential of one of the three elements is constant.
2 The fact that FeO is non-stoichiometric is neglected.

$$\Delta\mu_i^{o(s\to l)} = \Delta_{\text{fus}}H_i^o - T\Delta_{\text{fus}}S_i^o = \Delta_{\text{fus}}H_i^o\left(1 - \frac{T}{T_{\text{fus},i}}\right) \tag{4.18}$$

Substitution of eq. (4.18) into eqs. (4.15) and (4.16) gives

$$\ln\left(\frac{a_A^{\text{liq}}}{a_A^{ss}}\right) = -\frac{\Delta\mu_A^{o(s\to l)}}{RT} = -\frac{\Delta_{\text{fus}}H_A^o}{R}\left(\frac{1}{T} - \frac{1}{T_{\text{fus},A}}\right) \tag{4.19}$$

and

$$\ln\left(\frac{a_B^{\text{liq}}}{a_B^{ss}}\right) = -\frac{\Delta\mu_B^{o(s\to l)}}{RT} = -\frac{\Delta_{\text{fus}}H_B^o}{R}\left(\frac{1}{T} - \frac{1}{T_{\text{fus},B}}\right) \tag{4.20}$$

If we furthermore assume that the solid and liquid solutions are ideal the activities can be replaced by mole fractions and eqs. (4.19) and (4.20) rearrange to

$$x_A^{\text{liq}} = x_A^{ss}\exp\left[-\frac{\Delta_{\text{fus}}H_A^o}{R}\left(\frac{1}{T} - \frac{1}{T_{\text{fus},A}}\right)\right] \tag{4.21}$$

and

$$x_B^{\text{liq}} = x_B^{ss}\exp\left[-\frac{\Delta_{\text{fus}}H_B^o}{R}\left(\frac{1}{T} - \frac{1}{T_{\text{fus},B}}\right)\right] \tag{4.22}$$

Analytical equations for the solidus and liquidus lines can now be obtained from these equations by noting that $x_A^{\text{liq}} + x_B^{\text{liq}} = 1$ and $x_A^{ss} + x_B^{ss} = 1$, giving

$$x_A^{ss}\exp\left[-\frac{\Delta_{\text{fus}}H_A^o}{R}\left(\frac{1}{T} - \frac{1}{T_{\text{fus},A}}\right)\right] + x_B^{ss}\exp\left[-\frac{\Delta_{\text{fus}}H_B^o}{R}\left(\frac{1}{T} - \frac{1}{T_{\text{fus},B}}\right)\right] = 1 \tag{4.23}$$

and

$$x_A^{\text{liq}}\exp\left[\frac{\Delta_{\text{fus}}H_A^o}{R}\left(\frac{1}{T} - \frac{1}{T_{\text{fus},A}}\right)\right] + x_B^{\text{liq}}\exp\left[\frac{\Delta_{\text{fus}}H_B^o}{R}\left(\frac{1}{T} - \frac{1}{T_{\text{fus},B}}\right)\right] = 1 \tag{4.24}$$

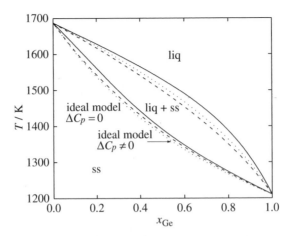

Figure 4.4 Phase diagram for the system Si–Ge at 1 bar. The solid lines represent experimental observations [2] while the dotted and dashed lines represent calculations assuming that the solid and liquid solutions are ideal with $\Delta C_p \neq 0$ and $\Delta C_p = 0$, respectively.

In this particular case of ideal solutions the phase diagram is defined solely by the temperature and enthalpy of fusion of the two components.

Using the analytical equations derived above, we are now able to consider the phase diagrams of the two nearly ideal systems mentioned above more closely. In the calculations we will initially use only the melting temperature and enthalpy of fusion of the two components as input parameters; both the solid and liquid solutions are assumed to be ideal. The observed (solid lines) and calculated (dashed lines) phase diagrams for the systems Si–Ge [2] and FeO–MnO [3] are compared in Figure 4.4 and 4.5. Although the agreement is reasonable, the deviation between the calculated and observed solidus and liquidus lines is significant.

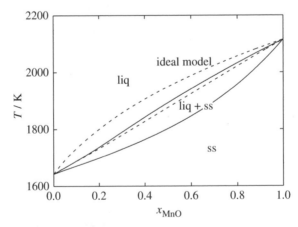

Figure 4.5 Phase diagram of FeO–MnO at 1 bar. The solid lines represent experimental observations [3]. The activity of iron is kept constant and equal to 1 by equilibration with liquid Fe. Dashed lines represent calculations assuming that the solid and liquid solutions are ideal.

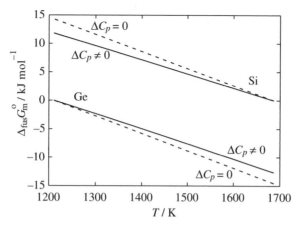

Figure 4.6 Gibbs energy of fusion of Ge and Si. The solid lines represent experimental data
[4] while the broken lines are calculated neglecting the heat capacity difference between
liquid and solid.

Let us now consider the effect of a difference between the heat capacity of pure
liquid i and pure solid i on the enthalpy and entropy of fusion and subsequently on
the phase diagram. This effect is easily taken into consideration by using eqs.
(1.24) and (1.54). $\Delta\mu_i^{o(s\rightarrow l)}$ is now given as

$$\Delta\mu_i^{o(s\rightarrow l)} = \Delta_{fus}H_i^o + \int_{T_{fus,i}}^{T}\Delta C_{p,i}^o\,dT - T\left(\Delta_{fus}S_i^o + \int_{T_{fus,i}}^{T}\frac{\Delta C_{p,i}^o}{T}\,dT\right) \quad (4.25)$$

where $\Delta C_p^o = C_p^{l,o} - C_p^{s,o}$. We will use the system Si–Ge as example. $\Delta\mu_i^{o(s\rightarrow l)}$ for
Si and Ge with (solid lines) and without (dashed lines) taking the heat capacity dif-
ference into consideration are shown in Figure 4.6 [4], while the effect of ΔC_p^o on
the calculated liquidus and solidus lines is shown in Figure 4.4 (dotted lines). In
this particular case, the liquids and solidus lines are shifted some few degrees up
and down in temperature, respectively, and the resulting two-phase field is only
slightly broader than that calculated without taking the heat capacity difference
between the liquid and the solid into consideration. The lack of quantitative agreement between the experimentally observed phase diagram and the calculated ones
shows that significant excess Gibbs energies of mixing are present for one or both
of the solution phases in the Si–Ge system. This indicates what is in general true:
non-ideal contributions to the solution energetics in general have a much larger
effect on the calculated phase diagrams than the heat capacity difference between
the liquid and solid. This is reflected in the phase diagram for the binary system
KCl–NaCl shown in Figure 4.7(a) [5]. This system is characterized by negative
deviation from ideal behaviour in the liquid state and positive deviation from
ideality in the solid state (see the corresponding G–x curves for the solid and liquid
solutions in Figure 4.7(b)). In general a negative excess Gibbs energy of mixing

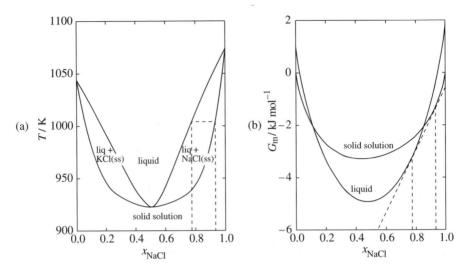

Figure 4.7 (a) Phase diagram of the system KCl–NaCl. (b) Gibbs energy curves for the solid and liquid solutions KCl–NaCl at 1002 K. Thermodynamic data are taken from reference [5].

corresponds to a stabilization of the solution and a deeper curvature of the G–x curve compared to ideal solution behaviour. Correspondingly, a positive deviation from ideal behaviour destabilizes the solution and the G–x curve becomes shallower. These features affect the resulting phase diagrams and the liquidus and solidus lines may show maxima or minima for intermediate compositions, as evident for the KCl–NaCl system in Figure 4.7(a).

Simple eutectic systems

Ag–Cu (Figure 4.1) and many other inorganic systems give rise to simple eutectic phase diagrams. In these systems the two solid phases have such different chemical and physical properties that the solid solubility is limited. The phases may have different structures and hence be represented by different Gibbs energy curves, or they may take the same structure but with a large positive enthalpic interaction giving rise to phase separation or immiscibility at low temperatures. The latter situation, where two solid solutions are miscible at high temperatures, is more usual for alloys and less usual in inorganic material systems. It is, however, a very useful situation for illustrating the link between thermodynamics and phase diagrams, as we will see in the next section on regular solution modelling. It is worth noting that two components that have different properties in the solid state still may form a near-ideal liquid solution.

The system MgO–Y_2O_3 [6] can be used to exemplify the link between Gibbs energy curves and the characteristic features of a simple eutectic phase diagram. The MgO–Y_2O_3 phase diagram is shown in Figure 4.8(a). MgO and Y_2O_3 have different crystal structures and the solid solubility of the two oxides is therefore limited. Furthermore, Y_2O_3 is found in both hexagonal and cubic polymorphs. Gibbs

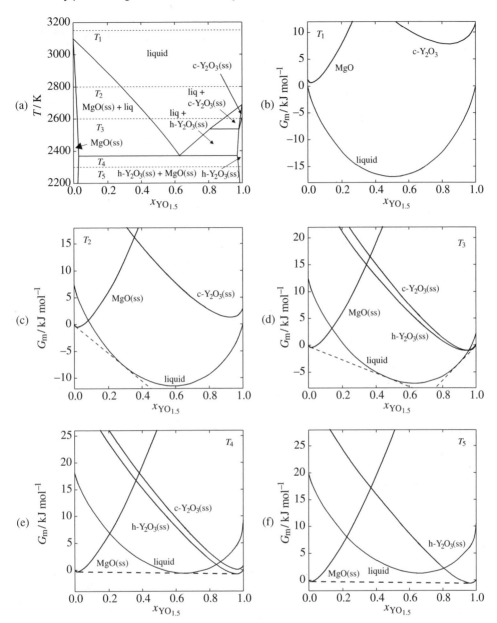

Figure 4.8 (a) Phase diagram of the binary system Y_2O_3–MgO at 1 bar defining the five temperatures for the Gibbs energy curves shown in (b) T_1; (c) T_2; (d) T_3; (e) T_4; (f) T_5. Thermodynamic data are taken from reference [6].

energy representations for the selected temperatures given in the phase diagram are shown in Figures 4.8(b)–(f). At T_1 the liquid is stable at all compositions (Figure 4.8(b)). At T_2 solid MgO(ss) has become stable for MgO-rich compositions (Figure 4.8(c)) and the two-phase field between MgO(ss) and the liquid is established by the common tangent construction. At T_3 solid MgO(ss) is stable for MgO-rich compositions, while the cubic polymorph of Y_2O_3(ss) is stable for Y_2O_3-rich compositions. At intermediate compositions the liquid is stable. The compositions of the liquid coexisting with MgO(ss) and Y_2O_3(ss) are again defined by common tangent constructions. Cubic Y_2O_3(ss) transforms to the hexagonal polymorph at the phase transition temperature given by the horizontal line at $T = 2540$ K. This transition will be further considered below. At the eutectic temperature, T_4, three phases are in equilibrium (see Figure 4.8(e)) according to

$$\text{liquid} = \text{h-}Y_2O_3(\text{ss}) + \text{MgO}(\text{ss}) \tag{4.26}$$

At an even lower temperature, T_5, a sample in equilibrium will consist of the crystalline phase h-Y_2O_3(ss), MgO(ss) or a two-phase mixture of these (see Figure 4.8(f)). The compositions of the two phases in equilibrium are again given by the common tangent construction.

Regular solution modelling

The examples focused on so far have demonstrated that phase diagrams contain valuable information about solution thermodynamics. We will illustrate this further by using the regular solution model, introduced in Section 3.4, to calculate a range of phase diagrams. Although the regular solution model may represent a very crude approximation for a large number of real solutions, it has proven to be very efficient in many respects.

The equilibrium conditions given by eqs. (4.15) and (4.16) can in general be expressed through the activity coefficients. Using a solid–liquid phase equilibrium as an example we obtain

$$\ln\left(\frac{a_A^{\text{liq}}}{a_A^{\text{ss}}}\right) = \ln\left(\frac{x_A^{\text{liq}}\gamma_A^{\text{liq}}}{x_A^{\text{ss}}\gamma_A^{\text{ss}}}\right) = -\frac{\Delta\mu_A^{o(s\to l)}}{RT} \tag{4.27}$$

and

$$\ln\left(\frac{a_B^{\text{liq}}}{a_B^{\text{ss}}}\right) = \ln\left(\frac{x_B^{\text{liq}}\gamma_B^{\text{liq}}}{x_B^{\text{ss}}\gamma_B^{\text{ss}}}\right) = -\frac{\Delta\mu_B^{o(s\to l)}}{RT} \tag{4.28}$$

These expressions can be simplified since the activity coefficient in the particular case of a regular solution can be expressed by the regular solution constant Ω through eqs. (3.84) and (3.85):

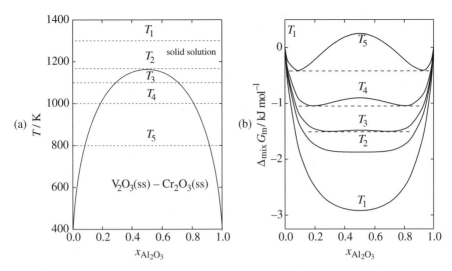

Figure 4.9 (a) Immiscibility gap of the binary solid solution V_2O_3–Cr_2O_3 as described by the regular solution model. (b) Gibbs energy of mixing curve of the solid solution at the temperatures marked in the phase diagram. Thermodynamic data are taken from reference [7].

$$\ln(1 - x_B^{liq}) + \frac{\Omega^{liq}}{RT}(x_B^{liq})^2 - \ln(1 - x_B^{ss}) - \frac{\Omega^{ss}}{RT}(x_B^{ss})^2 = -\frac{\Delta\mu_A^{o(s\rightarrow l)}}{RT} \qquad (4.29)$$

$$\ln x_B^{liq} + \frac{\Omega^{liq}}{RT}(1 - x_B^{liq})^2 - \ln x_B^{ss} - \frac{\Omega^{ss}}{RT}(1 - x_B^{ss})^2 = -\frac{\Delta\mu_B^{o(s\rightarrow l)}}{RT} \qquad (4.30)$$

These two simultaneous equations can then be solved numerically to calculate the solidus and liquidus lines.

It should be remembered that if we assume that a solution phase in a hypothetical A–B system is regular, a positive interaction parameter implies that the different types of atom interact repulsively and that if the temperature is not large enough phase separation will occur. Let us first consider a solid solution only. The immiscibility gap of the solid solution in the binary system V_2O_3–Cr_2O_3 [7] given in Figure 4.9(a) can be described by a regular solution model and thus may be used as an example. The immiscibility gap is here derived by using the positive interaction parameter reported for the solid solution [7]. There is no solubility at absolute zero. As the temperature is raised, the solubility increases with the solubility limits given by the interaction coefficient, Ω, and by temperature. Figure 4.9(b) show the Gibbs energy curves for the solid solution and the common tangent constructions defining the compositions of the coexisting solid solutions at different selected temperatures.

Let us now return to our hypothetical system A–B where we also consider the liquid and where the solid and liquid solutions are both regular (following Pelton and

Thompson [8]). Pure A and B are assumed to melt at 800 and 1000 K with the entropy of fusion of both compounds set to 10 J K^{-1} mol^{-1} (this is the typical entropy of fusion for metals, while semi-metals like Ga, In and Sb may take quite different values – in these three specific cases 18.4, 7.6 and 21.9 J K^{-1} mol^{-1}, respectively). The interaction coefficients of the two solutions have been varied systematically in order to generate the nine different phase diagrams given in Figure 4.10.

In the diagram in the middle (Figure 4.10(e)), both the solid and liquid solutions are ideal. Changing the regular solution constant for the liquid to –15 or +15 kJ mol^{-1}, while keeping the solid solution ideal evidently must affect the phase diagram. In the first case (Figure 4.10(d)), the liquid is stabilized relative to the solid solution. This is reflected in the phase diagram by a shift in the liquidus line to lower temperatures and in this particular case a minimum in the liquidus temperature is present for an intermediate composition. Correspondingly, the positive interaction energy for the liquid destabilizes the liquid relative to the solid solution and the liquidus is in this case shifted to higher temperatures: see Figure 4.10(f). For the composition corresponding to the maximum or minimum in the liquidus/solidus line, the melt has the same composition as the solid. A solid that melts and forms a liquid phase with the same composition as the solid is said to melt **congruently**. Hence the particular composition that corresponds to the maximum or minimum is termed a congruently melting solid solution.

Positive deviations from ideal behaviour for the solid solution give rise to a miscibility gap in the solid state at low temperatures, as evident in Figures 4.10(a)–(c). Combined with an ideal liquid or negative deviation from ideal behaviour in the liquid state, simple eutectic systems result, as exemplified in Figures 4.10(a) and (b). Positive deviation from ideal behaviour in both solutions may result in a phase diagram like that shown in Figure 4.10(c).

Negative deviation from ideal behaviour in the solid state stabilizes the solid solution. $\Omega^{sol} = -10$ kJ mol^{-1}, combined with an ideal liquid or a liquid which shows positive deviation from ideality, gives rise to a maximum in the liquidus temperature for intermediate compositions: see Figures 4.10(h) and (i). Finally, negative and close to equal deviations from ideality in the liquid and solid states produces a phase diagram with a shallow minimum or maximum for the liquidus temperature, as shown in Figure 4.10(g).

The mathematical treatment can be further simplified in one particular case, that corresponding to Figure 4.10(a). As we saw in the previous section, in some binary systems the two terminal solid solution phases have very different physical properties and the solid solubility may be neglected for simplicity. If we assume no solid solubility (i.e. $a_A^{ss} = a_B^{ss} = 1$) and in addition neglect the effect of the heat capacity difference between the solid and liquid components, eqs. (4.29) and (4.30) can be transformed to two equations describing the two liquidus branches:

$$\ln x_A^{liq} + \frac{\Omega^{liq}}{RT}(x_B^{liq})^2 = -\frac{\Delta_{fus}H_A^o}{R}\left(\frac{1}{T} - \frac{1}{T_{fus,A}}\right) \tag{4.31}$$

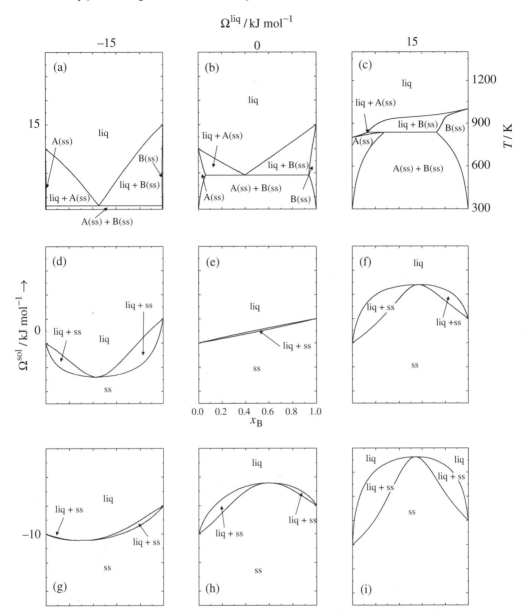

Figure 4.10 (a)–(i) Phase diagrams of the hypothetical binary system A–B consisting of regular solid and liquid solution phases for selected combinations of Ω^{liq} and Ω^{sol}. The entropy of fusion of compounds A and B is 10 J K^{-1} mol^{-1} while the melting temperatures are 800 and 1000 K.

$$\ln x_B^{liq} + \frac{\Omega^{liq}}{RT}(x_A^{liq})^2 = -\frac{\Delta_{fus} H_B^{\,o}}{R}\left(\frac{1}{T} - \frac{1}{T_{fus,B}}\right) \qquad (4.32)$$

The two branches intersect at the eutectic point and the phase diagram thus relies on a single interaction parameter, Ω^{liq}, only.

In the present section we have focused on the calculation of phase diagrams from an existing Gibbs energy model. We can turn this around and derive thermodynamic information from a well-determined phase diagram. Modern computational methods utilize such information to a large extent to derive consistent data sets for complex multi-components systems using both experimental thermodynamic data and phase diagram information. Still, it should be remembered that the phase diagram data does not give absolute values for the Gibbs energy, but rather relative values. A few well-determined experimental data points are, however, enough to 'calibrate the scale', and this allows us to deduce a large amount of thermodynamic data from a phase diagram.

Invariant phase equilibria

In the examples covered so far several invariant reactions defined by zero degrees of freedom have been introduced. For a two-component system at isobaric conditions, $F = 0$ corresponds to three phases in equilibrium. Eutectic equilibria have been present in several of the examples. Also, congruent melting for the solid solutions with composition corresponding to the maxima or minima in the liquidus lines present in Figures 4.10(f) and (d), for example, corresponds to invariant reactions. At the particular composition corresponding to the maximum or minimum, the system can be considered as a single-component system, since the molar ratio $n_A/n_B = x_A/x_B$ remains constant in both the solid and liquid solutions. The molar ratio between the two components is a stoichiometric restriction that reduces the number of components from two to one. A third invariant reaction is the hexagonal to cubic phase transition of pure Y_2O_3 represented in Figure 4.8(a). While pure Y_2O_3 is clearly a single-component system, the solid solubility of the component MgO in h-Y_2O_3(ss) and c-Y_2O_3(ss) increases the number of components by one relative to pure Y_2O_3. Two coexisting condensed phases give one degree of freedom and the solid–solid transition is no longer an invariant reaction according to the phase rule, but occurs over a temperature interval. The two-phase region is, however, narrow and not visible in Figure 4.8(a).

Several other invariant equilibria may take place. A **peritectic reaction** is defined as a reaction between a liquid and a solid phase under the formation of a second solid phase during cooling. Such an invariant reaction is seen in Figure 4.10(c), where the reaction

$$B(\text{ss}) + \text{liquid} \rightarrow A(\text{ss}) \tag{4.33}$$

takes place at $T = 837$ K. It is possible for a solid solution to play the role of the liquid in a similar reaction. Equilibria of this type between three crystalline phases are termed **peritectoid**. Similarly, eutectic reactions, where the liquid is replaced by a solid solution (hence involving only solid phases), are termed **eutectoid**.

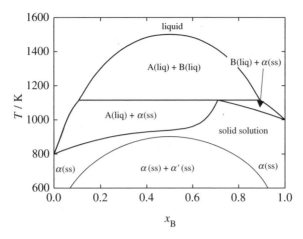

Figure 4.11 Phase diagram of the hypothetical binary system A–B consisting of regular solid and liquid solutions. $\Omega^{\text{liq}} = 20 \text{ kJ mol}^{-1}$ and $\Omega^{\text{sol}} = 15 \text{ kJ mol}^{-1}$. Thermodynamic data for components A and B as in Figure 4.10.

A miscibility gap in the liquid state in general results in another invariant reaction in which a liquid decomposes on cooling to yield a solid phase and a new liquid phase

$$\text{liq}_1 \rightarrow \beta(\text{ss}) + \text{liq}_2 \tag{4.34}$$

in a **monotectic reaction**.

Finally, a phase diagram showing phase separation in both the liquid and solid states is depicted in Figure 4.11. Here a **syntectic reaction** ($\text{liq}_1 + \text{liq}_2 \rightarrow \alpha(\text{ss})$) takes place at 1115 K.

Formation of intermediate phases

The binary systems we have discussed so far have mainly included phases that are solid or liquid solutions of the two components or end members constituting the binary system. **Intermediate phases**, which generally have a chemical composition corresponding to stoichiometric combinations of the end members of the system, are evidently formed in a large number of real systems. Intermediate phases are in most cases formed due to an enthalpic stabilization with respect to the end members. Here the chemical and physical properties of the components are different, and the new intermediate phases are formed due to the more optimal conditions for bonding found for some specific ratios of the components. The stability of a ternary compound like $BaCO_3$ from the binary ones (BaO and $CO_2(g)$) may for example be interpreted in terms of factors related to electron transfer between the two binary oxides; see Chapter 7. Entropy-stabilized intermediate phases are also frequently reported, although they are far less common than enthalpy-stabilized phases. Entropy-stabilized phases are only stable above a certain temperature,

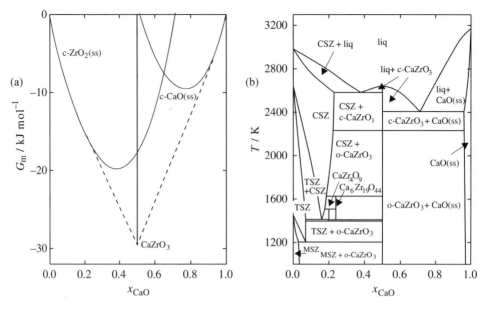

Figure 4.12 (a) Gibbs energy representation of the phases in the system ZrO_2–CaO at 1900 K. $\mu_{CaO}^o = \mu_{ZrO_2}^o = 0$. TSZ is not included for clarity. (b) Calculated phase diagram of the system ZrO_2–CaO. Thermodynamic data are taken from reference [9].

where the entropy contribution to the Gibbs energy exceeds the enthalpy difference between this phase and the phase or phase assemblage stable at lower temperatures. One example is wüstite, $Fe_{1-y}O$, which forms eutectoidally from Fe and Fe_3O_4 at 850 K. The formation of intermediate phases will naturally significantly influence the phase diagram of a given system.

Before we give some examples of phase diagrams involving intermediate phases, it is useful to discuss the compositional variation of the Gibbs energy of such phases. Some intermediate phases may be regarded as **stoichiometric**. Here the **homogeneity range** or the compositional width of the single-phase region is extremely narrow. This reflects the fact that the Gibbs energy curves rise extremely rapidly on each side of the minimum, which is located at exactly the stoichiometric composition of the phase. This is illustrated in Figure 4.12(a) for $CaZrO_3$, which may be seen as an intermediate phase of the system CaO–ZrO_2 [9].[3] The sharpness of the G–x curve implies that $CaZrO_3$ is represented by a vertical line in the CaO–ZrO_2 phase diagram shown in Figure 4.12(b). The fact that the solid solubility or non-stoichiometry of $CaZrO_3$ is negligible is understood by considering the crystal structures of the compounds involved; $CaZrO_3$ takes a perovskite-related crystal structure, while the two end members ZrO_2 and CaO have the fluorite and rock salt structures, respectively. In the perovskite structure there are

---

3   The system Ca–Zr–O is principally a ternary system. However, as long as the oxidation state of Zr and Ca are the same in all phases, the system can be redefined as a two-component system consisting of CaO and ZrO_2.

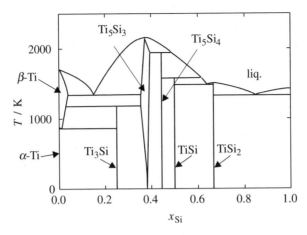

Figure 4.13 Phase diagram of the system Si–Ti [10].

unique lattice sites for both Ca and Zr, and neither interchange of Zr and Ca nor the generation of vacant sites are thermodynamically favourable for $CaZrO_3$.[4] In optimization of thermodynamic properties of stoichiometric compounds, the compositional variation is often neglected and the Gibbs energy is simply given as a function of temperature (and possibly pressure).

$CaZrO_3$ melts congruently, i.e. the coexisting liquid and solid phases have the exact same composition and $CaZrO_3$ may hence be considered as a single-component system. Here two phases present in equilibrium at constant pressure give zero degrees of freedom. The congruent melting of $CaZrO_3$ is therefore an invariant equilibrium. Correspondingly, an **incongruently** melting compound melts under the formation of a new solid phase and a liquid with composition different from the original compound.

The phase diagram of the binary system Si–Ti shown in Figure 4.13 [10] is even more complex. In this system several intermediate phases are formed. Solid solubility is present for the intermediate phase Ti_5Si_3, while the other intermediate phases Ti_3Si, Ti_5Si_4, $TiSi$ and $TiSi_2$ all have very narrow homogeneity ranges. The $G–x$ curve for 'Ti_5Si_3' should therefore display a shallow minimum at the Ti_5Si_3 composition, while the $G–x$ curve for the other intermediate phases should possess sharp minima at the exact composition of the phases. In the thermodynamic description of the Gibbs energy of the non-stoichiometric phase Ti_5Si_3, the variation of the Gibbs energy with composition must be taken into account explicitly in order to calculate the homogeneity range. In this particular case, the Gibbs energy model may contain several different sub-lattices (see Chapter 9) so that the distribution of different species on the relevant sub-lattices is represented.

---

4 It should be mentioned that oxygen vacancies are often formed in the perovskite-type structure ABO_3 in cases where the B atom is a transition metal that readily exists in more than one oxidation state.

The intermediate phases in the system Si–Ti display also a variety of other features. While $TiSi_2$ and Ti_5Si_3 are congruently melting phases, Ti_5Si_4 and $TiSi$ melt incongruently. Finally, Ti_3Si decomposes to β-Ti and Ti_5Si_3 at $T = 1170$ K, in a peritectoid reaction while β-Ti decomposes eutectoidally on cooling forming α-Ti + Ti_3Si at $T = 862$ K.

Melting temperature: depression or elevation?

While until now we have considered relatively simple phase diagrams and the fundamentals of the connection between phase diagrams and thermodynamics, we are here going to consider a somewhat more complex example, but only briefly.

The calculation of phase diagrams is possible if the equilibrium between the different phases can be evaluated as a function of the variables of the system. A relatively simple case is obtained by considering the effect of impurities on the melting temperature of a 'pure metal' following Lupis's treatment of the calculation of the phase boundaries in the vicinity of invariant points [11]. The impurity may be solved both in the solid and liquid phases and the presence of impurities in a 'pure' metal leads to interval freezing. The fusion interval is generally offset towards higher or lower temperatures depending on the nature of the impurity. These alternatives are discussed here with reference to binary phase diagrams exemplifying either a eutectic-type or peritectic-type behaviour in the composition range adjoining the pure metal. The term eutectic-like behaviour is used for all diagrams with the liquidus line sloping downwards, and peritectic-like is used for all those with the liquidus line sloping upwards. Monotectic diagrams, as well as those that include intermediate phases decomposing peritectically below the fusion temperature of the pure metal (e.g. SnSb in the phase diagram of the Sn–Sb system in Figure 4.14) are presently classed in the eutectic-like category. Nearly ideal

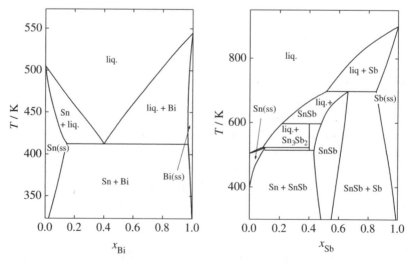

Figure 4.14 Phase diagrams of the systems (a) Sn–Bi [13] and (b) Sn–Sb [14]. Reprinted from [12]. Copyright (1999), with permission from Elsevier.

systems, with complete solid and liquid solubility, are categorized as peritectic if the impurity fuses at a higher temperature than the pure metal.

The equilibrium compositions of an impurity B, x_B^α and x_B^β, in the two phases α and β at a given temperature are given by eqs. (4.27) and (4.28), which may be rewritten as

$$
\ln\left(\frac{1 - x_B^\alpha}{1 - x_B^\beta}\right) + \ln \gamma_A^\alpha - \ln \gamma_A^\beta - \frac{\Delta \mu_A^{o(\alpha \to \beta)}}{RT} = 0 \tag{4.35}
$$

$$
\ln\left(\frac{x_B^\alpha}{x_B^\beta}\right) + \ln \gamma_B^\alpha - \ln \gamma_B^\beta - \frac{\Delta \mu_B^{o(\alpha \to \beta)}}{RT} = 0 \tag{4.36}
$$

Here Raoultian standard states are used for both the pure metal and the impurity. The slope dx_B/dT of the phase boundaries can now be derived by differentiation with respect to temperature. Let $f(x_B)$ denote the left-hand side of eq. (4.35) or (4.36); then (see Lupis, Further reading)

$$
\begin{aligned}
\left(\frac{df(x_B)}{dT}\right)_{eq} &= \left(\frac{\partial f(x_B)}{\partial x_B^\alpha}\right)_T \left(\frac{dx_B^\alpha}{dT}\right) + \left(\frac{\partial f(x_B)}{\partial x_B^\beta}\right)_T \left(\frac{dx_B^\beta}{dT}\right) + \left(\frac{\partial f(x_B)}{\partial T}\right)_{x_B^\alpha, x_B^\beta} \\
&= 0
\end{aligned}
\tag{4.37}
$$

where dx_B^α/dT and dx_B^β/dT are the slopes of the two phase boundaries [11]. Equation 4.37 is identical to zero since $f(x_B)$ is zero along the boundaries. The slope of the phase boundaries can now be evaluated using eqs. (4.35) and (4.36). Further treatment [11] gives the following equations for the slopes of the phase boundaries – using a solid–liquid transition (melting) as an example:

$$
\left(\frac{dx_B^{ss}}{dT}\right)_{x_B^{ss} \to 0} = \frac{\gamma_2^{\infty,liq}(\Delta_{fus} S_A^o / RT_{fus,A})}{\gamma_B^{\infty,liq} - \gamma_B^{\infty,ss} \exp(-\Delta_{fus} G_B^o / RT_{fus,A})} \tag{4.38}
$$

and

$$
\left(\frac{dx_B^{liq}}{dT}\right)_{x_B^{liq} \to 0} = \frac{\gamma_B^{\infty,ss}(\Delta_{fus} S_A^o / RT_{fus,A}) \exp(-\Delta_{fus} G_B^o / RT_{fus,A})}{\gamma_B^{\infty,liq} - \gamma_B^{\infty,ss} \exp(-\Delta_{fus} G_B^o / RT_{fus,A})} \tag{4.39}
$$

Here $\gamma_B^{\infty,\text{liq}}$ and $\gamma_B^{\infty,\text{ss}}$ are the activity coefficients of component B in the liquid and solid solutions at infinite dilution with pure solid and liquid taken as reference states. $\Delta_{\text{fus}} S_A^o$ is the standard molar entropy of fusion of component A at its fusion temperature $T_{\text{fus},A}$ and $\Delta_{\text{fus}} G_B^o$ is the standard molar Gibbs energy of fusion of component B with the same crystal structure as component A at the melting temperature of component A.

The melting temperature depression or enhancement may now be expressed in terms of the melting temperature and the entropy of fusion of component A, the activity coefficients of impurity B in the liquid and solid solutions at infinite dilution, and the total concentration of the impurity B, x_B [12]:

$$\Delta T = T - T_{\text{fus},A} = -\frac{x_B R T_{\text{fus},A}}{\Delta_{\text{fus}} S_A^o}\left[\frac{1}{F-K}\right] \tag{4.40}$$

where F is the fraction of the sample melted at ΔT departure from $T_{\text{fus},A}$, and K is an interaction coefficient. If the liquid standard state is used for the activity coefficient of component B in both solid and liquid solutions:

$$K = \frac{\gamma_B^{\infty,\text{liq}}}{\gamma_B^{\infty,\text{liq}} - \gamma_B^{\infty,\text{ss}}} \tag{4.41}$$

In this case the equations are greatly simplified and the ratio of the slopes of the two phase boundaries at $x_A = 1$ is given by the activity coefficients of B at infinite dilution in the liquid and solid phases [11]:

$$\frac{(dT / dx_B^{\text{ss}})_{x_B^{\text{ss}} \to 0}}{(dT / dx_B^{\text{liq}})_{x_B^{\text{liq}} \to 0}} = \frac{\gamma_B^{\infty,\text{ss}}}{\gamma_B^{\infty,\text{liq}}} \tag{4.42}$$

Eutectic behaviour persists for $0 > K > -\infty$, that is for $\gamma_B^{\infty,\text{liq}} - \gamma_B^{\infty,\text{ss}} < 0$. Peritectic behaviour is obtained for $1 < K < \infty$, that is for $\gamma_B^{\infty,\text{liq}} - \gamma_B^{\infty,\text{ss}} > 0$.

The phase diagrams of the Sn–Bi and Sn–Sb systems are shown in Figure 4.14, and they illustrate the effect of Bi and Sb on the melting temperature of pure Sn. Experimental thermodynamic data for the Sn–Bi system gives $\lambda_{Bi}^{\infty,\text{ss}} = 7.5$ and $\lambda_{Bi}^{\infty,\text{liq}} = 1.3$ [13]; the slope of the Sn solidus is steeper than for the Sn liquidus, $K = -0.2$, and a eutectic type behaviour is expected. For the Sn–Sb system $\lambda_{Sb}^{\infty,\text{ss}} = 0.12$ and $\lambda_{Sb}^{\infty,\text{liq}} = 0.27$ [14], the slope of the Sn liquidus is steeper than that for the Sn solidus, $K = 1.8$, and a peritectic-type behaviour is suggested. Both results are in agreement with the experimental phase diagrams.

Activity coefficients at infinite dilution are in general very important and frequently used in thermodynamic analyses. Examples are analyses of trace element partitioning between solids and melts in geological systems [15] and analyses of the distribution of long-lived chemicals throughout the environment [16].

Minimization of Gibbs energy and heterogeneous phase equilibria

The heterogeneous phase equilibria considered in the preceding sections have implicitly been derived by finding the phase or phase assemblage with the lowest Gibbs energy. Heterogeneous phase equilibria in general are calculated by minimizing the Gibbs energy, and computer software has been available for several decades to perform similar calculations in multi-component systems consisting of any number of components and phases. Generally, the Gibbs energy in a system consisting of the components A, B, C, ..., the stoichiometric phases $\alpha, \beta, \gamma, ...$ and the solution phases 1, 2, 3, ... can be expressed as

$$G = n_\alpha \Delta_f G_\alpha^o + n_\beta \Delta_f G_\beta^o + n_\gamma \Delta_f G_\gamma^o + ...$$
$$+ \sum_{i=1,2,3,...} \sum_{j=A,B,C,...} n_j (\mu_j^o + RT \ln a_j) \tag{4.43}$$

For a given set of constraints (for example temperature, pressure and overall composition), the algorithm identifies the phases present and the relative amounts of these phases, as well as the mole fraction of all the components in all phases. The global minimum evidently must obey the Gibbs phase rule, and not all phases need to be present at the global equilibrium.

A thorough description of strategies and algorithms for minimization of Gibbs energy in multi-component systems is outside the scope of the present text. The monograph by Smith and Missen (see *Further reading*) gives an excellent overview of the topic.

4.2 Multi-component systems

Ternary phase diagrams

For three-component ($C = 3$) or ternary systems the Gibbs phase rule reads $Ph + F = C + 2 = 5$. In the simplest case the components of the system are three elements, but a ternary system may for example also have three oxides or fluorides as components. As a rule of thumb the number of independent components in a system can be determined by the number of elements in the system. If the oxidation state of all elements are equal in all phases, the number of components is reduced by 1. The Gibbs phase rule implies that five phases will coexist in invariant phase equilibria, four in univariant and three in divariant phase equilibria. With only a single phase present $F = 4$, and the equilibrium state of a ternary system can only be represented graphically by reducing the number of intensive variables.

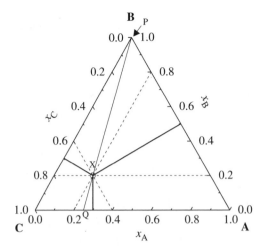

Figure 4.15 Equilateral composition triangle for defining composition in a ternary system A–B–C.

It is sometimes convenient to fix the pressure and decrease the degrees of freedom by one in dealing with condensed phases such as substances with low vapour pressure. The Gibbs phase rule for a ternary system at isobaric conditions is $Ph + F = C + 1 = 4$, and there are four phases present in an invariant equilibrium, three in univariant equilibria and two in divariant phase fields. Finally, three dimensions are needed to describe the stability field for the single phases; e.g. temperature and two compositional terms. It is most convenient to measure composition in terms of mole fractions also for ternary systems. The sum of the mole fractions is unity; thus, in a ternary system A–B–C:

$$x_A + x_B + x_C = 1 \tag{4.44}$$

and there are two independent compositional variables. A representation of composition, symmetrical with respect to all three components, may be obtained from the equilateral composition triangle as shown for the system A–B–C in Figure 4.15. The three corners of the triangle correspond to the three pure components. Along the three edges are found the compositions corresponding to the three binary systems A–B, B–C and A–C. Lines of constant mole fraction of component A are parallel to the B–C edge (exemplified by the broken line for $x_A = 0.2$), while lines of constant mole fraction of B and C are parallel to the A–C and A–B binary edges respectively (exemplified by broken lines for $x_B = 0.2$ and $x_C = 0.6$). The three lines intersect at the point marked X, which thus have the composition $x_A = 0.2$, $x_B = 0.2$, $x_C = 0.6$. Note that the sum of the lengths of the perpendiculars from any composition point to the three edges is constant (using the point X as an example once more, the perpendiculars are given by bold lines in the figure). It is upon this property that the representation is based and the three perpendiculars are measures of the mole fractions of the three components.

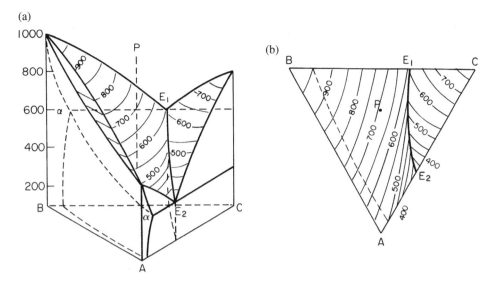

Figure 4.16 (a) Triangular prism phase diagram for a ternary system A–B–C with the equilateral triangular base giving composition. Temperature is given along the vertical axis [17]. (b) Projection of the liquidus surface onto the ternary composition triangle. The bold line is the intersection between the primary crystallization fields of C and the solid solution α. The dashed line represents the extension of the solid solution α. Reprinted with permission of The American Ceramic Society, www.ceramics.org. Copyright [1984]. All rights reserved.

To represent a ternary system at constant pressure completely, the effect of temperature must be incorporated in the phase diagram. Such a ternary temperature–composition phase diagram at constant pressure may be plotted as a three-dimensional space model using a triangular prism. The ternary composition triangle would then form the base while temperature is given along the vertical axes as shown in Figure 4.16(a) for a hypothetical ternary system A–B–C [17]. On the three faces of the prism we find the phase diagrams of the three binary systems A–B, B–C and A–C. In the hypothetical system illustrated in Figure 4.16(a) complete solid solubility is present in the close to ideal binary system A–B (the solid solution phase is denoted α), while the two other binary systems B–C and A–C are eutectic systems with a limited solubility of C in the solid solution phase α.

Recall that the Gibbs phase rule gives $F + Ph = C + 1 = 4$ for a ternary system at constant pressure. Within the prism two liquidus surfaces are shown: one descending from the melting temperature of pure C and the other from the liquidus of α in the binary A–B system. Compositions on the two surfaces corresponds to compositions of the liquid in equilibrium with one of the two solid phases, C or α. For an equilibrium between a solid and the liquid, $Ph = 2$ and thus $F = 2$; the two surfaces are divariant. The two **liquidus surfaces** intersect along the univariant line ($F = 1$) starting from one of the binary eutectics (E_1) and ending in the other (E_2). The intersection of adjoining liquidus surfaces in a ternary phase diagram is generally termed a **common boundary line**. Along the univariant common boundary line three phases (liq, C and α) are in equilibrium.

In the present case there are no ternary invariant equilibria in the system, partly due to the complete solid solubility of the A–B system. In a ternary system composed from three binary eutectic sub-systems, three univariant lines would meet in a ternary eutectic equilibrium:

$$\text{liq.} \leftrightarrow \text{A(ss)} + \text{B(ss)} + \text{C(ss)} \tag{4.45}$$

which is an invariant equilibrium at isobaric conditions.

Phase diagrams based on the triangular prism give an illustrative representation of isobaric ternary systems, but the construction of the diagram is very time-consuming and of less convenience. A more convenient two-dimensional representation of the ternary liquidus surface may be obtained by an orthogonal projection upon the base composition triangle. This is shown for the system A–B–C in Figure 4.16(b) [17]. The lines of constant temperature are called **liquidus isotherms**. In Figure 4.16(b) the bold line shows the common boundary line of the two liquidus surfaces descending from the melting temperature of pure C and from the liquidus of α in the binary A–B system discussed above. Often an arrow is used to indicate the direction of decreasing temperature along univariant lines.

We will now apply Figure 4.16 to find the equilibrium behaviour of a sample with overall composition marked P in the diagrams, when it is cooled from above the liquidus surface to below the solidus temperature for the given overall composition. The point marked P lies within the **primary crystallization field** of α. That is, it lies within the composition region in which α will be the first solid to precipitate during cooling. When cooling the liquid, α will start to precipitate when the liquidus temperature is reached just below 700 °C; see Figure 4.16(b). During further cooling the amount of solid α will increase at the expense of the amount of liquid. It is important to note that the composition of α is not constant during the crystallization.

Two selected **isothermal sections** of the phase diagram that show relevant two-phase equilibria are given in Figure 4.17 [17]. The thin lines illustrate the tielines between the compositions of two phases in equilibrium (α + liq.) or (C + liq). The tieline going through the overall composition point P in Figure 4.17(a) defines the composition of the two conjugate phases, α and liquid, at that particular temperature. During cooling the composition of α is enriched on A and also the composition of the liquid changes. The two phases remain in equilibrium until the liquid reaches the intersection of the primary crystallization fields of α and phase C. At this temperature, the second solid phase, denoted phase C, will start to precipitate in addition to α. On further cooling, the composition of the liquid is defined by the common boundary line from E_1 to E_2 in Figure 4.16, where the liquid is in equilibrium with α and C. The compositions of the three phases in equilibrium are given by a triangle in an isothermal section. This is illustrated for the temperature corresponding to that where the liquid phase disappears, i.e. when P reaches the edge of

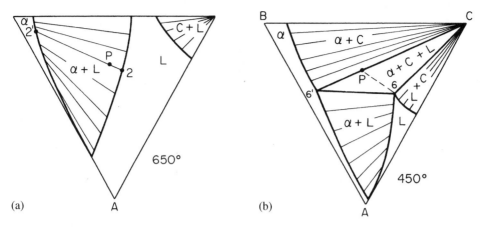

Figure 4.17 Isothermal sections of the ternary phase diagram A–B–C shown in Figure 4.16 at (a) 650 °C and (b) 450 °C [17]. Here L denotes liq. Reprinted with permission of The American Ceramic Society, www.ceramics.org. Copyright [1984]. All rights reserved.

the three-phase triangle in Figure 4.17(b). On further cooling the two solid phases remain in equilibrium. Since the solubility limit of α decreases with decreasing temperature, the relative amounts of the two phases in equilibrium also change.

A sample in the primary crystallization field of phase C will behave differently during crystallization. Here phase C precipitates with composition identical to C (no solid solubility) during cooling keeping the A:B ratio in the melt constant until the melt hits the intersection of the two primary crystallization fields. At this temperature α will start to precipitate together with further C and from this point on the cooling process corresponds to that observed for the sample with overall composition P after this sample reaches the same stage of the crystallization path.

The relative amount of the different phases present at a given equilibrium is given by the lever rule. When the equilibrium involves only two phases, the calculation is the same as for a binary system, as considered earlier. Let us apply the lever rule to a situation where we have started out with a liquid with composition P and the crystallization has taken place until the liquid has reached the composition 2 in Figure 4.17(a). The liquid with composition 2 is here in equilibrium with α with composition 2′. The relative amount of liquid is then given by

$$y_{\text{liq}} = \frac{2'P}{2'2} \tag{4.46}$$

where 2′P denotes the distance between 2′ and P and 2′2 that from 2′ to 2. The amount of solid (α) y_α is given by $1 - y_{\text{liq}}$. With three phases in equilibrium, the relative amounts of the three phases are also given by the lever rule, but its use is slightly more complex. In this case the relative amounts of the three phases are determined in terms of a triangle defined by the composition of the three phases in equilibrium. A line from each of the three corners is drawn through the point representing the overall composition of the sample to the opposite edge. The relative

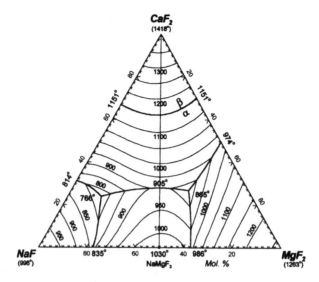

Figure 4.18 Phase diagram for the ternary system NaF–MgF$_2$–CaF$_2$ [18]. Reprinted with permission © 2001 by ASM International and TMS (The Minerals, Metals and Materials Society).

amount of a given phase (represented as a corner in the diagram) is defined as the length of the line from the overall composition to the opposite edge divided by the total length from the corner to the edge. This is illustrated in Figure 4.15. Here the relative amount of phase B for a sample with overall composition X is equal to QX/QP. Similar procedures for the two other components give the relative amount of all three phases in equilibrium.

Let us now consider two real ternary systems to illustrate the complexity of ternary phase diagrams in some detail. While the first is a system in which the solid state situation is rather simple and attention is primarily given to the liquidus surfaces, the solid state is the focus of the second example.

The phase diagram of the ternary system NaF–MgF$_2$–CaF$_2$ is shown in Figure 4.18 [18]. Of the three binary sub-systems NaF–CaF$_2$ and MgF$_2$–CaF$_2$ are simple eutectic systems, while an intermediate phase, NaMgF$_3$, is formed in the third system, NaF–MgF$_2$. The latter can however be divided into two simple eutectic subsystems: NaF–NaMgF$_3$ and NaMgF$_3$–MgF$_2$. The overall system consists of the four solid phases described above, all with their own primary crystallization field and all four phases melt congruently. The borderline between the primary crystallization fields of the phases are shown as bold lines. Two ternary eutectics are shown with the eutectic compositions within the two ternary subsystems NaF–CaF$_2$–NaMgF$_3$ and CaF$_2$–MgF$_2$–NaMgF$_3$. The binary join between CaF$_2$ and NaMgF$_3$ is termed a true **Alkemade line** defined as a join connecting the compositions of two primary phases having a common boundary line. This Alkemade line intersects the boundary curve separating the primary phases CaF$_2$ and NaMgF$_3$. The point of intersection represents the temperature maximum on the

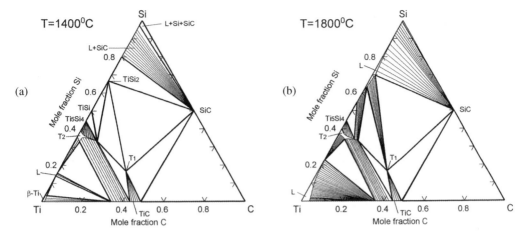

Figure 4.19 Isothermal sections of the ternary system C–Si–Ti at (a) 1400 °C and (b) 1800°C [10]. Reprinted with permission of The American Ceramic Society, www.ceramics.org. Copyright [2000]. All rights reserved.

boundary curve, while the liquidus along the Alkemade line has a minimum at the same composition; hence the intersection represents a saddle point. If the Alkemade line does not intersect the boundary curve, then the maximum on the boundary curve is represented by that end which if prolonged would intersect the Alkemade line. The binary join NaF–MgF$_2$ is not an Alkemade line since these two solid phases are not coexistent. Finally, the α–βCaF$_2$ phase transition at 1151 °C is shown as a bold line in the figure.

For ternary systems with complex phase behaviour in the solid state it is more convenient to use only isothermal sections. This is shown for two temperatures for the ternary system Ti–Si–C in Figure 4.19 [10]. In this system several binary and ternary intermediate phases are stable, and the system is divided into several ternary sub-systems. Tielines for two-phase equilibria are also shown in the two isothermal sections.

Quaternary systems

In a quaternary system, three dimensions are required to represent composition and a fourth dimension is needed for the temperature if the temperature dependence is to be displayed. Since we live in a three-dimensional world this is awkward. The dilemma is partly overcome by constructing a diagram, which is analogous to the plane projection made for ternary systems. This is shown for the system A–B–C–D in Figure 4.20. The phase diagram is a tetrahedron, and the four corners of the tetrahedron correspond to the four components. The four faces of the tetrahedron correspond to the plane projections of the four limiting ternary systems. The Gibbs phase rule for the quaternary system at isobaric conditions is $Ph + F = C + 1 = 5$. With only a single phase present and for a given temperature, three composition variables may be varied (i.e. $F = 3$), and the stability field for each phase is thus a

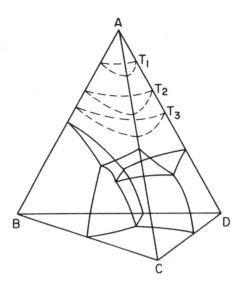

Figure 4.20 Tetrahedron space model for the phase diagram of the quaternary system A–B–C–D. The isotherms T_1, T_2, T_3 are shown for the primary phase volume of component A [17]. Reprinted with permission of The American Ceramic Society, www.ceramics.org. Copyright [1984]. All rights reserved.

volume in the tetrahedron. Two phases are in equilibrium along surfaces, three phases are present in univariant equilibria and finally there are four phases in invariant equilibria.

The addition of further components makes the presentation of the phase diagrams increasingly complex. The principles are general, however, and calculation of a vertical section in a quinternary system like Fe–Cr–Mo–W–C [19], for example, is fairly easily done by the use of large computer programs for calculation of phase diagrams based on thermodynamics.

Ternary reciprocal systems

A ternary reciprocal system is a system containing four components, but where these components are related through a reciprocal reaction. One example is the system LiCl–LiF–KCl–KF. Solid LiCl, LiF, KCl and KF are highly ionic materials and take the rock salt crystal structure, in which the cations and anions are located on separate sub-lattices. It is therefore convenient to introduce ionic fractions (X_i) for each sub-lattice as discussed briefly in Section 3.1. The ionic fractions of the anions and cations are not independent since electron neutrality must be fulfilled:

$$X_{F^-} + X_{Cl^-} = X_{Li^+} + X_{K^+} = 1 \tag{4.47}$$

For this reason, the system is defined by the four neutral components LiCl, LiF, KCl and KF, which in addition can be related by the **reciprocal reaction**

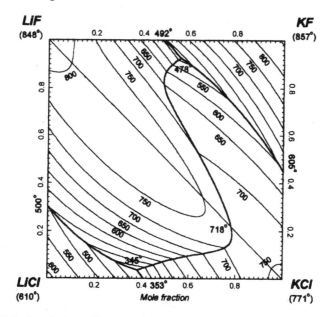

Figure 4.21 Calculated phase diagram of the ternary reciprocal system LiCl–LiF–KCl–KF [20]. Reprinted with permission © 2001 by ASM International and TMS (The Minerals, Metals and Materials Society).

$$LiCl + KF = LiF + KCl \qquad (4.48)$$

The sign of the Gibbs energy of this reciprocal reaction determines which of the two pairs of compounds are coexistent at a given temperature (and pressure). In our specific case, the Gibbs energy of the reciprocal reaction is negative and the products are coexistent phases, while the reactants are not. A reciprocal ternary phase diagram is in general constructed by the combination of two ternary systems that both contain the two coexistent phases. Thus in the present case the ternary phase diagrams of the systems LiF–KCl–LiCl and LiF–KCl–KF are combined. The calculated phase diagram of the ternary reciprocal system considered is shown in Figure 4.21 [20]. Here the sign of the reciprocal reaction is reflected in that the stable diagonal in the system is LiF–KCl and not LiCl–KF.

Both the ternary systems are simple eutectic ones and the composition of the system is represented by the ionic fraction of one of the cations and one of the anions. In Figure 4.21 the ionic fraction of Li^+ is varied along the X-axis, while the ionic fraction of F^- is varied along the Y-axis.

4.3 Predominance diagrams

In the preceding sections the phase diagrams have been represented in terms of composition. Alternatively, the chemical potential of one or more of the components may be used as variables. This gives rise to a range of similar diagrams that

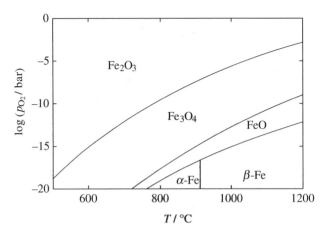

Figure 4.22 $\log p_{O_2}$ versus T predominance phase diagram for the binary system Fe–O. Thermodynamic data are taken from reference [21].

have many applications in materials science. These will here be termed **predominance diagrams**. They are of great importance for understanding materials' stability under hostile conditions (for example hot corrosion) and in planning the synthesis of materials; for example for chemical vapour deposition. In addition to being of great practical value, they also further illustrate the principles of Gibbs energy minimization and the Gibbs phase rule.

In predominance diagrams one or more **base elements** are defined which must be present in all the condensed phases. A predominance diagram for the binary system Fe–O is shown in Figure 4.22. The diagram is divided into areas or domains of stability of the various solid phases of the Fe–O system. In this simple binary case the base element is iron, which is present in all five condensed phases in the system: three oxides and two solid modifications of Fe. The Gibbs phase rule reads $Ph + F = C + 2 = 4$ if the pressure of oxygen is considered as the mechanical potential p. Alternatively, p_{tot} may be considered to be constant e.g. 1 bar. In the latter case, a third component, an inert gas, must be added to the system to maintain the isobaric condition. Thus $Ph + F = C + 1$, which for $C = 3$ again gives $Ph + F = 4$. In conclusion, we may have a maximum of four phases in equilibrium: three condensed phases and a gas phase. A univariant line ($F = 1$) is for this two component system a phase boundary separating the domains of two condensed phases, for instance Fe_3O_4 and Fe_2O_3. These univariant lines are defined by heterogeneous phase equilibria like

$$4/3\,Fe_3O_4(s) + 1/3\,O_2(g) = 2\,Fe_2O_3(s) \tag{4.49}$$

The stability fields for the condensed phases correspond to $F = 2$, which means that both temperature and the partial pressure of O_2 can be varied independently.

In order to derive the phase boundaries in Figure 4.22 we need the Gibbs energy of formation of the oxides. This type of data is conveniently given in an **Ellingham**

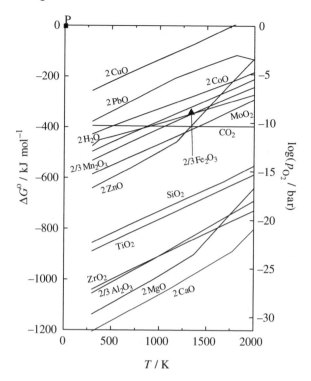

Figure 4.23 Ellingham diagram for various metal oxides. Thermodynamic data are taken from reference [21].

diagram [22,23] where the Gibbs energy 'of formation' of various oxides is plotted versus the temperature, as shown in Figure 4.23. Note that the Gibbs energy of formation is given per mole O_2, which is not in accordance with the definition of the energies of formation given in Chapter 1 and used frequently thereafter.

For a binary oxide like Fe_2O_3 the reaction in question is

$$4/3\ Fe(s) + O_2(g) = 2/3\ Fe_2O_3(s) \tag{4.50}$$

Assuming that the metal and oxygen are in their standard states, the equilibrium constant corresponding to reaction (4.50) is given as

$$K = 1/p_{O_2} = \exp(-\Delta_r G^{\circ}/RT) \tag{4.51}$$

If the oxygen partial pressure is lower than $p_{O_2} = 1/K$ the reactant (in our case Fe) is stable. If it is higher, the product (in our case Fe_2O_3) is formed.

The slopes of the lines in the Ellingham diagram are given by the entropy change of the formation reaction: $-\Delta_r S^{\circ}$. The entropy changes are in general negative due to the consumption of gas molecules with higher entropy and the slopes are thus positive. In the large scale of the plot the lines appear to be linear, suggesting constant entropies of the reactions (or in other words that the difference in heat capacity between the reactants and products is zero). On a different scale the curvature of the Gibbs energy curves is visible. Furthermore, it should be noted that the breaks in the slopes of the curves are due to first-order phase transitions of the metal or the oxide.

The Ellingham diagram contains a lot of useful information. By drawing a line from the point P that intersects the Gibbs energy curve of a particular compound at a temperature of interest, the partial pressure that corresponds to decomposition/formation of the oxide from pure metal and gas at that particular temperature can be derived. The partial pressure of oxygen is obtained by extrapolation to the $\log p_{O_2}$ scale on the right-hand side of the diagram. For example, at 1000 K the partial pressure of oxygen corresponding to equilibrium between Zn and its monoxide, ZnO, is 10^{-26} bar. From the diagram in Figure 4.23 it is evident that the oxides with the more negative Gibbs energies of formation have the highest stability and are harder to reduce to the elemental state.

In materials science, the controlled partial pressure of oxygen is often obtained by using gas mixtures. Here the ratio of the partial pressures of e.g. $H_2(g)$ and $H_2O(g)$ or $CO(g)$ and $CO_2(g)$ are varied to give the desired p_{O_2} at a given temperature. The ratios p_{H_2}/p_{H_2O} and p_{CO}/p_{CO_2} are related to the partial pressure of O_2 by the reactions

$$2H_2O(g) = 2H_2(g) + O_2(g) \tag{4.52}$$

$$2CO_2(g) = 2CO(g) + O_2(g) \tag{4.53}$$

Calculated ratios p_{CO}/p_{CO_2} and p_{H_2}/p_{H_2O} for selected partial pressures of oxygen at a total pressure of 1 bar are given in Figure 4.24.

The equilibrium between a metal and an oxide in a $CO–CO_2$ atmosphere can then be obtained by combining the formation reaction of the oxide with reaction (4.53). As an example the equilibrium between Co, O_2 and CoO combined with reaction (4.53) gives

$$Co(s) + CO_2(g) = CO(g) + CoO(s) \tag{4.54}$$

After finding the partial pressure of oxygen at a given temperature through Figure 4.23, the composition of the gas mixture is obtained from Figure 4.24(a).

Let us now include an additional component to the Fe–O system considered above, for instance S, which is of relevance for oxidation of FeS and for hot corrosion of Fe. In the Fe–S–O system iron sulfides and sulfates must be taken into consideration in addition to the iron oxides and 'pure' iron. The number of components C is now 3 and the Gibbs phase rule reads $Ph + F = C + 2 = 5$, and we may have a maximum of four condensed phases in equilibrium with the gas phase. A two-dimensional illustration of the heterogeneous phase equilibria between the pure condensed phases and the gas phase thus requires that we remove one degree of

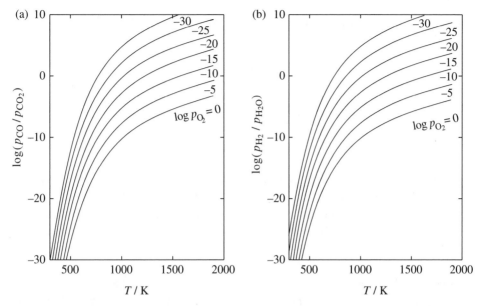

Figure 4.24 Relationship between the partial pressure of oxygen and the composition of CO–CO$_2$ and H$_2$–H$_2$O gas mixtures at 1 bar. (a) p_{CO}/p_{CO_2} versus temperature and (b) p_{H_2}/p_{H_2O} versus temperature at selected partial pressures of oxygen. Thermodynamic data are taken from reference [21].

freedom. This can be done by keeping either the temperature or a chemical potential constant. To exemplify the former choice, the isothermal predominance diagram for the Fe–S–O system at 800 K is shown in Figure 4.25. Here the partial pressures of SO$_2$ and O$_2$ are used as variables. An univariant line ($F = 1$) or phase boundary separates domains of two different condensed phases. For Fe$_2$O$_3$ and FeSO$_4$ this line is defined by the heterogeneous phase equilibrium

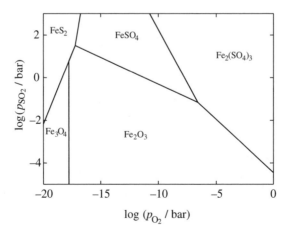

Figure 4.25 log p_{SO_2} versus log p_{O_2} predominance diagram for the system Fe–S–O at 527 °C. Thermodynamic data are taken from reference [21].

$$\tfrac{1}{2}Fe_2O_3(s) + SO_2(g) + \tfrac{1}{4}O_2(g) = FeSO_4(s) \tag{4.55}$$

Since

$$\Delta_r G^\circ = -RT \ln K = -RT \ln\left(\frac{1}{p_{SO_2}\, p_{O_2}^{1/4}} \right) \tag{4.56}$$

the phase boundary is given by the line $\log K = -\log p_{SO_2} - \tfrac{1}{4}\log p_{O_2}$, where K is the equilibrium constant for reaction (4.55). The domain of stability for each of the condensed phases corresponds to two degrees of freedom ($F = 2$), while three condensed phases are in equilibrium with the gas phase at an invariant point ($F = 0$). Three lines corresponding to three different univariant phase equilibria meet at an invariant point.

As indicated above, rather than keeping the temperature constant, we can replace the partial pressure of one of the gas components with the temperature as a variable. Figure 4.26 is a diagram of the Fe–S–O system in which $\ln p_{O_2}$ is plotted versus temperature. Here p_{SO_2} is fixed in order to allow a two-dimensional representation.

There are in principle no restrictions on the number of components in a predominance diagram and examples of four- and five-component systems are shown in Figure 4.27. In Figure 4.27(a) the predominance diagram of the system Si–C–O–N at 1500 K is given as a function of $\log p_{O_2}$ and $\log p_{N_2}$. The system has four components, and Si and C are the two base elements. The amount of Si is assumed to be in excess relative to SiC. At constant temperature, the Gibbs phase rule gives $Ph + F = C + 1 = 5$. Thus at an invariant point four condensed phases are in equilibrium with the gas phase. Two such invariant points are evident in Figure 4.27(a). Three condensed phases are in equilibrium along the lines in the diagram ($F = 1$), whereas

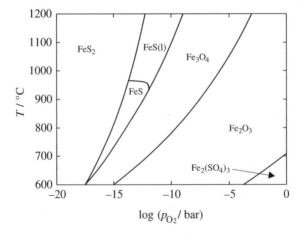

Figure 4.26 T versus $\log p_{O_2}$ predominance diagram for the Fe–S–O system at $p_{SO_2} = 0.1$ bar. Thermodynamic data are taken from reference [21].

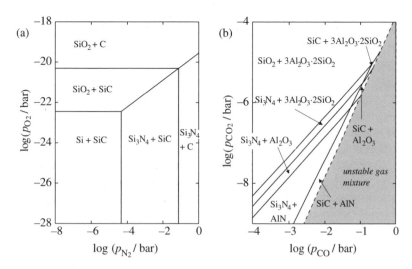

Figure 4.27 (a) Predominance diagram for the system Si–C–O–N at 1500 K, $N_{Si} > N_C$. (b) Predominance diagram for the system Al–Si–C–O–N at 1700 K and $p_{N_2} = 0.5$ bar, $N_{Si} > 0.25 N_{Al}$. Thermodynamic data are taken from reference [21].

two condensed phases are coexistent in the two-dimensional phase fields. Si_3N_4 and SiC, which are often present in ceramic composites, are only thermodynamically coexistent in a narrow $\log p_{N_2}$ range at low partial pressures of O_2. Thermodynamic data for the oxynitride phase Si_2ON_2 are not available, but at 1500 K it has a narrow stability region between SiO_2 and Si_3N_4.

In Figure 4.27(b) the predominance diagram of the five-component system Si–Al–C–O–N is shown as a function of $\log p_{CO_2}$ and $\log p_{CO}$ at a constant partial pressure of N_2 equal to 0.5 bar at 1700 K. The two base elements of the plot are Si and Al ($n_{Si} > 0.25 n_{Al}$). The Gibbs phase rule reads $Ph + F = C + 2 = 7$, which at constant temperature and constant partial pressure of N_2 gives $Ph + F = 5$. The predominance diagram shown in Figure 4.27(b) is therefore analogous to the one shown in Figure 4.27(a), in that the same number of phases is present for a certain degree of freedom. Aluminium nitride is only stable at low partial pressures of CO and CO_2.

Note that the gas mixture in the lower right corner in Figure 4.27(b) is unstable due to the **Boudouard reaction**

$$C(s) + CO_2(g) = 2\,CO(g) \tag{4.57}$$

At 1700 K the equilibrium constant for this reaction is

$$K = 6946 = p_{CO}^2 / p_{CO_2} \tag{4.58}$$

The CO–CO_2 gas mixture is therefore unstable at conditions below the line defined by eq. (4.57) and will here lead to formation of graphite. It may be useful to note

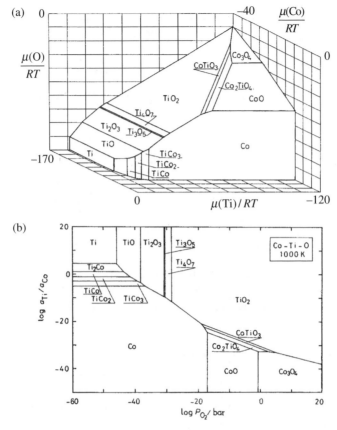

Figure 4.28 (a) Three-dimensional chemical potential diagram for the system Co–Ti–O at 1000 K. (b) Two-dimensional chemical potential diagram at the same conditions [24]. Reprinted with permission of The American Ceramic Society, www.ceramics.org. Copyright [1989]. All rights reserved.

that in many commercial thermodynamic software packages it is not prohibited to calculate phase equilibria for unstable gas compositions, and care should be taken.

In cases where ternary compounds, e.g. oxides, are being investigated, other related types of diagrams may be more efficient. The thermodynamic stability of ternary oxides at constant pressure, for example, is visually well represented in three-dimensional chemical potential diagrams [24]. In Figure 4.28(a) the phase relations in the system Co–Ti–O are plotted as a function of the chemical potential of the three elements. At constant temperature, the Gibbs phase rule gives $Ph + F = 3 + 1 = 4$, and an invariant point corresponds to three condensed phases in equilibrium with the gas phase. The stability field of each single phase is given as a plane, while two phases are in equilibrium along univariant lines. The same phase equilibria may also be represented in two dimensions, as exemplified by Figure 4.28(b). Here the stability of the metallic elements and their binary oxides and double oxides are presented as areas in a $\log(a_A / a_B)$ versus $\log p_{O_2}$ plot (at constant temperature). Complex phase relations for double oxides are in this way visualized in a clear and compact manner.

References

[1] P. R. Subramanian and J. H. Perepezko, Ag–Cu (silver–copper) system, in *Phase Diagrams of Binary Copper Alloys* (P. R. Subramanian, D. J. Chakrabarti and D. E. Laughlin, eds.). Materials Park, OH: ASM International, 1994.

[2] C. Bergman, R. Chastel and R. Castanet, *J. Phase Equilibria* 1992, **13**, 113.

[3] P. Wu, G. Eriksson and A. D. Pelton, *J. Am. Ceram. Soc.* 1993, **76**, 2065.

[4] *NIST-JANAF Thermochemical Tables*, 4th edn (M. W. Chase Jr, ed.). *J. Phys. Chem. Ref. Data Monograph* No. 9, 1998.

[5] A. D. Pelton, A. Gabriel and J. Sangster, *J. Chem. Soc. Faraday Trans.* 1, 1985, **81**, 1167.

[6] Y. Du and Z. P. Jin, *J. Alloys and Comp.* 1991, **176**, L1.

[7] S. S. Kim and T. H. Sanders Jr, *J. Am. Ceram. Soc.* 2001, **84**, 1881.

[8] A. D. Pelton and W. T. Thompson, *Progr. Solid State Chem.* 1975, **10**, 119.

[9] Y. Du, Z. P. Jin and P. Y. Huang, *J. Am. Ceram. Soc.* 1992, **75**, 3040.

[10] Y. Du, J. C. Schuster, H. J. Seifert and F. Aldinger, *J. Am. Ceram. Soc.*, 2000, **83**, 197.

[11] C. H. P. Lupis, *Chemical Thermodynamics of Materials*, New York: North-Holland, 1983, p. 221.

[12] S. Stølen and F. Grønvold, *J. Chem. Thermodyn.* 1999, **31**, 379.

[13] H. Ohtani and K. Ishida, *J. Electron. Mater.* 1994, **23**, 747.

[14] B. Jönsson and J. Ågren, *Mater. Sci. Techn.* 1986, **2**, 913.

[15] N. L. Allan, J. D. Blundy, J. A. Purton, M. Yu. Laurentiev and B. J. Wood, *EMU Notes in Mineralogy*, 2001, **3**(11), 251.

[16] S. I. Sandler, *Pure Appl. Chem.* 1999, **71**, 1167.

[17] C. G. Bergeron and S. H. Risbud, *Introduction to Phase Equilibria in Ceramics*, The American Ceramic Society, Westerville, Ohio, 1984.

[18] P. Chartrand and A. D. Pelton, *Metal. Mater. Trans.* 2001, **32A**, 1385.

[19] B. Jönsson and B. Sundman, *High Temp. Sci.* 1990, **26**, 263.

[20] P. Chartrand and A. D. Pelton, *Metal. Mater. Trans.* 2001, **32A**, 1417.

[21] C. W. Bale, A. D. Pelton and W. T. Thompson, *F*A*C*T* (Facility for the Analysis of Chemical Thermodynamic), Ecole Polytechnique, Montreal, 1999, http://www.crct.polymtl.ca/.

[22] H. T. T. Ellingham, *J. Soc. Chem. Ind.* 1944, **63**, 125.

[23] F. D. Richardson and J. H. E. Jeffes, *J. Iron Steel Inst.* 1948, **160**, 261.

[24] H. Yokokawa, T. Kawada and M. Dokiya, *J. Am. Ceram. Soc.* 1989, **72**, 2104.

Further reading

J. W. Gibbs, *The Collected Works, Volume I Thermodynamics*. Harvard: Yale University Press, 1948.

M. Hillert, *Phase Equilibria, Phase Diagrams and Phase Transformations*. Cambridge: Cambridge University Press, 1998.

L. Kaufman and H. Bernstein, *Computer Calculation of Phase Diagrams*. New York: Academic Press, 1970.

C. H. P. Lupis, *Chemical Thermodynamics of Materials*, New York: North-Holland, 1983.

A. Prince, *Alloy Phase Equilibria*. Amsterdam: Elsevier Science, 1966.

A. D. Pelton, Phase diagrams, in *Physical Metallurgy*, 4th edn (R. W. Cahn and P. Haasen, eds.). Amsterdam: Elsevier Science BV. Volume 1, Chapter 6, 1996.

N. Saunders, and A. P. Miodownik, *CALPHAD: Calculation of Phase Diagrams*. Oxford: Pergamon, Elsevier Science, 1998.

W. R. Smith and R. W. Missen, *Chemical Reaction Equilibrium Analysis: Theory and Algorithms*. New York: Wiley, 1982.

5

Phase stability

When referring to a phase as stable in thermodynamics we usually mean the phase that has the lowest Gibbs or Helmholtz energy at the given conditions. In Section 1.1 the concept of metastability was introduced. Both stable and metastable phases are in **local equilibrium**, but only the thermodynamically stable phase is in **global equilibrium;** a metastable state has higher Gibbs energy than the true equilibrium state. We may also have **unstable phases**, and here, as will be described further below, the nature of the instability is reflected in the second derivative of the Gibbs energy with regard to the thermodynamic potentials defining the system.

Both stable and metastable states are in **internal equilibrium** since they can explore their complete phase space, and the thermodynamic properties are equally well defined for metastable states as for stable states. However, there is a limit to how far we can extend the metastable region with regard to temperature, pressure and composition. If we use temperature as a variable, there is a limit to super-heating a crystal above its melting temperature or cooling a liquid below its freezing temperature. A supercooled liquid will either crystallize or transform to a glass. Glasses are materials out of equilibrium or in other words **non-ergodic states**; glasses cannot explore their complete phase space and some degrees of freedom are frozen in.

An analogous situation is obtained if we consider pressure as variable instead of temperature. Some crystals may exist, as metastable phases, far above the pressure where thermodynamically they should transform to a denser high-pressure polymorph. However, there is a limit for 'superpressurizing' a crystal above its transformation pressure. The phase will either recrystallize (in a non-equilibrium transition) to the more stable phase, or transform to an amorphous state with higher density. To make the analogy with superheating and supercooling complete, high-pressure phases may remain as metastable states when the pressure is released.

Chemical Thermodynamics of Materials by Svein Stølen and Tor Grande
© 2004 John Wiley & Sons, Ltd ISBN 0 471 492320 2

However, at some specific pressure the high-density polymorph becomes mechanically unstable. This low-pressure limit is seldom observed, since it often corresponds to negative pressures. When the mechanical stability limit is reached the phase becomes unstable with regard to density fluctuations, and it will either crystallize to the low-pressure polymorph or transform to an amorphous phase with lower density.

Phases may also become unstable with regard to compositional fluctuations, and the effect of compositional fluctuations on the stability of a solution is considered in Section 5.2. This is a theme of considerable practical interest that is closely connected to spinodal decomposition, a diffusion-free decomposition not hindered by activation energy.

Since the formation of a stable phase may be kinetically hindered, it is of interest to calculate phase diagrams without the presence of a particular phase. This is an exercise easily done using thermodynamic software for phase diagram analysis, but the general effects can be understood based on Gibbs energy rationalizations. Closely related to this topic is the thermal evolution of metastable states with time. The reactivity of a metastable phase is governed by both thermodynamic and kinetic factors. Although the transformation toward equilibrium is irreversible, the direction is given, and the rate of transformation influenced, by the Gibbs energy associated with the transformation. Finally, kinetic factors are also of great importance in many other applications of materials and kinetic demixing, and decomposition of materials in potential gradients are briefly described in the last section of the chapter.

5.1 Supercooling of liquids – superheating of crystals

It is well known that a liquid can be cooled below its equilibrium freezing temperature. The crystallization of the stable crystalline phase is hindered due to an activation barrier caused by the surface energy of the crystal nuclei. In some cases, such as B_2O_3, stable crystals barely form, and the supercooled liquid turns into a glass even at very slow cooling rates. In other cases high cooling rates are needed to produce glasses, notably metallic glasses where cooling rates of the order of 10^6 K s^{-1} might be needed. The supercooled liquid passes through a transition to a glass at the glass transition temperature, T_g, which is typically $\frac{2}{3}$ of the melting temperature, T_{fus}. At this transition some degrees of freedom are frozen in and the sample becomes non-ergodic. Since, the transition is an out-of-equilibrium transition, the properties of the resulting glass depend on its thermal history (see Section 8.5).

The entropy difference between the supercooled liquid and the crystal is given by

$$\Delta_{fus} S_m^o (T) = \Delta_{fus} S_m^o (T_{fus}) + \int_{T_{fus}}^{T} \frac{\Delta C_p^o}{T} dT \tag{5.1}$$

where $\Delta_{fus} S_m^o (T_{fus})$ is the standard entropy of fusion at the melting temperature and ΔC_p^o is the difference between the standard heat capacity of the supercooled liquid and the stable crystalline phase. Many supercooled liquids possess heat capacities that substantially exceed those of the corresponding crystals, as in the case of selenium shown in Figure 5.1(a) [1–3]. The entropy difference between a crystal and the corresponding liquid, which is positive at the fusion temperature, is reduced with decreasing temperature and become zero at some temperature below the equilibrium freezing temperature. Although helium-3 melts exothermally [4], a negative entropy of fusion is in general considered to be a paradox since the entropy of the disordered phase then becomes lower than the entropy of the ordered phase. This argument was first put forward by Kauzman [5] and is often referred to as the **Kauzmann paradox**. By extrapolation of the heat capacity of the super-cooled liquid below its T_g, the temperature at which the entropy of fusion becomes zero can be calculated. This temperature is called the **Kauzmann temperature**, T_K, or the **ideal glass transition temperature**. The entropy of crystalline and liquid selenium is shown as an example in Figure 5.1(b). Here the entropy of the supercooled liquid crosses the entropy of crystalline Se at around $T = 180$ K. Kauzmann proposed that this paradox is avoided through a non-equilibrium transition above the ideal glass transition temperature where a glass is formed. Experiments have confirmed this prediction and all known glass-forming liquids display a glass transition at temperatures above T_K. For our example, Se, the glass transition temperature is approximately 120 K above the Kauzmann temperature.

Unlike supercooling of liquids, superheating of crystalline solids is difficult due to nucleation of the liquid at surfaces. However, by suppressing surface melting, superheating to temperatures well above the equilibrium melting temperature has

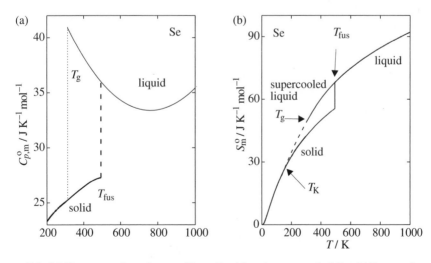

Figure 5.1 (a) Heat capacity of crystalline, liquid and supercooled liquid Se as a function of temperature [1–3]. (b) Entropy of crystalline, liquid and supercooled liquid Se as a function of temperature.

been achieved. As for many other phenomena in physical sciences, superheating is discussed using both kinetic and thermodynamic arguments. Of the early models, those by Lindemann [6] and Born [7] are the most important. Lindemann [6] proposed that bulk melting is caused by a vibrational instability in the crystal lattice when the root mean displacement of the atoms reaches a critical fraction of the distance between them. Somewhat later, Born [7] proposed that a 'rigidity catastrophe' caused by a vanishing elastic modulus determines the melting temperature of the bulk crystal in the absence of surfaces.

The conditions for mechanical instability can be derived from a set of criteria for the stability of equilibrium systems put forward by Gibbs [8]. Considering instability with regard to temperature and pressure, the criteria for stability are

$$\left(\frac{\partial^2 G}{\partial p^2}\right)_T = \left(\frac{\partial V}{\partial p}\right)_T < 0 \qquad (5.2)$$

$$\left(\frac{\partial^2 G}{\partial T^2}\right)_p = -\left(\frac{\partial S}{\partial T}\right)_p < 0 \qquad (5.3)$$

Equation (5.2) requires that the bulk modulus is positive.

$$K_T = \frac{1}{\kappa_T} = -\frac{V}{(\partial V / \partial p)_T} > 0 \qquad (5.4)$$

When this criterion is fulfilled the compound is stable with respect to the spontaneous development of inhomogeneities in the average atomic density. The phase is in other words stable with regard to infinitesimal density fluctuations. Equation (5.3) requires that the heat capacity is positive.

Equation (5.2) also implies that a crystalline solid becomes mechanically unstable when an elastic constant vanishes. Explicitly, for a three-dimensional cubic solid the stability conditions can be expressed in terms of the elastic stiffness coefficients of the substance [9] as

$$C_{11} + 2C_{12} > 0 \qquad (5.5)$$

$$C_{44} > 0 \qquad (5.6)$$

$$C_{11} - C_{12} > 0 \qquad (5.7)$$

The complexity of the stability conditions increases the lower the symmetry of the crystal. For an isotropic condensed phase, such as a liquid or fluid the criteria can be simplified. Here, $C_{11} - C_{12} = 2C_{44}$ and the stability conditions reduce to

$$3K_T = 2C_{44} + 3C_{12} > 0 \tag{5.8}$$

$$C_{44} > 0 \tag{5.9}$$

where K_T is the bulk modulus and C_{44} is the shear modulus.

The temperature dependences of the isothermal elastic moduli of aluminium are given in Figure 5.2 [10]. Here the dashed lines represent extrapolations for $T > T_{fus}$. Tallon and Wolfenden found that the shear modulus of Al would vanish at $T = 1.67T_{fus}$ and interpreted this as the upper limit for the onset of instability of metastable superheated aluminium [10]. Experimental observations of the extent of superheating typically give $1.1T_{fus}$ as the maximum temperature where a crystalline metallic element can be retained as a metastable state [11]. This is considerably lower than the instability limits predicted from the thermodynamic arguments above.

In recent years other types of thermodynamic arguments for the upper limit for superheating a crystal have also been proposed. One argument is based on the fact that the heat capacity of the solid increases rapidly with temperature above the melting temperature due to vacancy formation. Inspired by the Kauzmann paradox, Fecht and Johnson [12] argued that the upper limit for superheating is defined by the isoentropic temperature, at which the entropies for a superheated crystal and the corresponding liquid become equal. The argument is thus the superheating equivalent of the Kauzmann paradox. The temperature corresponding to this 'entropy catastrophe' is again calculated using eq. (5.1) except that we now have to extrapolate the heat capacity of the solid above the melting temperature. The resulting entropies for liquid and solid aluminium [12] are shown in Figure 5.3. Here, the temperature at which the entropy of supercooled liquid aluminium reaches that of crystalline aluminium, the ideal glass transition temperature, is $0.24T_{fus}$. Correspondingly, the temperature at which the entropy of the crystal on

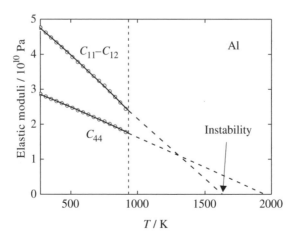

Figure 5.2 Temperature dependence of the isothermal elastic stiffness constants of aluminium [10].

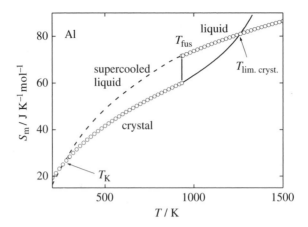

Figure 5.3 Entropy of liquid and crystalline aluminium in stable, metastable and unstable temperature regions [12]. The temperatures where the entropy of liquid and crystalline aluminium are equal are denoted T_K and $T_{\text{lim cryst}}$, respectively.

heating again becomes as large as that of the liquid is $1.38T_{\text{fus}}$. The latter is far lower than $1.67T_{\text{fus}}$, obtained from the Born stability criteria [10]. The vacancy concentration at the stability limit is approximately 10% and the volume effect of this amount of vacancies corresponds to the volume change of melting at T_{fus}. It has therefore been argued that the isentropic temperature in general may coincide with the temperature at which the volume of the superheated crystal becomes equal to the volume of the liquid.

Finally, Tallon [13] has suggested another instability point where the entropy of the superheated crystal becomes equal to that for a superheated diffusionless liquid (a glass) rather than that of the liquid. Since the glass has lower entropy than the liquid, this instability temperature is lower than that predicted by Fetch and Johnson [12].

5.2 Fluctuations and instability

The driving force for chemical reactions: definition of affinity

The equilibrium composition of a reaction mixture is the composition that corresponds to a minimum in the Gibbs energy. Let us consider the simple chemical equilibrium A $\leftrightarrow$ B, where A and B could for example be two different modifications of a molecule. The changes in the mole numbers dn_A and dn_B are related by the stoichiometry of the reaction. We can express this relation as $-dn_A = dn_B = d\xi$ where the parameter $d\xi$ represents an infinitesimal change in the extent of the reaction and expresses the changes in mole numbers due to the chemical reaction. The rate of reaction is the rate at which the extent of the reaction changes with time. The driving force for a chemical reaction is called affinity and is defined as the slope of the Gibbs energy versus the extent of reaction, ξ. The differential of the Gibbs

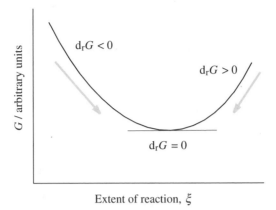

Figure 5.4 Gibbs energy as a function of the extent of reaction.

energy at constant T and p is (taking into consideration the Gibbs–Duhem equation, eq. (1.93))

$$d_r G = \mu_A \, dn_A + \mu_B dn_B = -\mu_A \, d\xi + \mu_B d\xi = (\mu_B - \mu_A) d\xi \qquad (5.10)$$

The **affinity** of the reaction, A_k, is defined as the difference between the chemical potential of the reactant and the product at a particular composition of the reaction mixture:

$$A_k = \mu_A - \mu_B \qquad (5.11)$$

Since the chemical potential varies with the fraction of the two molecules, the slope of the Gibbs energy against extent of reaction changes as the reaction proceeds. The reaction A $\rightarrow$ B is spontaneous when $\mu_A > \mu_B$, whereas the reverse reaction, B $\rightarrow$ A, is spontaneous when $\mu_B > \mu_A$. The different situations are illustrated in Figure 5.4. The slope of the Gibbs energy versus the extent of the reaction is zero when the reaction has reached equilibrium. At this point we have

$$\mu_A = \mu_B \qquad (5.12)$$

and the equilibrium criteria for a system at constant temperature and pressure given by eq. (1.84) are thus fulfilled.

Stability with regard to infinitesimal fluctuations

In general, the first derivative of the Gibbs energy is sufficient to determine the conditions of equilibrium. To examine the stability of a chemical equilibrium, such as the one described above, higher order derivatives of G are needed. We will see in the following that the Gibbs energy versus the potential variable must be upwards convex for a stable equilibrium. Unstable equilibria, on the other hand, are

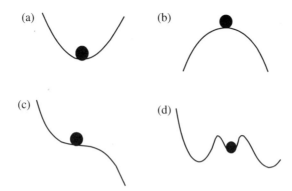

Figure 5.5 Ball in a gravitational field; illustration of (a) stable, (b) unstable, (c) spinodal and (d) metastable equilibria.

characterized by a downward convex Gibbs energy versus the potential variable. This is illustrated in Figure 5.5 where we have used a ball in a gravitational field as an example. In example (a) the ball is in a stable equilibrium and is stable against fluctuations in both directions. In (b) the ball is unstable towards fluctuations in both directions and it follows that this is an unstable equilibrium. In (c) the ball is stable for fluctuations to the left but unstable for fluctuations to the right. This is defined as a **spinodal equilibrium**. Finally, in (d) the ball is located in a locally stable but globally metastable equilibrium.

Let us assume the existence of a Taylor series for the Gibbs energy at the equilibrium point. This implies that the Gibbs energy and all its derivatives vary continuously at this point. The Taylor series is given as

$$(-A_k)_{\zeta \to 0} = \left(\frac{\partial G}{\partial \zeta} \right)_{\zeta=0} \zeta + \frac{1}{2!} \left(\frac{\partial^2 G}{\partial \zeta^2} \right)_{\zeta=0} \zeta^2 + \dots + \frac{1}{n!} \left(\frac{\partial^n G}{\partial \zeta^n} \right)_{\zeta=0} \zeta^n \quad (5.13)$$

where ζ is an infinitesimal fluctuation. In principle, the fluctuation could be a fluctuation in concentration, temperature or pressure. Equilibrium is identified when the affinity is zero, which means that the first derivative $(\partial G / \partial \zeta)_{\zeta=0} = 0$. If $(\partial^2 G / \partial \zeta^2)_{\zeta=0} \neq 0$ the sign of $(-A_k)_{\zeta \to 0}$ is the sign of $(\partial^2 G / \partial \zeta^2)_{\zeta=0} \zeta^2$. Since ζ^2 is always positive, the equilibrium is stable if

$$\left(\frac{\partial^2 G}{\partial \zeta^2} \right)_{\zeta=0} > 0 \quad (5.14)$$

The equilibrium is unstable if this second derivative is negative. If

$$\left(\frac{\partial^2 G}{\partial \zeta^2} \right)_{\zeta=0} = 0 \quad (5.15)$$

Table 5.1 The criteria for stability of solutions with regard to infinitesimal fluctuations.

Criteria	Equilibrium state	Comment
$\left(\dfrac{\partial^2 G}{\partial \zeta^2}\right)_{\zeta=0} > 0$	Stable or metastable	
$\left(\dfrac{\partial^2 G}{\partial \zeta^2}\right)_{\zeta=0} < 0$	Unstable	
$\left(\dfrac{\partial^2 G}{\partial \zeta^2}\right)_{\zeta=0} = 0$	Spinodal point	Separates a stable region from an unstable region.

we have to examine higher order derivatives. The affinity is then given as

$$(-A_k)_{\zeta \to 0} = \frac{1}{3!}\left(\frac{\partial^3 G}{\partial \zeta^3}\right)_{\zeta=0} \zeta^3 + \frac{1}{4!}\left(\frac{\partial^4 G}{\partial \zeta^4}\right)_{\zeta=0} \zeta^4 + \ldots \qquad (5.16)$$

If $(\partial^3 G / \partial \xi^3)_{\xi=0} \neq 0$, it is possible to choose the sign of ζ so that $(-A_k)_{\zeta \to 0}$ is negative. Hence the equilibrium is unstable since small compositional fluctuations can have any sign. The stability criteria are summarized in Table 5.1.

The correspondence with a ball in a gravitational field illustrated in Figure 5.5 is evident. The stable and unstable regions are defined as the regions where the second derivative of the Gibbs energy with regard to ζ are positive and negative, respectively, and correspond to upward and downward convexity of the Gibbs energy with respect to ζ. When the second derivative is zero we have a situation corresponding to the inflection point which separates the regions of instability and stability with regard to small fluctuations. This inflection point represents a spinodal equilibrium and is called a **spinodal point**.

Compositional fluctuations and instability

The criterion given in Table 5.1 may be used to consider the stability of different compositions of a liquid or solid solution by looking at the variation of the Gibbs energy with composition. As discussed in Section 4.1, the miscibility gap of a solution is usually due to a positive enthalpy of mixing balanced by the entropy increment obtained when a disordered solution is formed. The enthalpy is here a segregation force, whereas the entropy is an opposing mixing force. At low temperatures the $T\Delta S$ term of the Gibbs energy is less important than $\Delta_{mix}H$, and segregation occurs. At high temperatures, the entropy gained by distributing different species on a given lattice is large and complete solubility is obtained. If we start from the absolute zero, the miscibility gap decreases with increasing temperature until a certain temperature, called the **critical temperature**, T_c. Above the critical temperature complete solubility in the liquid or solid state is obtained.

Let us initially consider the Gibbs energy of the solid solution Al_2O_3–Cr_2O_3 at 1200 K [14] given in Figure 5.6. The solution is partly miscible and the composition of the two coexisting solutions α and β is given by the equilibrium condition

$$\mu^{\alpha}_{Al_2O_3} = \mu^{\beta}_{Al_2O_3} \quad \text{and} \quad \mu^{\alpha}_{Cr_2O_3} = \mu^{\beta}_{Cr_2O_3} \tag{5.17}$$

The phase boundaries at this specific temperature are given by the points x_1 and x_2 in Figure 5.6(a), defined by the common tangent (the dotted line). Three different situations for the variation of the Gibbs energy of the solution with composition are marked by the points A, B and C/C'. A solution with composition A is stable with regard to fluctuations in composition, while one with composition B is unstable. At the two spinodal points (C and C') the second derivative of the Gibbs energy with regard to composition changes sign (see Figure 5.6(b)) and in general

$$\left(\frac{\partial^2 G_m}{\partial x_B^2} \right)_{x_B = x_B^s} = 0 \tag{5.18}$$

where x_B^s is the composition at the spinodal point. The samples with composition C and C' are stable with regard to fluctuations in composition in one direction, but not with regard to fluctuations in composition in the other direction. The compositions of the two spinodal points vary with temperature and approach each other as the temperature is raised, and the two points finally merge at the critical temperature, where

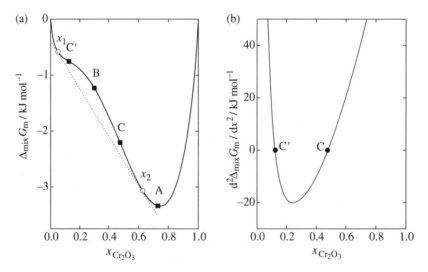

Figure 5.6 (a) Gibbs energy of mixing of the system Al_2O_3–Cr_2O_3 at 1200 K [14]. A, B and C/C' correspond to stable, unstable and spinodal points. The points x_1 and x_2 give the compositions of the two coexisting solutions. (b) The compositional dependence of $(d^2\Delta_{mix}G_m / d^2 x)$.

$$\left(\frac{\partial^3 G_{\mathrm{m}}}{\partial x_{\mathrm{B}}^3}\right)_{x_{\mathrm{B}}=x_{\mathrm{B}}^{\mathrm{c}}} = 0 \tag{5.19}$$

Here $x_{\mathrm{B}}^{\mathrm{c}}$ is the composition at the critical point. Let us now calculate the immiscibility gap and the spinodal line for a regular solution A–B:

$$G_{\mathrm{m}} = x_{\mathrm{A}}\mu_{\mathrm{A}}^{\mathrm{o}} + x_{\mathrm{B}}\mu_{\mathrm{B}}^{\mathrm{o}} + RT(x_{\mathrm{A}}\ln x_{\mathrm{A}} + x_{\mathrm{B}}\ln x_{\mathrm{B}}) + \Omega x_{\mathrm{A}} x_{\mathrm{B}} \tag{5.20}$$

Let us for simplicity assume that $\mu_{\mathrm{B}}^{\mathrm{o}} = \mu_{\mathrm{A}}^{\mathrm{o}} = 0$, for which case the immiscibility gap is given by

$$\frac{\partial G_{\mathrm{m}}}{\partial x_{\mathrm{B}}} = RT(\ln x_{\mathrm{B}} - \ln x_{\mathrm{A}}) + \Omega(1 - 2x_{\mathrm{B}}) = 0 \tag{5.21}$$

An analytical expression that defines the compositions of the two coexisting solutions is easily derived:

$$\ln \frac{x_{\mathrm{B}}}{x_{\mathrm{A}}} = -\frac{\Omega}{RT}(1 - 2x_{\mathrm{B}}) \tag{5.22}$$

The spinodal line is correspondingly given using eq. (5.18) as

$$\frac{\partial^2 G_{\mathrm{m}}}{\partial x_{\mathrm{B}}^2} = RT\left(\frac{1}{x_{\mathrm{A}}} + \frac{1}{x_{\mathrm{B}}}\right) - 2\Omega = 0 \tag{5.23}$$

and it follows that the spinodal line for a regular solution is a parabola:

$$x_{\mathrm{A}} x_{\mathrm{B}} = (1 - x_{\mathrm{B}})x_{\mathrm{B}} = \frac{RT}{2\Omega} \tag{5.24}$$

At the critical point $x_{\mathrm{A}} = x_{\mathrm{B}} = 0.5$ and the critical temperature and the interaction coefficient are related through

$$T_{\mathrm{c}} = \Omega / 2R \tag{5.25}$$

Both the **binodal line**, defining the immiscibility gap, and the spinodal line are for a regular solution symmetrical about $x_{\mathrm{A}} = x_{\mathrm{B}} = 0.5$. This is shown in Figure 5.7(a), where theoretical predictions of the miscibility gaps in selected semiconductor systems are given [15].

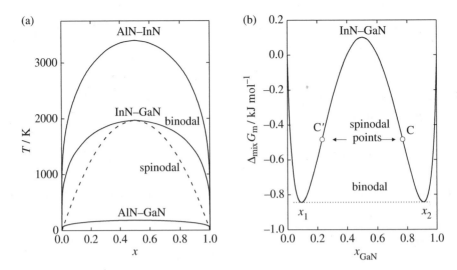

Figure 5.7 (a) Theoretical predictions of the unstable regions (miscibility gap) of the solid solutions in the systems AlN–GaN, InN–GaN and AlN–InN [15]. For the system InN–GaN both the phase boundary (binodal) and spinodal lines are shown. (b) Gibbs energy of mixing for the solid solution InN–GaN at 1400 K.

Physically, the spinodal lines separate two distinct regions in a phase diagram. Between the binodal and the spinodal lines (i.e. the compositional regions between x_1 and C′ and between C and x_2 in Figure 5.7(b)) the solution is in a metastable state. Between the spinodal points or in the spinodal region (i.e. the compositional regions between C′ and C in Figure 5.7(b)), the solution is unstable. Samples with overall composition within the metastable regions and within spinodal regions are expected to behave differently on cooling from high temperatures where complete solid solubility prevails.

Let us consider a portion of the Gibbs energy curve for a composition within and outside the spinodal region in some detail. In Figure 5.8(a), we consider the Gibbs

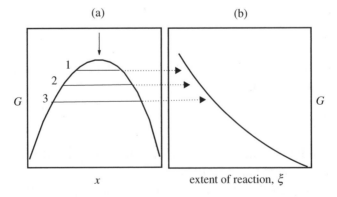

Figure 5.8 (a) Gibbs energy curve for an unstable system. (b) Gibbs energy of the unstable system as a function of the extent of reaction during spinodal decomposition of a sample with composition indicated by the arrow in figure (a).

energy in the vicinity of a composition that is well within the spinodal region. Assume that a sample with the overall composition marked with an arrow is cooled to below the critical temperature. Below the critical temperature, small fluctuations in composition lead to a continuous decrease in the Gibbs energy, as illustrated in Figure 5.8(b), and the separation of the original homogeneous solution occurs without nucleation of a new phase. Instead, two different regions with different composition emerge. As the system approaches equilibrium the difference in the composition increases and approaches the difference between the two equilibrium compositions. The decomposition occurs without any thermal activation and with continuous changes in composition. The decomposition of a homogeneous solution resulting from infinitesimal concentration fluctuations is called **spinodal decomposition**.

The decomposition of a solution with composition outside the spinodal region but within the metastable region can be analyzed in a similar way. Let us assume that a sample with composition in this region is cooled to low temperatures. Small fluctuations in composition now initially lead to an increase in the Gibbs energy and the separation of the original homogeneous solution must occur by nucleation of a new phase. The formation of this phase is thermally activated. Two solutions with different composition appear, but in this case the composition of the nucleated phase is well defined at all times and only the relative amount of the two phases varies with time.

Both decomposition mechanisms are used actively in the design of materials. The two important commercial glasses in the system $Na_2B_8O_{13}$–SiO_2, Pyrex and Vycor, represent striking examples. Figure 5.9 shows the relevant phase diagram for the system $Na_2B_8O_{13}$–SiO_2 [16]. Here the liquid shows immiscibility below the liquidus temperature of the system. On supercooling liquids with compositions given by the arrows the immiscibility dome at some point is reached. The supercooled liquids are here still in internal equilibrium since the glass transition

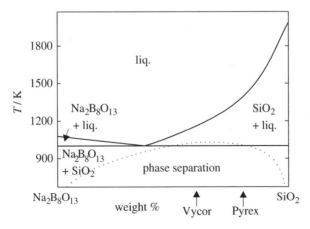

Figure 5.9 Phase diagram showing liquid immiscibility in the $Na_2B_8O_{13}$–SiO_2 system below the liquidus [16].

temperatures are even lower. On cooling the liquid further partial immiscibility is obtained and the composition of the two liquids in metastable equilibrium at different temperatures are given by the dotted line in the figure.[1]

While Vycor has a composition within the spinodal region of the system, Pyrex lies outside the spinodal region but within the binodal region. In Vycor, the spinodal mechanism secures a complete connectivity of the two metastable liquid/glass phases throughout the matrix. The $Na_2B_8O_{13}$-rich phase can be etched out with acid and a 'close to pure' silica glass is produced. Since the original $Na_2B_8O_{13}$–rich melt can be homogenized at relatively low temperatures compared with pure SiO_2 this secures a rather low production cost.

For Pyrex the composition of the melt is outside the spinodal region and the $Na_2B_8O_{13}$ phase is formed by nucleation and growth. Complete connectivity is not obtained and spherical particles of an $Na_2B_8O_{13}$-rich melt forms a minority phase within the SiO_2-rich matrix. A glass with low softening and melting temperatures is thus produced.

The van der Waals theory of liquid–gas transitions

In Section 2.2 we introduced the van der Waals equation of state for a gas. This model, which provides one of the earliest explanations of critical phenomena, is also very suited for a qualitative explanation of the limits of mechanical stability of a homogeneous liquid. Following Stanley [17], we will apply the van der Waals equation of state to illustrate the limits of the stability of a liquid and a gas below the critical point.

The van der Waals equation of state for one mole of gas is expressed in terms of the critical pressure, temperature and volume by eq. (2.40) as

$$\left(\bar{p} + \frac{3}{\bar{V}^2} \right)(3\bar{V} - 1) = 8\bar{T} \tag{5.26}$$

where $\bar{p} = p/p_c$, $\bar{T} = T/T_c$ and $\bar{V} = V/V_c$.

For a simple gas comprised of spherical molecules, eq. (5.26) is fairly well obeyed at low density or high temperature. At higher density the equation of state become less accurate in describing real systems. Some p–V isotherms of the van der Waals equation of state for H_2O are shown in Figure 5.10(a). At high temperatures the volume of the gas falls near asymptotically with increasing pressure, as expected from the ideal gas law. At a particular temperature, corresponding to the temperature at the critical point, the first derivative $(\partial p / \partial V)_T = 0$ becomes zero for a given value of $\bar{V}$. This is an inflection point where not only $(\partial^2 G/\partial p^2)_T = 0$,

---

1 In practice the glasses are made by first quenching the liquid. The phase separation takes place by reheating and annealing the glass between the glass transition and the critical temperature.

but also $(\partial^3 G/\partial p^3)_T = 0$. This critical point is analogous to the critical point for the miscibility gap of solutions that phase separate into two phases of different composition. In the present case, the supercritical fluid separates into a liquid and a gas with different density.

For sub-critical isotherms $(T < T_c)$, the parts of the isotherm where $(\partial p/\partial V)_T < 0$ become unphysical, since this implies that the thermodynamic system has negative compressibility. At the particular reduced volumes where $(\partial p/\partial V)_T = 0$, $(\partial^2 G/\partial p^2)_T = 0$ and we have spinodal points that correspond to those discussed for solutions in the previous section. This breakdown of the van der Waals equation of state can be bypassed by allowing the system to become heterogeneous at equilibrium. The two phases formed at $T < T_c$, liquid and gaseous H_2O, must have the same temperature and pressure in order to obey the equilibrium criteria.

The variation of the Helmholtz energy of the van der Waals equation of state for H_2O with volume can be calculated by

$$\left(\frac{\partial A}{\partial V}\right)_T = -p \tag{5.27}$$

Integration gives

$$A(V) = -\int_{V_i}^{V_f} p\,dV \tag{5.28}$$

The Helmholtz energy curves for three different isotherms are given in Figure 5.10(b). The volume of the two phases in equilibrium at a given temperature can be derived in at least two different ways. In the first approach, we can apply a common tangent construction to the Helmholtz energy curve as shown for $\overline{T} = 0.8$ in Figure 5.10(b). The dashed line is tangent to the Helmholtz energy curve at the 'high-density' point A and at the 'low-density' point B. These volumes define the volumes of the two phases (liquid and gas) in equilibrium. At a specific temperature, the pressure is implicitly defined by these two volumes. Alternatively, the equilibrium pressure for the coexistence of liquid and gas can be determined using what is called the **Maxwell equal-area construction**. In practice, this is done by adjusting the horizontal line in Figure 5.10(a), so that the areas marked C and D become equal.

The equilibrium pressure for which liquid and gas are in equilibrium is given as a function of temperature for the van der Waals equation of state for H_2O in Figure 5.11(a). The corresponding equilibrium densities of the coexisting liquid and gas are given in Figure 5.11(b). In these two figures the spinodal lines defining the mechanical stability limits of the liquid and gas phases are shown as dotted curves. The stable regions of the potential space for the liquid and gas phases are separated by the equilibrium line for the heterogeneous phase equilibrium, while the

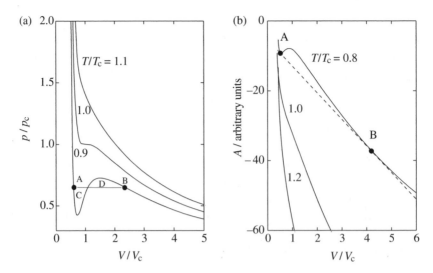

Figure 5.10 (a) p-V isotherms of the van der Waals equation of state for H_2O at $T/T_c = 1.1$, 1.0 and 0.9. (b) The Helmholz energy $A(T,V)$ for the van der Waals equation of state of H_2O as a function of V/V_c at $T/T_c = 1.2$, 1.0 and 0.8.

spinodal lines define the maximum extension of the metastable regions for the liquid and gas phase, respectively. For example, the liquid can be extended to negative pressure (liquid under tension) if nucleation of gas can be avoided. However, below the spinodal line the liquid becomes mechanically unstable.

The T–ρ plot shown in Figure 5.11(b) resembles the T–x plot of a binary solution. The equilibrium between the two phases is, as we have seen above, given by a similar set of equilibrium conditions in both cases. Within the spinodal regions of the

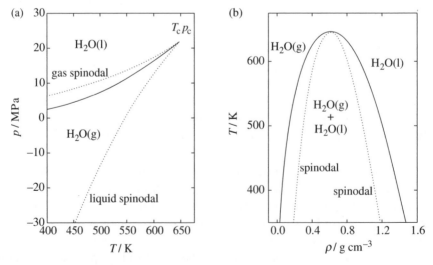

Figure 5.11 The p-T (a) and the T-ρ (b) phase diagrams of H_2O calculated using the van der Waals equation of state.

potential space, phase separation in the binary solution system is driven by concentration fluctuations, while the formation of the stable state in the single-component system is driven by density fluctuations.

Pressure-induced amorphization and mechanical instability

Pressure-induced amorphization of solids has received considerable attention recently in physical and material sciences, although the first reports of the phenomenon appeared in 1963 in the geophysical literature (actually amorphization on reducing the pressure [18]). During isothermal or near isothermal compression, some solids, instead of undergoing an equilibrium transition to a more stable high-pressure polymorph, become amorphous. This is known as **pressure-induced amorphization**. In some systems the transition is sharp and mimics a first-order phase transition, and a discontinuous drop in the volume of the substance is observed. Occasionally it is strictly not an amorphous phase that is formed, but rather a highly disordered denser nano-crystalline solid. Here we are concerned with the situation where a true amorphous solid is formed.

The report of the pressure-induced amorphization and amorphous–amorphous transition of porous Si captures most of the current understanding of this phenomenon [19]. In Figure 5.12(a) the p, T phase diagram of Si is shown. Three phases are present: the four-coordinated low-pressure modification (diamond-type) and the six-coordinated high-pressure modification (β–Sn-type) of crystalline Si, as well as liquid Si. The behaviour of crystalline Si under compression is most easily understood by considering the melting line, which is extended into the metastable pressure region in Figure 5.12(a). This melting line has been extrapolated by using a simple two-species lattice model for the liquid, first introduced by Rapoport [20, 21]. In this model the liquid is seen to consist of atoms that are all in one of two different possible states. These two states are described as two different species A and B that, for the given liquid, are assumed to be in chemical equilibrium at any given temperature and pressure. Thus,

$$Si_A \rightleftharpoons Si_B \tag{5.29}$$

The Gibbs energy of reaction (5.29) is

$$\Delta_r G = G_{Si,B} - G_{Si,A} = \Delta_r H - T\Delta_r S + p\Delta_r V \tag{5.30}$$

Here $G_{Si,A}$ and $G_{Si,B}$ are the Gibbs energy of the 'pure' species A and B. $\Delta_r H = H_{Si,B} - H_{Si,A}$, $\Delta_r S = S_{Si,B} - S_{Si,A}$ and $\Delta_r V = V_{Si,B} - V_{Si,A}$ are the corresponding enthalpy, entropy and volume changes of reaction (5.29). The integral Gibbs energy of the liquid is determined by the relative population of the two states or in other words by the equilibrium constant for the reaction. We will deduce an expression for this below.

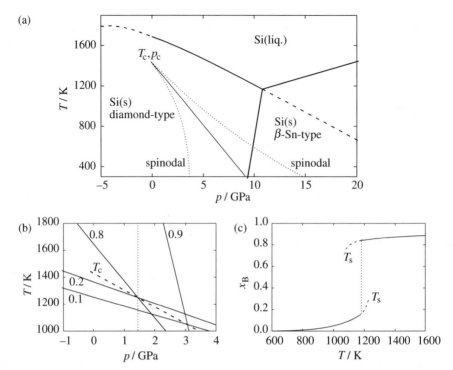

Figure 5.12 (a) The p-T phase diagram of Si. The melting lines for the low-pressure polymorph of Si and the liquid–liquid phase transition are calculated by using the two-state model and the parameters given in Table 5.2. (b) Iso-concentration lines for species B in the p-T plane. (c)) The fraction of species B as a function of temperature at constant pressure p = 2 GPa.

In the two-state model [20, 21] the two different species interact and the interaction can be expressed using the regular solution model. Thus the Gibbs energy of the liquid is

$$G^{\mathrm{liq}} = x_{\mathrm{Si,A}} G_{\mathrm{Si,A}} + x_{\mathrm{Si,B}} G_{\mathrm{Si,B}} + RT[x_{\mathrm{Si,A}} \ln x_{\mathrm{Si,A}} + x_{\mathrm{Si,B}} \ln x_{\mathrm{Si,B}}] \quad (5.31)$$
$$+ \Omega x_{\mathrm{Si,A}} x_{\mathrm{Si,B}}$$

Table 5.2 Model parameters used in the thermo-dynamic description of liquid Si.

$\Delta_r V = V_{\mathrm{Si,B}} - V_{\mathrm{Si,A}} = -2.6 \text{ cm}^3 \text{ mol}^{-1}$

$\Delta_r H = H_{\mathrm{Si,B}} - H_{\mathrm{Si,A}} = 30 \text{ kJ mol}^{-1}$

$\Delta_r S = S_{\mathrm{Si,B}} - S_{\mathrm{Si,A}} = 21 \text{ J K}^{-1} \text{ mol}^{-1}$

$\Omega = 24 \text{ kJ mol}^{-1}$

$\Delta_{\mathrm{fus}} H_{\mathrm{Si,A}} = H_{\mathrm{Si,A}} - H^{\mathrm{sol}} = 26.4 \text{ kJ mol}^{-1}$

$\Delta_{\mathrm{fus}} S_{\mathrm{Si,A}} = S_{\mathrm{Si,A}} - S^{\mathrm{sol}} = 9 \text{ J K}^{-1} \text{ mol}^{-1}$

$\Delta_{\mathrm{fus}} V_{\mathrm{Si,A}} = V_{\mathrm{Si,A}} - V^{\mathrm{sol}} = 0.95 \text{ cm}^3 \text{ mol}^{-1}$

Here Ω is the regular solution constant and $x_{Si,B}$ the fraction of Si atoms in silicon state B. By noting that $x_{Si,A} = 1 - x_{Si,B}$, equation (5.31) becomes

$$
\begin{aligned}
G^{liq} = G_{Si,A} + x_{Si,B}(G_{Si,B} - G_{Si,A}) \\
+ RT[(1 - x_{Si,B})\ln(1 - x_{Si,B}) \\
+ x_{Si,B}\ln x_{Si,B}] + \Omega(1 - x_{Si,B})x_{Si,B}
\end{aligned}
\tag{5.32}
$$

Equation (5.32) looks like the Gibbs energy for a regular binary solution. However, it is important to note that $x_{Si,B}$ for the two-state model has a slightly different interpretation than x_B for a binary regular solution. In the latter case, x_B is an external parameter that describes the composition of the solution. For the two-state model, on the other hand, $x_{Si,B}$ is an internal parameter describing the relative population of the two states present in the single-component system. The equilibrium value of $x_{Si,B}$ is determined by minimizing G^{liq} with respect to $x_{Si,B}$:

$$
\begin{aligned}
\frac{\partial G^{liq}}{\partial x_{Si,B}} &= G_{Si,B} - G_{Si,A} + \Omega(1 - 2x_{Si,B}) + RT \ln \frac{x_{Si,B}}{1 - x_{Si,B}} \\
&= \Delta_r G + \Omega(1 - 2x_{Si,B}) + RT \ln \frac{x_{Si,B}}{1 - x_{Si,B}} = 0
\end{aligned}
\tag{5.33}
$$

Equation (5.33) can alternatively be written in terms of the equilibrium constant for reaction (5.29) as

$$
K_{5.29} = \exp\left(-\frac{\Delta_r G}{RT}\right) = \frac{x_{Si,B}}{1 - x_{Si,B}} \exp\left[\frac{\Omega}{RT}(1 - 2x_{Si,B})\right]
\tag{5.34}
$$

The relative populations of the two states vary with temperature and pressure. The species A which dominates at low temperature and low pressure has the larger molar volume, while the denser species B becomes increasingly more favoured at high pressures.

We are now able to use this model for the Gibbs energy of the liquid to calculate the melting line for four-coordinated Si by using the Clapeyron equation (eq. 2.10):

$$
\frac{dT}{dp} = \frac{\Delta_{fus}V}{\Delta_{fus}S}
\tag{5.35}
$$

When Ω is assumed to be independent of temperature and pressure the entropy and volume of the liquid is given as

$$S^{\text{liq}} = -\left(\frac{\partial G^{\text{liq}}}{\partial T}\right)_p = x_{\text{Si,A}} S_{\text{Si,A}} + x_{\text{Si,B}} S_{\text{Si,B}}$$

$$- R[(1 - x_{\text{Si,B}}) \ln(1 - x_{\text{Si,B}}) + x_{\text{Si,B}} \ln x_{\text{Si,B}}]$$

(5.36)

and

$$V^{\text{liq}} = \left(\frac{\partial G^{\text{liq}}}{\partial p}\right)_T = x_{\text{Si,A}} V_{\text{Si,A}} + x_{\text{Si,B}} V_{\text{Si,B}}$$

(5.37)

The slope of the melting curve follows:

$$\frac{dT}{dp} = \frac{V^{\text{liq}} - V^{\text{sol}}}{S^{\text{liq}} - S^{\text{sol}}}$$

$$= \frac{x_{\text{Si,A}} V_{\text{Si,A}} + x_{\text{Si,B}} V_{\text{Si,B}} - V^{\text{sol}}}{x_{\text{Si,A}} S_{\text{Si,A}} + x_{\text{Si,B}} S_{\text{Si,B}} - R[(1 - x_{\text{Si,B}}) \ln(1 - x_{\text{Si,B}}) + x_{\text{Si,B}} \ln x_{\text{Si,B}}] - S^{\text{sol}}}$$

$$= \frac{\Delta_{\text{fus}} V_{\text{Si,A}} + x_{\text{Si,B}} \Delta_r V}{\Delta_{\text{fus}} S_{\text{Si,A}} + x_{\text{Si,B}} \Delta_r S - R[(1 - x_{\text{Si,B}}) \ln(1 - x_{\text{Si,B}}) + x_{\text{Si,B}} \ln x_{\text{Si,B}}]}$$

(5.38)

The parameters of the model, given in Table 5.2, are obtained by fitting expression (5.38) to the experimental melting line. The extensions of the melting line to negative pressures and beyond the triple point between the liquid and the two solid polymorphs are given by dotted lines in Figure 5.12(a). Note that the model is predicting a melting temperature maximum for Si at negative pressure (−3 GPa).

We are now in a position to analyze the behaviour of Si under compression. Compression of crystalline Si at temperatures above that of the triple point will lead to conventional melting of four-coordinated Si. Compression at temperatures between that of the triple point and the glass transition temperature will also cause melting of four-coordinated Si, but this melting is not an equilibrium reaction. The reason is that the transformation of four-coordinated Si to the six-coordinated polymorph (denoted β-Sn type in the figure) is kinetically hindered. Following this argument, four-coordinated Si melts at the pressure corresponding to the extrapolated melting line for the melting of four-coordinated Si, but with subsequent crystallization of the stable six-coordinated high-pressure polymorph. Under compression at temperatures below the glass transition, an ergodic liquid can no longer be produced, since the thermal energy is too low for the needed structural rearrangements. Still, at some given pressure four-coordinated crystalline Si becomes unstable towards density fluctuations and an amorphous, non-ergodic solid is formed. It follows that the pressure-induced amorphization at ambient temperature occurs at pressures exceeding those of the extension of the melting line for four-coordinated crystalline Si. The amorphization is therefore not a two-phase melting process, but occurs at the mechanical stability limit for the low-pressure modification of crystalline Si.

On decompression, amorphous Si formed by mechanical amorphization undergoes an amorphous–amorphous transition. The volume change associated with the transition is large, reflecting a change in coordination of Si from six to four. Transitions of this type between two different amorphous states have been reported in other cases as well. One of the more studied transitions is that between low- and high-pressure amorphous ice [22]. The existence of more than one amorphous modification for a given substance has been given the name **polyamorphism** [23], by analogy with polymorphism used for crystalline compounds.

A closely related topic is that of liquid–liquid transitions. The possible coexistence of two modifications of a given liquid with the same chemical composition but with different densities has been much discussed. While two coexisting chemically identical liquids with different densities have been reported in quenched Y_2O_3–Al_2O_3 melts [24], a first-order-like liquid–liquid transition has been reported in an *in situ* high-temperature–high-pressure study of liquid phosphorous [25]. At ambient pressure molten phosphorus is a molecular liquid consisting of P_4 molecules, while at high pressures molecular units with low density become unstable and a denser liquid is formed [25].

The regular solution-type two-state model for the liquid induces two coexisting liquids under certain p,T conditions when the interaction parameter, Ω, is positive. Transitions between two amorphous states have for that reason been rationalized in terms of the model. One example is the analysis of the transition between low and high-density amorphous ice by Ponyatovsky *et al.* [26]. We will now consider the transition between low- and high-density amorphous Si in a similar analysis. In the following discussion we disregard the fact that the liquid becomes a glass below the glass transition and thereby transforms to a non-ergodic state.

Below the critical temperature, $G^{\text{liq}}(x_{\text{Si,B}})$ has two minima that are interpreted as representing two different liquids. The deeper minimum in Gibbs energy corresponds to the equilibrium phase at a given temperature and pressure. Hence the two different phases are stable in different parts of the p,T potential space. At the particular conditions where the Gibbs energies of the two minima are equal, the two liquids coexist. Using the two-state model based on the regular solution expression, the relative populations of state B in the two coexisting phases are related by $x_{\text{Si,B}}^{\text{liq.1}} + x_{\text{Si,B}}^{\text{liq.2}} = 1$ since the regular solution expression is symmetrical about $x = 0.5$.

Let us now return to the relative population of the two states of the liquid given by eq. (5.34). Iso-concentration lines ($x_{\text{Si,B}}$ = constant) in the p-T potential space are shown in Figure 5.12(b) (data are taken from Table 5.2). At low temperature and pressure (or even negative pressure) species A with high molar volume and low entropy relatively to the denser polymorph is favoured and $x_{\text{Si,B}}$ is low. At increasing pressure species B become increasingly more favoured and $x_{\text{Si,B}}$ is high at high pressures irrespective of the temperature.

Furthermore, Figure 5.12(b) illustrates the temperature–pressure conditions where two liquids coexist in equilibrium. The temperature and pressure of both phases must be equal at the transition point. In addition, the relative population of state B of the two liquids in equilibrium must be related through $x_{Si,B}^{liq.1} + x_{Si,B}^{liq.2} = 1$ since we use a regular solution-type two-state model. This implies (using a specific example) that two liquids coexist at the intersection point of the iso-concentration lines for $x_{Si,B} = 0.2$ and 0.8. On heating the liquid at the pressure corresponding to this intersection point (along the dotted line in the figure), the population of state B, $x_{Si,B}$, increases continuously until reaching the temperature where the two liquids coexist in equilibrium. Here, $x_{Si,B}$ jumps discontinuously from the value in the low temperature phase, $x_{Si,B} = 0.2$, to that of the high temperature phase, $x_{Si,B} = 0.8$. Further heating results in a slow increase in $x_{Si,B}$. The variation of the population of state B with temperature for $p = 2$ GPa is given in Figure 5.12(c). The dashed lines in the figure give the concentration of B for the two liquids in the metastable regions limited by the spinodals (see below).

The equilibrium line for the liquid–liquid transition given in Figure 5.12(a) is also given in Figure 5.12(b). This line goes through the intersections of all pairs of lines that satisfy $x_{Si,B}^{liq.1} + x_{Si,B}^{liq.2} = 1$, e.g. the intersection of the lines for $x_{Si,B} = 0.1$ and 0.9 and that for $x_{Si,B} = 0.2$ and 0.8. These two specific examples represent two discrete points on the equilibrium curve.

It can be shown that the equilibrium temperature for the liquid–liquid transition is given as

$$T_{trs} = \frac{\Omega(1 - 2x_{B,Si})/R}{\ln(x_{B,Si}/(1 - x_{B,Si}))} \tag{5.39}$$

This equilibrium line terminates at the critical point where the critical temperature is given by the two-state regular model as $T_c = \Omega/2R$. The pressure corresponding to a given equilibrium temperature is given by

$$p = \frac{T_{trs}\Delta_r S - \Delta_r H}{\Delta_r V} \tag{5.40}$$

We have now derived the phase boundary between the two liquids. By analogy with our earlier examples, the two phases may exist as metastable states in a certain part of the p,T potential space. However, at some specific conditions the phases become mechanically unstable. These conditions correspond to the spinodal lines for the system. An analytical expression for the spinodals of the regular solution-type two-state model can be obtained by using the fact that the second derivative of the Gibbs energy with regards to $x_{Si,B}$ is zero at spinodal points. Hence,

$$\left(\frac{\partial^2 G^{liq.}}{\partial x_{Si,B}^2}\right)_T = -2\Omega + \frac{RT}{x_{Si,B}(1 - x_{Si,B})} = 0 \tag{5.41}$$

The calculated spinodal lines are given together with the equilibrium phase boundary in Figure 5.12(a). All three lines terminate at the critical point. The amorphous–amorphous transition observed during decompression of amorphous Si can now be understood in terms of crossing the spinodal line for the high-pressure amorphous phase on reducing pressure. Finally, it should be noted that a transition similar to that shown in Figure 5.12(c) may illustrate what is observed during flash heating of amorphous Si at ambient pressure [27]. Amorphous Si shows a 'first-order-like' transition to supercooled liquid Si when crystallization of amorphous Si is suppressed by a high heating rate. In spite of the apparent success of the two state model, the model used gives a very simplistic description of a liquid and further experimental and theoretical evidence is needed to confirm the very existence of liquid–liquid transitions.

5.3 Metastable phase equilibria and kinetics

Metastable materials are becoming increasingly important as the use of extreme far-from-equilibrium conditions during preparation is expanding. The formation of a large fraction of these phases cannot be rationalized using thermodynamic arguments. Most zeolites for example are kinetically stabilized through the use of templates during synthesis. Another and more specific example is $YMnO_3$, which takes the perovskite structure under equilibrium conditions, but crystallizes in a different structure as a thin film or powder prepared from precursors [28, 29]. While thermodynamic arguments may fail in cases like this, thermodynamic analyses can in other cases be used to predict synthesis routes to new compounds, and there are numerous examples of metastable alloys formed when the nucleation of the stable phases is suppressed. In these cases the metastable phase equilibria can be analysed thermodynamically, and may even be represented in phase diagrams.

Phase diagrams reflecting metastability

One of the classical examples of metastable heterogeneous phase equilibria occurs in the system Fe–C. The eutectic between γ-Fe and graphite shown in Figure 5.13(a) is important for many cast irons. If the C-level is low, as in steels, solidification directly to δ- or γ-Fe may occur. On cooling, these steels become unstable with regard to the formation of graphite. Although there is a driving force for precipitation of graphite, the volume change for this precipitation reaction in the solid state is high and nucleation becomes difficult. For many conditions the metastable phase Fe_3C, cementite, is formed instead. Hence for practical purposes the metastable $Fe–Fe_3C$ phase diagram is more important than the equilibrium Fe–C phase diagram. In the section of the binary phase diagram Fe–C shown in Figure 5.13(a), both the stable phase boundaries and phase boundaries corresponding to metastable phase equilibria involving cementite, Fe_3C, are given. The stability field of γ-Fe when γ-Fe is in metastable equilibrium with cementite is slightly larger than when γ-Fe is in the stable equilibrium with graphite [30]. This can be

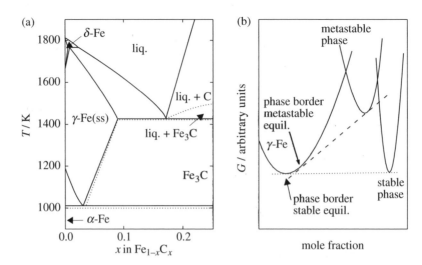

Figure 5.13 (a) Phase diagram of the Fe–C system. Solid lines represent the stable phase diagram where γ-Fe is in equilibrium with graphite. Dotted lines represent the metastable phase diagram where γ-Fe is in equilibrium with cementite, Fe_3C. (b) Schematic Gibbs energy rationalization of the effect of a metastable equilibrium on solid solubility.

rationalized for a general situation through the schematic Gibbs energy diagram given in Figure 5.13(b). The Gibbs energy of a metastable phase is higher than that of the mixture of γ-Fe and the stable phase. The result is that the common tangent construction that determines the extension of the phase field of γ-Fe touches the Gibbs energy curve of γ-Fe at a higher mole fraction of carbon for the metastable case than is the case for the stable phase equilibrium. Cementite is marginally metastable, and it follows that the effect on the phase boundaries is small.

Thermodynamic representations of phase diagrams may reveal such and more complex metastable situations. The phase relations in the Sn–Sb system are shown in Figure 5.14 [31]. The diagram on the right-hand side represent a metastable situation where the non-stoichiometric phase SnSb is considered as kinetically hindered from formation. The terminal solid solubility of Sn in Sb and of Sb in Sn increases; furthermore, Sn_3Sb_2 will in this case be apparently stable even at low temperatures. Diagrams of this type may correspond to situations high in Gibbs energy that are of little practical importance. In other situations they correspond to phase equilibria close in Gibbs energy to the stable situation. Such situations may be observed as metastable equilibria under certain conditions.

Thermal evolution of metastable phases

Although the formation of a large number of metastable materials that are far from equilibrium cannot be explained thermodynamically, thermodynamics predicts that they will with time transform to the stable phase or phase mixture, often via intermediate phases. More than one hundred years ago, Ostwald pointed out that

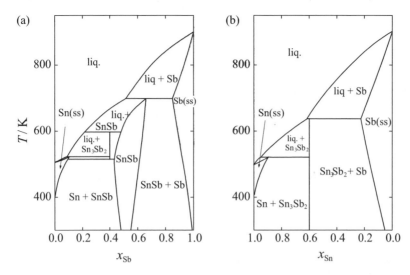

Figure 5.14 Calculated (a) stable and (b) metastable phase diagrams for the Sn–Sb system.

non-equilibrium thermodynamic systems appear to evolve through a sequence of states of progressively lower Gibbs energy [32]. In this rationalization scheme, it is not the most stable form of the material with the lowest Gibbs energy that is obtained as the initial product, but the least stable that is nearest to the original in Gibbs energy. If several metastable phases, or mixtures of phases, are possible, they will follow one another in the order of a stepwise decrease in Gibbs energy. Ostwald's step rule has considerable value in materials science. This will be illustrated here by considering crystallization of metallic glasses in the system Fe–B [33] and diffusional amorphization in the system Ni–Zr [34].

The glass formation ability in the system Fe–B is largest for alloys close to the eutectic where a melt with $x_B = 0.20$ on cooling solidifies under the formation of Fe and Fe_2B. The crystallization sequence observed depends on the composition of the glass and often involves the metastable compound Fe_3B [33]. A large portion of the experimental observations can be rationalized using the schematic Gibbs energy representation given in Figure 5.15. For an alloy with composition given by the point A, primary crystallization of Fe(ss) is observed. The glass/supercooled liquid near T_g is simultaneously enriched in B. On further thermal evolution, the metastable phase equilibrium between Fe(ss) and Fe_3B is reached. Only in a third and final stage is the stable phase equilibrium between Fe(ss) and Fe_2B obtained. The three stages of the crystallization are marked in Figure 5.15 by arrows. In this simple rationalization of the experimental observations, the fact that Fe can crystallize in different structures which are close in Gibbs energy is not taken into consideration. It has for example been shown that molten droplets of certain Fe–Ni alloys crystallize in a bcc-type structure before transforming to the more stable fcc structure [35]. Although the Gibbs energy rationalization can often predict the

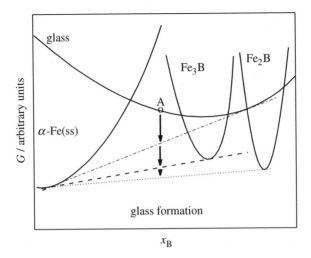

Figure 5.15 Schematic Gibbs energy rationalization of the crystallization of metallic glasses in the system Fe–B.

phase sequence formed, some phases may be kinetically hindered from formation, as in the case of graphite in the system Fe–C described above.

Thermodynamic rationalization of diffusional amorphization similarly relies on Ostwald's step rule. One example is the formation of amorphous Ni–Zr alloys on interdiffusion in stacks of thin –Ni–Zr–Ni–Zr– foils [34]. Since, the kinetics is slow, the formation of intermetallic phases is prevented. The thermodynamic stability of the liquid determines whether or not a glass is formed. The Gibbs energy of formation of the liquid must be intermediate in Gibbs energy between the Gibbs energy of the mixture of the elements and the mixture of different intermetallic compounds. Glass formation and growth in diffusion couples like this was first observed for Ni–Zr in 1983, but has subsequently also been seen in a large number of other intermetallic systems [36].

Materials in thermodynamic potential gradients

A last example of kinetic effects is given by the behaviour of materials in thermodynamic potential gradients [37]. Materials are often applied in situations where they are not in equilibrium with their immediate surroundings. Gradients in temperature, chemical or electrical potential act as driving forces on atoms in a crystalline material, and fluxes of atoms across an initially homogeneous solid solution result. This effect tends to separate the components if they have different mobilities and is a phenomenon of practical significance. Materials subject to thermodynamic potential gradients in general may demix or even decompose. One example is engineering components subject to large temperature gradients like turbine blades (Ludvig–Soret effect).

Let us initially look at a semiconducting binary oxide $A_{1-\delta}O$ in a chemical gradient; an oxygen potential gradient. Reduction takes place on the low oxygen

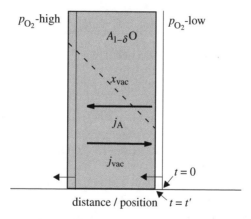

Figure 5.16 Schematic illustration of demixing in a binary oxide $A_{1-\delta}O$ in a gradient in the chemical potential of O_2.

activity side with the result that the oxide loses oxygen at the surface. Oxygen molecules leave the oxide while the cations left behind move toward the higher oxygen potential side; see Figure 5.16. Oxidation takes place at the high oxygen activity side were cations are recombined with oxygen atoms from the gas and the oxide grows. Both the oxide/gas surfaces move in the direction of the high oxygen activity side (indicated by arrows in Figure 5.16) relative to the immobile oxygen lattice. The composition, the vacancy concentration, of the binary oxide will vary across the oxide due to the oxygen potential gradient.

For ternary mixed cation oxides like $(A,B)_{1-\delta}O$ more pronounced effects may be encountered in addition to growth at the high p_{O_2} surface at the expense of the low p_{O_2} surface [37, 38]. Demixing of the different cations will occur in an applied oxygen potential gradient in cases where the two cations have different mobility. The result of these transport processes is concentration gradients and a material is usually enriched in the more mobile cation species at the high oxygen potential side. The degree of demixing increases with increasing difference in mobility between the cations and with decreasing thickness of the material.

The segregation or demixing is a purely kinetic effect and the magnitude depends on the cation mobility and sample thickness, and is not directly related to the thermodynamics of the system. In some specific cases, a material like a spinel may even decompose when placed in a potential gradient, although both potentials are chosen to fall inside the stability field of the spinel phase. This was first observed for Co_2SiO_4 [39]. Formal treatments can be found in references [37] and [38].

References

[1]   F. Grønvold, *J. Chem. Thermodyn.* 1973, **5**, 525.

[2]   S. S. Chang and A. B. Bestul, *J. Chem. Thermodyn.* 1974, **6**, 325.

[3]   S. Stølen, H. B. Johnsen, C. S. Bøe, T. Grande and O. B. Karlsen, *J. Phase Equil.* 1999, **20**, 17.
[4]   J. C. Wheatley, *Rev. Mod. Phys.* 1975, **47**, 415.
[5]   W. Kauzmann, *Chem. Rev.* 1948, **43**, 219.
[6]   F. A. Lindemann, *Z. Phys.* 1910, **11**, 609.
[7]   M. Born, *J. Chem. Phys.* 1931, **7**, 591.
[8]   J. W. Gibbs, *The Collected Work, Volume I Thermodynamics*. Harvard: Yale University Press, 1948.
[9]   M. Born and K. Huang, *Dynamic Theory of Crystal Lattices*. Oxford: Oxford University Press, 1954.
[10] J. L. Tallon and A. Wolfenden, *J. Phys. Chem. Solids* 1979, **40**, 831.
[11] K. Lu and Y. Li, *Phys. Rev. Lett.* 1998, **80**, 4474.
[12] H. J. Fecht and W. L. Johnson, *Nature* 1988, **334**, 50.
[13] J. L. Tallon, *Nature*, 1989, **342**, 658.
[14] S. S. Kim and T. H. Sanders Jr, *J. Am. Ceram. Soc.* 2001, **84**, 1881.
[15] T. Takayama, M. Yuri, K. Itoh, and J. S. Harris, Jr, *J. Appl. Phys.* 2001, **90**, 2358.
[16] T. J. Rockett and W. R. Foster, *J. Am. Ceram. Soc.* 1966, **49**, 30.
[17] H. E. Stanley, *Introduction to Phase Transitions and Critical Phenomena*. Oxford: Oxford University Press, 1971.
[18] B. J. Skinner and J. J. Fahey, *J. Geophys. Res.* 1963, **68**, 5595.
[19] S. K. Deb, M. Wilding, M. Somayazulu and P. F. McMillan, *Nature*, 2001, **414**, 528.
[20] E. Rapoport, *J. Chem. Phys.* 1967, **46**, 2891.
[21] E. Rapoport, *J. Chem. Phys.* 1968, **48**, 1433.
[22] O. Mishima, L. D. Calvert and E.Whelley, *Nature* 1985, **314**, 76.
[23] P. H. Poole, T. Grande, C. A. Angell and P. F. McMillan, *Science*, 1996, **275**, 322.
[24] S. Aasland and P. F. McMillan, *Nature* 1994, **369**, 633.
[25] Y. Katayama, T. Mizutani, W. Utsumi, O. Shimomura, M. Yamakata and K. Funakoshi, *Nature* 2000, **403**, 170.
[26] E. G. Ponyatovsky, V.V. Sinitsyn and T. A. Pozdnyakova, *J. Chem. Phys.* 1998, **109**, 2413.
[27] M. O. Thompson, G. J. Galvin, J. W. Mayer, P. S. Peercy, J. M. Poate, D. C. Jacobson, A. G. Cullis and N. G. Chew, *Phys. Rev. Lett.* 1984, **52**, 2360.
[28] H. W. Brinks, H. Fjellvåg and A. Kjekshus, *J. Solid State Chem.* 1997, **129**, 334.
[29] P. A. Salvador, T. D. Doan, B. Mercey and B. Raveau, *Chem. Mater.* 1998, **10**, 2592.
[30] R. Hultgren, P. D. Desai, D. T. Hawkins, M. Gleiser and K. K. Kelley, *Selected Values of the Thermodynamic Properties of Binary Alloys*. Metals Park, OH: Am. Soc. Metals, 1973.
[31] B. Jönsson and J. Ågren, *Mater. Sci. Techn.* 1986, **2**, 913.
[32] W. Ostwald, *Z. Phys. Chem.* 1897, **22**, 289.
[33] K. Köster and K. Herold, *Topics in Applied Physics*, Vol. 46. Berlin: Springer-Verlag, 1981.
[34] M. Atzmon, J. R. Verhoeven, E. D. Gibson and W. L. Johnson, *Appl. Phys. Lett.* 1984, **45**, 1052.
[35] R. E. Cech, *Trans. Met. Soc. AIME* 1956, **206**, 585.
[36] W. L. Johnson, *Progr. Mater. Sci.* 1986, **30**, 81.
[37] H. Schmalzried, in *Chemical Kinetics of Solids*. Weinheim: VCH, 1995, Chapter 8.
[38] M. Martin, *Mater. Sci. Rep.* 1991, **7**, 1.

[39] H. Schmalzried, W. Laqua and P. L. Lin, Z. *Naturforsch.* 1979, **A34**, 192.

Further reading

J. W. Gibbs, *The Collected Work, Volume I Thermodynamics*. Harvard: Yale University Press, 1948.

D. Kondepudi and I. Prigogine, *Modern Thermodynamics: from Heat Engines to Dissipative Structures*. New York: John Wiley & Sons, 1998.

H. Schmalzried, *Chemical Kinetics of Solids*. Weinheim: VCH, 1995.

H. E. Stanley, *Introduction to Phase Transitions and Critical Phenomena*. Oxford: Oxford University Press, 1971.

Surfaces, interfaces and adsorption

In Chapter 1, heterogeneous systems were described as a set of homogeneous regions separated by surfaces or interfaces. The surface or interface is chemically different from the bulk material and the surface or interface energy represents an excess energy of the system relative to the bulk. When considering the macroscopic thermodynamic properties of a system the surface/interface contribution can be neglected as long as the homogeneous regions are large. 'Large' in this context can be identified from Figure 6.1. Here the enthalpy of formation of NaCl is shown as a function of the size of single crystals formed as cubes where a is the

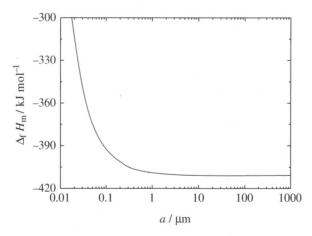

Figure 6.1 The molar enthalpy of formation of NaCl as a function of the cube edge a of the NaCl crystal cubes.

Chemical Thermodynamics of Materials by Svein Stølen and Tor Grande
© 2004 John Wiley & Sons, Ltd ISBN 0 471 492320 2

length of a cube edge. The enthalpy of formation of NaCl becomes less negative as the cube size decreases because the surface energy is positive (energy is required to form surfaces), and the relative contribution from the surface increases with decreasing size of the cubes. However, the number of surface atoms is significant only for very small cubes and the contribution from the surface energy becomes measurable only when the cubes are smaller than ~1 μm. A thermodynamic system can therefore be analyzed in terms of the bulk properties of the system when the homogeneous regions are larger than roughly 1 μm.

Although the surface energy may be neglected in considering macroscopic systems, it is still very important for the kinetics of atomic mobility and for kinetics in heterogeneous systems. Nucleation and crystal growth in solid or liquid phases and sintering or densification in granular solids are largely influenced by the surface or interface thermodynamics. In these cases the complexity of the situation further increases since the curvature of the surfaces or interfaces is a key parameter in addition to surface energy. Moreover, materials science is driven towards smaller and smaller dimensions, and the thermodynamics of surfaces and interfaces are becoming a key issue for materials synthesis and for understanding the properties of nano-scale materials. For a cube containing only 1000 atoms, as many as 50% of the atoms are at the surface and the surface energy is of great importance.

The purpose of this chapter is to introduce the effect of surfaces and interfaces on the thermodynamics of materials. While *interface* is a general term used for solid–solid, solid–liquid, liquid–liquid, solid–gas and liquid–gas boundaries, *surface* is the term normally used for the two latter types of phase boundary. The thermodynamic theory of interfaces between isotropic phases were first formulated by Gibbs [1]. The treatment of such systems is based on the definition of an isotropic **surface tension**, σ, which is an excess surface stress per unit surface area. The Gibbs surface model for fluid surfaces is presented in Section 6.1 along with the derivation of the equilibrium conditions for curved interfaces, the Laplace equation.

Surfaces of crystals, which are inherently anisotropic in nature, are also briefly treated. Gibbs' treatment of interfaces was primarily related to fluid surfaces, and the thermodynamic treatment of solid surfaces was not fully developed before the second half of the 20th century [2]. While the thermodynamics of surfaces and interfaces in the case of isotropic systems are defined in terms of the surface tension, surface energy is the term used for non-isotropic systems. The **surface energy**, γ, is defined as the energy of formation of a new equilibrium surface of unit area by cutting a crystal into two separate parts. The surface energy according to this definition cannot be isotropic since the chemical bonds broken due to the cleavage depend on the orientation of the crystal. The consequences of surface energy anisotropy for the crystal morphology are discussed. Trends in surface tension and average surface energy of the elements and some salt systems are reviewed and finally the consequences of differences in surface energy/tension between different phases in equilibrium on the morphology of the interface are considered generally.

In the last two sections the formal theory of surface thermodynamics is used to describe material characteristics. The effect of interfaces on some important heterogeneous phase equilibria is summarized in Section 6.2. Here the focus is on the effect of the curvature of the interface. In Section 6.3 adsorption is covered. Physical and chemical adsorption and the effect of interface or surface energies on the segregation of chemical species in the interfacial region are covered. Of special importance again are solid–gas or liquid–gas interfaces and adsorption isotherms, and the thermodynamics of physically adsorbed species is here the main focus.

6.1 Thermodynamics of interfaces

Gibbs surface model and definition of surface tension

A real interface region between two homogeneous phases α and β is schematically illustrated in Figure 6.2(a). A hypothetical geometric surface termed the **Gibbs dividing surface**, Σ, is constructed lying in the region of heterogeneity between the two phases α and β, as shown in Figure 6.2(b). In the Gibbs surface model [1], Σ has no thickness and only provides a geometrical separation of the two homogeneous phases. At first sight this simple description may seem to be inadequate for a real interface, but in the following we will show the usefulness of the model. The energetic contribution of the interface is obtained by assigning to the bulk phases the values of these properties that would pertain if the bulk phases continued uniformly up to the dividing surface. The value of any thermodynamic property for the system as a whole will then differ from the sum of the values of the thermodynamic properties for the two bulk phases involved. These excess thermodynamic properties, which may be positive or negative, are assigned to the interface.

Let us now consider an interface between two isotropic multi-component phases. The number of moles of a component i in the two phases adjacent to the interface are given as n_i^{α} and n_i^{β}. Since the mass balance of the overall system must be obeyed, it is necessary to assume that the dividing surface contains a certain

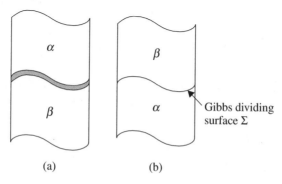

(a)   (b)

Figure 6.2 (a) Illustration of a real physical interface between two homogeneous phases α and β. (b) The hypothetical Gibbs dividing surface Σ.

number of moles of species i, n_i^σ, such that the total number of moles of i in the real system, n_i, is equal to

$$n_i = n_i^\alpha + n_i^\beta + n_i^\sigma \qquad (6.1)$$

The **surface excess moles** or the number of moles of species i adsorbed or present at the surface is then defined as

$$n_i^\sigma = n_i - n_i^\alpha - n_i^\beta \qquad (6.2)$$

n_i^σ divided by the area A_s of Σ yields the **adsorption** of i:

$$\Gamma_i = n_i^\sigma / A_s \qquad (6.3)$$

Γ_i may become positive or negative, depending on the particular interface in question. Other **surface excess properties**, such as the surface internal energy and surface entropy, are defined similarly:

$$U^\sigma = U - U^\alpha - U^\beta \qquad (6.4)$$

$$S^\sigma = S - S^\alpha - S^\beta \qquad (6.5)$$

Recall that the Gibbs dividing surface is only a geometrical surface with no thickness and thus has no volume:

$$V^\sigma = V - V^\alpha - V^\beta = 0 \qquad (6.6)$$

It follows that the surface excess properties are macroscopic parameters only.

In order to define the surface tension we will consider the change in internal energy connected with a reversible change in the system. For an open system dU is given by eq. (1.79) as

$$dU = TdS - pdV + \sum_i \mu_i dn_i \qquad (6.7)$$

For a reversible process, where the interfaces remain fixed, the volumes of the two phases remain constant and eq. (6.7) becomes

$$dU = TdS + \sum_i \mu_i dn_i \qquad (6.8)$$

An infinitesimal change in the surface internal energy

$$dU^\sigma = dU - dU^\alpha - dU^\beta \qquad (6.9)$$

can be expressed in terms of the changes in internal energy of the two homogeneous phases separated by the fixed boundary

$$dU^\alpha = TdS^\alpha + \sum_{i=1}^{C} \mu_i dn_i^\alpha \qquad (6.10)$$

$$dU^\beta = TdS^\beta + \sum_{i=1}^{C} \mu_i dn_i^\beta \qquad (6.11)$$

Here C is the number of components in the system. Combination of eqs. (6.9), (6.10) and (6.11) yields

$$dU^\sigma = T(dS - dS^\alpha - dS^\beta) + \sum_{i=1}^{C} \mu_i (dn_i - dn_i^\alpha - dn_i^\beta) \qquad (6.12)$$

The expressions in the two parentheses can be identified as the surface excess moles and surface excess entropy defined by eqs. (6.2) and (6.5). Equation (6.12) thus reduces to

$$dU^\sigma = TdS^\sigma + \sum_{i=1}^{C} \mu_i dn_i^\sigma \qquad (6.13)$$

The exact position of the geometrical surface can be changed. When the location of the geometrical surface Σ is changed while the form or topography is left unaltered, the internal energy, entropy and excess moles of the interface vary. The thermodynamics of the interface thus depend on the location of the geometrical surface Σ. Still, eq. (6.13) will always be fulfilled.

The effect of variations in the form of the geometrical interface on the energy can be deconvoluted into two contributions: changes in energy related to changes in the area of the interface and changes in energy related to changes in the curvatures of the interface [3]. The **two principal curvatures** c_1 and c_2 at a point Q on a arbitrary surface are indirectly illustrated in Figure 6.3. Two planes normal to the surface at Q are defined by the normal at point Q and the unit vectors in the two principal directions, **u** and **v**. A circle can be constructed in each of the two planes which just touches the surface at point Q. The radii r_1 and r_2 of the two circles are the **two principal radii** at point Q and the two principle curvatures are defined as the reciprocal radii $c_1 = 1/r_1$ and $c_2 = 1/r_2$. For systems where the thickness of the real physical interface is much smaller than the curvature of the interface, Gibbs [1] showed that the dividing surface could be positioned such that the contribution from the curvature of the interface is negligible. Assuming the surface to have such a position, only the term related to a change in the interfacial area needs to be considered. An infinitesimal change in the surface internal energy is

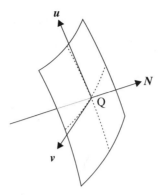

Figure 6.3 Illustration of the curvature of a geometrical surface.

$$dU^\sigma = TdS^\sigma + \sum_i \mu_i dn_i^\sigma + \sigma dA_s \qquad (6.14)$$

where σ is the partial derivative of U with respect to the area A_s. We are now going to investigate the significance of the variable σ. For a reversible process dU is

$$dU = dU^\sigma + dU^\alpha + dU^\beta \qquad (6.15)$$

The change in internal energy for the two phases adjacent to the interface is now

$$dU^\alpha = TdS^\alpha + \sum_{i=1}^{C} \mu_i dn_i^\alpha - p^\alpha dV^\alpha \qquad (6.16)$$

$$dU^\beta = TdS^\beta + \sum_{i=1}^{C} \mu_i dn_i^\beta - p^\beta dV^\beta \qquad (6.17)$$

Incorporating these two equations in eq. (6.15) yields the following expression for the change in internal energy for the system:

$$dU = T(dS^\sigma + dS^\alpha + dS^\beta)$$
$$+ \sum_{i=1}^{C} \mu_i (dn_i^\sigma + dn_i^\alpha + dn_i^\beta) - p^\alpha dV^\alpha - p^\beta dV^\beta + \sigma dA_s \qquad (6.18)$$

or

$$dU = TdS + \sum_{i=1}^{C} \mu_i dn_i - p^\alpha dV^\alpha - p^\beta dV^\beta + \sigma dA_s \qquad (6.19)$$

where the **surface tension**, σ, is

$$\sigma = \left(\frac{\partial U}{\partial A_s} \right)_{S, V^\alpha, V^\beta, n_i} \tag{6.20}$$

This is the definition of the surface tension according to the Gibbs surface model [1]. According to this definition, the surface tension is related to an interface, which behaves mechanically as a membrane stretched uniformly and isotropically by a force which is the same at all points and in all directions. The surface tension is given in J m^{-2}. It should be noted that the volumes of both phases involved are defined by the Gibbs dividing surface Σ that is located at the position which makes the contribution from the curvatures negligible.

Equilibrium conditions for curved interfaces

The equilibrium conditions for systems with curved interfaces [3] are in part identical to those defined earlier for heterogeneous phase equilibria where surface effects where negligible:

$$T^\alpha = T^\beta = T^\sigma \tag{6.21}$$

and

$$\mu_i^\alpha = \mu_i^\beta = \mu_i^\sigma \tag{6.22}$$

Note that the chemical potential of a given component at the interface is equal to that in the two adjacent phases. This is important since this implies that adsorption can be treated as a chemical equilibrium, as we will discuss in Section 6.3.

To establish the **equilibrium conditions for pressure** we will consider a movement of the dividing surface between the two phases α and β. The dividing surface moves a distance dl along its normal while the entropy, the total volume and the number of moles n_i are kept constant. An infinitesimal change in the internal energy is now given by

$$dU = -p^\alpha dV^\alpha - p^\beta dV^\beta + \sigma dA_s \tag{6.23}$$

The changes in the volume of the two phases are related by

$$dV^\alpha = A_s dl = -dV^\beta \tag{6.24}$$

and also the change in area of the surface is related to dl. dA_s can be expressed in terms of the two principal curvatures c_1 and c_2 of the interface [3]:

$$dA_s = (c_1 + c_2)A_s \, dl \tag{6.25}$$

Substitution of eqs. (6.24) and (6.25) into eq. (6.23) yields

$$dU = (p^\beta - p^\alpha)A_s \, dl + \sigma(c_1 + c_2)A_s \, dl = [(p^\beta - p^\alpha) + \sigma(c_1 + c_2)]A_s \, dl \tag{6.26}$$

At equilibrium $(dU)_{S,V,n_i} = 0$, which leads to the equilibrium condition for pressure expressed in terms of the two principal curvatures or alternatively in terms of the two principal radii of curvature:

$$p^\beta - p^\alpha = \sigma(c_1 + c_2) = \sigma\left(\frac{1}{r_1} + \frac{1}{r_2}\right) \tag{6.27}$$

Equation (6.27) is the **Laplace equation**, or Young–Laplace equation, which defines the equilibrium condition for the pressure difference over a curved surface. In Section 6.2 we will examine the consequences of surface or interface curvature for some important heterogeneous phase equilibria.

For planar surfaces the pressure difference over the interface becomes zero and the equilibrium condition for pressure, eq. (6.27) reduces to

$$p^\beta = p^\alpha \tag{6.28}$$

The surface tension for a planar surface thus is

$$\sigma = \left(\frac{\partial U}{\partial A_s}\right)_{S,V,n_i} \tag{6.29}$$

and here only the total volume needs to be kept constant. The position of the geometrical surface Σ no longer affects the definition of σ, as for curved surfaces.

The surface energy of solids

The surface tension defined above was related to an interface that behaved mechanically as a membrane stretched uniformly and isotropically by a force which is the same at all points on the surface. A surface property defined this way is not always applicable to the surfaces of solids and the **surface energy** of planar surfaces is defined to take anisotropy into account. The surface energy is often in the literature interchanged with surface tension without further notice. Although this may be useful in practice, it is strictly not correct.

The surface energy can be derived by an alternative treatment. Let us initially consider a large homogeneous crystal that contains N atoms and that has a planar surface. The change in energy on forming solid surfaces is often deconvoluted into two contributions. The first contribution is due to a change in the surface area that does not disturb the structural arrangement of the atoms and which thus leaves the surface structure identical to that of the bulk. The second contribution is elastic in nature, and relates to the deformation of the surface when relaxed or reconstructed.

To create a new surface we have to break bonds and remove the superfluous atoms. At equilibrium at constant pressure and temperature the work demanded to increase the surface area of a one-component system by an amount dA_s is given as

$$dW_{T,p} = \gamma dA_s \qquad (6.30)$$

where γ is the surface energy (J m^{-2}). This energy is the excess energy relative to the bulk and depends on the number of bonds per unit area and the strength of these bonds. The reversible work is equal to the change in Gibbs energy due to the formation of a surface, and the change in the Gibbs energy of a one-component system can now be written as

$$dG = -SdT + Vdp + \gamma dA_s \qquad (6.31)$$

where

$$\gamma = \left(\frac{\partial G}{\partial A_s} \right)_{T,p} \qquad (6.32)$$

For an isotropic phase there are no differences between surface energy and surface tension. However, for crystals, which are anisotropic in nature, the relationship between these two quantities is significant and also theoretically challenging, see e.g. the recent review by Rusanov [2].

It is important to note that the formation of a surface always leads to a positive Gibbs energy contribution. This implies that smaller particles are unstable relative to larger particles and that the equilibrium shape of crystals is determined by the tendency for surfaces of higher energy to be sacrificed while those of lower Gibbs energy grow. This is the topic of the next section.

Anisotropy and crystal morphology

Basically, the surface energy is given by the number of bonds per unit surface area and by the bond strength. Different crystal surfaces have different numbers of bonds per unit surface area and the measured surface energies for crystals are often an average value over many different crystal surfaces. Using a face-centred cubic structure as an example, the density of atoms in specific planes generally decreases with increasing Miller indices [hkl]. The exception is the close-packed [111] plane. For a [111] plane there are six nearest neighbours in the plane, three above and three below the plane. Hence three bonds are broken for each surface atom when the crystal is cut in two along [111]. For the [100] and [110] planes there are four and six broken bonds respectively, and thus taking only nearest neighbours into consideration the surface energies for these planes are larger. The different surface energies for different types of crystal surfaces control the equilibrium shape of a crystal, as first discussed by Wolf [11]. This important phenomenon occurs not only for solid–gas interfaces but also for all other interfaces. For liquid–gas interfaces the surface tension is independent of orientation and the equilibrium shape is a sphere. This represents the smallest surface area for a body of a given size. Experimental studies indicate that spheres become energetically favourable also for solids at high temperatures. Hence the difference in surface energy between different surfaces is less important at high temperatures.

The equilibrium shape of a crystal can be constructed using the Wulff construction [4]. Consider a one-component system in which only the solid and gas phases are present. Assume that the phases have their equilibrium volumes and can only change their shape. Hence we need to be able to describe the volume of the crystal. Let us start looking at a single crystal in the form of a polyhedron of some kind. This is shown for a two-dimensional case in Figure 6.4. From some point O in the interior of a crystal, normals to all crystal faces are drawn. The distance between O and the face v is h_v. If a straight line is drawn from O to each corner of the body, the crystal will be divided into N pyramids of height h_v, base A_v and volume $1/2A_vh_v$. Using a similar analysis, the volume of a three-dimensional crystal can be expressed as

$$V = \frac{1}{3}\sum_{v=1}^{N} A_v h_v \qquad (6.33)$$

For a reversible change at constant temperature and volume of both phases and for a constant number of moles of the components, the equilibrium shape can be

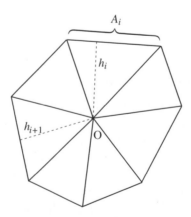

Figure 6.4 Geometric parameters describing a two-dimensional crystal.

found by minimization of the Helmholtz energy of the system. It can be shown that the equilibrium morphology of a single crystal is given by [3]

$$\frac{\gamma_1}{h_1} = \frac{\gamma_2}{h_2} = \ldots = \frac{\gamma_N}{h_N} \tag{6.34}$$

Here γ_i is the surface energy of the crystal surface i. The equilibrium shape of a crystal is thus a polyhedron where the area of the crystal facets is inversely proportional to their surface energy. Hence the largest facets are those with the lowest surface energy.

Equation (6.34), defining the equilibrium shape of crystals, is only relevant for crystals of a certain size. For large crystals changes in shape involve diffusion of large numbers of atoms and the driving force may not be sufficient, since the surface contribution is small compared with the bulk. Hence metastable crystal shapes are more likely to be reached. But even for small crystals the Wulff relationship may break down. Here twinning may lead to configurations which lower the Gibbs energy of the crystal, and this results in a different crystal morphology. Herring [5, 6] and Mullins [7] give extended discussions of the topic.

The Laplace equation (eq. 6.27) was derived for the interface between two isotropic phases. A corresponding Laplace equation for a solid–liquid or solid–gas interface can also be derived [3]. Here the pressure difference over the interface is given in terms of the factor that determines the equilibrium shape of the crystal:

$$p^\alpha - p^\beta = 2\frac{\gamma_1}{h_1} = 2\frac{\gamma_v}{h_v} = \ldots = 2\frac{\gamma_N}{h_N} \tag{6.35}$$

Comparing this expression with eq. (6.27), we see that γ_v/h_v for each crystal face represents σ divided by the radius of curvature for an isotropic spherical phase. As a first approximation we may replace γ_v/h_v with γ/r for near-spherical crystals. In this case γ represents an average surface energy of all possible crystal faces.

In the remaining part of the chapter we will use the term γ for interfaces that involve solids. It should then implicitly be understood that we are here considering bulk solids that are treated as isotropic systems and that the surface energy thus defined is the average value of the surface energies for different crystal surfaces. Furthermore, we will consistently use superscripts to denote the phases adjacent to the interface in the rest of Section 6.1 and in Section 6.2.

Trends in surface tension and surface energy

Periodic variations in the surface tension of liquid metals, σ^{lg}, are shown in Figure 6.5. The much higher surface tension of d-block metals compared to the s- and p-block metals suggests that the surface tension relates to the strength of interatomic bonding. Similar periodic trends can be found also for the melting temperature and the enthalpy of vaporization, and the surface tension of liquid metals is strongly

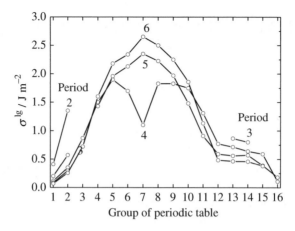

Figure 6.5 Periodic trends in the surface tension of selected liquid elements in periods 2–6 at their melting temperature [8].

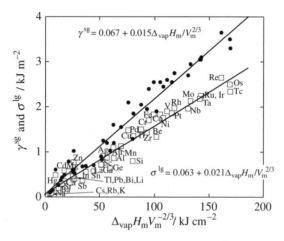

Figure 6.6 Surface tension of the liquid elements at T_{fus}, σ^{lg} (open squares), and surface energy of the solid elements at 0 K, γ^{sg} (filled circles), versus $\Delta_{fus}H_m V_m^{-2/3}$ [8].

correlated with these and other physical properties that depend on the strength of the interatomic bond. The correlation between the surface tension of the liquid metals and their enthalpy of vaporization, in the form of $\Delta_{vap}H_m^{o} \cdot V_m^{-2/3}$, first discussed by Shapski [9] and Grosse [10], is shown in Figure 6.6. Data for the average surface energy of solid metals, γ^{sg}, are also included in this figure. It can be noted that the average surface energy of solids has also been shown to correlate with other cohesion-related properties like Young's modulus and the Debye temperature [11]. Surface tension and average surface energies for selected inorganic compounds are given in Table 6.1.

In Figure 6.7, different interfacial tensions or energies of metals are correlated with the fusion temperature in the form $T_{fus} \cdot V_m^{-2/3}$. In general the ratio of the average surface energy of the solid to the surface tension of the liquid is around 1.2

Table 6.1 Surface tension or average surface energy of some solid and liquid substances [12, 13].

Substance	σ^{lg} /J m^{-2}
NaCl(l) (1000 °C)	0.098
Al$_2$O$_3$(l) (2050 °C)	0.69
SiO$_2$(l) (1800 °C)	0.307
P$_2$O$_5$(l) (100 °C)	0.06
Cu$_2$S(l) (1200 °C)	0.4
NiS(l) (1200 °C)	0.577
PbS(l)(1200 °C)	0.2
Sb$_2$S$_3$(l) (1200 °C)	0.094
H$_2$O(l) (25 °C)	0.072
	γ^{sg}/J m^{-2}
LiF(s) (25 °C)	0.34
CaF$_2$(s) (25 °C)	0.45
NaCl(s) (25 °C)	0.227
MgO(s) (25 °C)	1.2

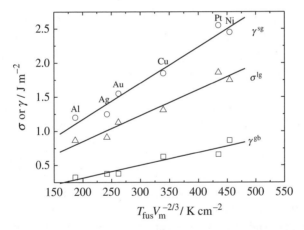

Figure 6.7 Average grain boundary energy, γ^{gb}, surface energy of crystals at 0 K, γ^{sg}, and surface tension σ^{lg} of liquid Al, Ag, Au, Ni and Pt as a function of melting temperature $T_{fus}V_m^{-2/3}$ [8, 11].

at a given temperature. γ^{sg} for Al and Pt at 0 K are 1.2 and 2.55 J m^{-2} respectively, while σ^{lg} at the melting temperature of the metals are 0.865 and 1.86 J m^{-2} [8]. Although less pronounced, similar trends can be found also for molten salts, as shown in Figure 6.8 [14, 15].

γ^{sl} is as a first estimate proportional to the enthalpy of fusion and separate proportionality coefficients are reported for metallic and semi-metallic elements [16].

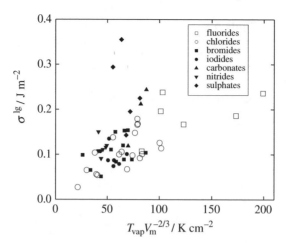

Figure 6.8 Surface tension of fused salts as a function of melting temperature normalized with the molar volume $T_{vap}V_m^{-2/3}$ [14, 15].

The average interfacial energy between the solid and liquid forms of a given element is generally lower than γ^{sg} and σ^{lg}, e.g. 0.093 and 0.24 J m^{-2} for Al and Pt. For materials which change coordination number upon melting γ^{sl} is larger.

Solid–solid interface energies are normally termed grain boundary energies, $\gamma^{ss} = \gamma^{gb}$, and they are comparable to solid–liquid interfacial energies. Average grain boundary energies for some f.c.c. metals are given in Figure 6.7. Again a correlation with $T_{fus} \cdot V_m^{-2/3}$ is observed.

We have now treated surface and interfacial energies without considering the effect of temperature. The excess Gibbs energy of a surface is expected to decrease with temperature since the excess entropy of the surface compared to the bulk is expected to be positive. Intuitively, the surface atoms have more degrees of freedom than atoms in the bulk and thus higher vibrational entropy. In addition, the formation of vacancies and disorder in general at the surface gives a positive configurational contribution to the entropy. Typically $(d\gamma^{sg}/dT)$ is -45 mJ m^{-2} K^{-1} for pure solid elements and slightly less for liquid elements. A semi-empirical equation for predicting the temperature variation of the surface energy of liquids was proposed by van der Waals [17] and Guggenheim [18]. The surface tension is here given as

$$\sigma^{lg} = \sigma_o^{lg}\left(1 - \frac{T}{T_c}\right)^n \tag{6.36}$$

The equation implies that the surface tension becomes zero at the critical temperature, T_c, where the two phases become indistinguishable. The exponent n has been determined to be around 1.2 for metals [11].

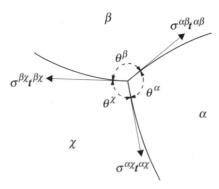

Figure 6.9 Two-dimensional projection of equilibrium at a plane of contact between three phases α, β and χ where the angles between the three two-phase boundaries meeting in a line of contact are denoted θ^α, θ^β and θ^χ.

Morphology of interfaces

The equilibrium shape of a crystal is, as described above, a polyhedron where the size of the crystal facets is inversely proportional to their surface energy, γ^{sg}. In the present section we will consider other types of interfaces as well and we will show that the interface energies determine the equilibrium morphology of interfaces in general.

A two-dimensional illustration of three phases α, β and χ in equilibrium is shown in Figure 6.9. Two phases coexist in equilibrium in planes perpendicular to the lines indicated in the two-dimensional figure and all three phases coexist along a common line also perpendicular to the plane of the drawing. Each of the three two-phase boundaries, which meet at the point of contact, has a characteristic interfacial tension, e.g. $\sigma^{\alpha\beta}$ for the α–β interface, which tends to reduce the area of the boundary. Here we assume the interfacial tensions to be independent of the orientation and that the surface forces are the only ones present. The three forces for the three boundaries are in mechanical equilibrium if

$$\sigma^{\alpha\beta} t^{\alpha\beta} + \sigma^{\beta\chi} t^{\beta\chi} + \sigma^{\alpha\chi} t^{\alpha\chi} = 0 \tag{6.37}$$

where t^{ij} is a unit vector tangent to the i–j boundary at the point of contact. Using the three angles defined in Figure 6.9, the equilibrium condition becomes

$$\frac{\sigma^{\alpha\beta}}{\sin\theta^\chi} = \frac{\sigma^{\beta\chi}}{\sin\theta^\alpha} = \frac{\sigma^{\alpha\chi}}{\sin\theta^\beta} = 0 \tag{6.38}$$

If the three interfacial tensions are equal, the three angles are also equal: $\theta^i = 120°$.

The conditions for mechanical equilibrium can now be applied to a simple case of great practical importance. Let us consider the interfaces that occur when a liquid phase is brought into equilibrium with a solid surface in a gaseous

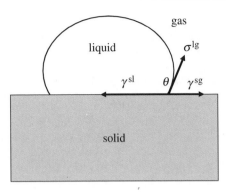

Figure 6.10 Contact angle θ of a liquid drop resting on a solid surface. The definition of the forces used in the figure eliminates the contribution from gravity.

atmosphere. The **wetting** of a solid by a liquid drop is characterized by the **contact angle**, θ, defined in Figure 6.10. The condition for mechanical equilibrium can be rearranged to give the surface energy balance:

$$\gamma^{sg} = \gamma^{sl} + \sigma^{lg} \cos\theta = 0 \tag{6.39}$$

Equation (6.39) was first derived by Young, and is often referred to as the **Young–Dupré** equation. We usually distinguish between full ($\theta < 90°$) and partial wetting ($\theta > 90°$) and an alternative measure of the same property is given by the **wetting coefficient**:

$$k = \frac{\gamma^{sg} - \gamma^{sl}}{\sigma^{lg}} = \cos\theta \tag{6.40}$$

A solid is not wetted if $k \leq -1$, partly wetted for $-1 < k < 1$ and fully wetted for $k > 1$. Wetting is favoured when the difference ($\gamma^{sg} - \gamma^{sl}$) approaches and becomes larger than σ^{lg}. In this case the interaction between the droplet and the substrate increases and the contact angle decreases. It follows that materials with high surface energy are better substrates for deposition of another phase than substrates with low surface energy. One consequence is that metal surfaces are often readily wetted while polymeric surfaces often are not.

The **sessile drop technique** for determination of interfacial energies is based on the configuration shown in Figure 6.10. If the surface tension of the liquid is known, the difference between the interfacial energies of the solid–gas and solid–liquid interfaces can be determined directly by measurement of the contact angle. Experimentally it is difficult to obtain reproducible data on wetting because of two factors: the influence of the surface roughness and the sensitivity of the interfacial energies to the presence of surface-active species. These species may be introduced through contaminations from the surrounding atmosphere or they may be present in the materials used.

Interfacial tensions also play an important role in the distribution of phases in polycrystalline solids. The presence of secondary phases is quite typical, since as grain growth proceeds non-soluble impurities accumulate, usually at the grain boundaries. In other cases the secondary phases may become embedded in the majority phase if the grain boundaries are not pinned at the inclusions. In powder metallurgy and ceramic technology secondary phases are often introduced on purpose in order to enhance sintering or inhibit grain growth. Generally, an equilibrium situation is difficult to obtain due to slow kinetics when governed by solid (bulk or grain boundary) diffusion. When a secondary liquid phase is present the kinetics is governed by liquid diffusion and equilibrium situations are more likely to be reached.

The distribution of the liquid is determined by the interfacial energy between the liquid and the solid matrix relatively to the grain boundary energy. An example is shown in Figure 6.11(a), where an important characteristic of grain boundaries, the

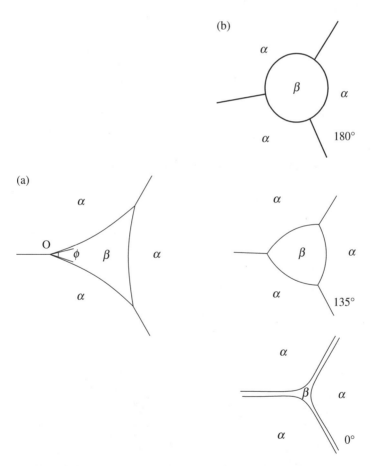

Figure 6.11 (a) Definition of the dihedral angle, ϕ, at a junction of three grain boundaries in a polycrystalline solid. (b) Schematic illustration of the shape of an inclusion phase for different dihedral angles.

dihedral angle ϕ, is defined. The dihedral angle at the point O is at equilibrium defined through

$$\gamma^{\alpha\alpha} - 2\gamma^{\alpha\beta} \cos\left(\frac{\phi}{2}\right) = 0 \qquad (6.41)$$

Depending on the value of the two interfacial energies the dihedral angle can take any value from 0 to 180°. The shapes taken by the secondary phase for different dihedral angles are illustrated in Figure 6.11(b). Since the dihedral angle governs the distribution of secondary phases, the mechanical properties of polycrystalline solids are to a great extent determined by the interfacial energies, although it is also evident that the amount of the secondary phase is important. In composites the strength of the material can be modified by changing the interfacial tension and thereby the distribution of phases. The fracture toughness of the material is here to a large degree determined by the mechanical strength of the interface between the two phases.

Our final example of the effect of interfacial energies relates to microscopic studies of grain sizes and grain distributions in sintered materials. The materials are typically polished and then thermally etched (annealing in air or an inert atmosphere at temperatures significantly below the sintering temperature). After polishing, the grain boundaries are not easily seen by electron microscopy. On thermal annealing the surface relaxes and the surface microstructure become modified. A schematic illustration of the effect of thermal etching on a grain boundary is given in Figure 6.12. The dihedral angle defines the microstructure after etching and the relaxed surface microstructure is much more visible; the grain boundaries are easily seen. The surface of the ceramic material $La_{0.5}Sr_{0.5}Fe_{0.5}Co_{0.5}O_3$ shown in Figure 6.13 constitutes an excellent example.

In cases where the interfacial energy is dependent on orientation, the equilibrium condition (6.41) does not hold [19]. Some grain boundaries will then represent higher Gibbs energies than others, and if kinetics allow for reorientation, certain grain boundaries will become dominant. However, in most cases the kinetics of

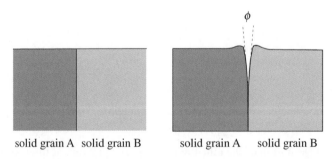

solid grain A   solid grain B   solid grain A   solid grain B

Figure 6.12 Grain boundary after polishing (a) and after the subsequent thermal etching (b).

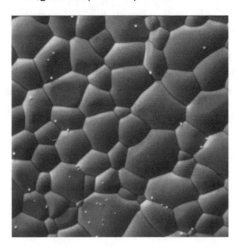

Figure 6.13 Surface of thermally etched $La_{0.5}Sr_{0.5}Fe_{0.5}Co_{0.5}O_3$, a polycrystalline ceramic material .

reorientation of the crystal lattice in order to reduce the interfacial energy is slow and the distribution of grain boundaries is thus not determined by thermodynamics.

6.2 Surface effects on heterogeneous phase equilibria

For small particles with large curvature the surface has, as previously stated, a significant effect on the thermodynamics, and the concepts developed apply to all types of interfaces between solid, liquid and gas.

The fact that the curvature of the surface affects a heterogeneous phase equilibrium can be seen by analyzing the number of degrees of freedom of a system. If two phases α and β are separated by a planar interface, the conditions for equilibrium do not involve the interface and the Gibbs phase rule as described in Chapter 4 applies. On the other hand, if the two coexisting phases α and β are separated by a curved interface, the pressures of the two phases are no longer equal and the Laplace equation (6.27) (eq. 6.35 for solids), expressed in terms of the two principal curvatures of the interface, defines the equilibrium conditions for pressure:

$$p^\alpha - p^\beta = \sigma^{\alpha\beta}(c_1 + c_2) \tag{6.42}$$

Equation (6.42) introduces a new independent variable of the system: the **mean curvature** $c = \frac{1}{2}(c_1 + c_2)$. This variable must be taken into account in the Gibbs phase rule, which now reads $F + Ph = C + 2 + 1$. The number of degrees of freedom (F) of a two-phase system ($Ph = 2$) with a curved interface is given by

$$F = C + 1 \tag{6.43}$$

Effect of particle size on vapour pressure

In this first example, a single-component system consisting of a liquid and a gas phase is considered. If the surface between the two phases is curved, the equilibrium conditions will depart from the situation for a flat surface used in most equilibrium calculations. At equilibrium the chemical potentials in both phases are equal:

$$\mu^l = \mu^g \tag{6.44}$$

For any reversible change

$$d\mu^l = d\mu^g \tag{6.45}$$

At constant temperature $d\mu = Vdp$ and eq. (6.45) becomes

$$V^g dp^g = V^l dp^l \tag{6.46}$$

Here V^g and V^l are the molar volume of the two phases, but the subscript m is not used for simplicity. The pressure of the two phases is related by the Laplace equation (6.27), which for a spherical liquid droplet surrounded by its own vapour becomes, in differential form,

$$d(p^g - p^l) = d\left(\frac{2\sigma^{lg}}{r}\right) \tag{6.47}$$

Here $2/r = [(1/r_1) + (1/r_2)]$ since $r_1 = r_2$. Combining eqs. (6.46) and (6.47) yields

$$\frac{V^g - V^l}{V^l} dp^g = d\left(\frac{2\sigma^{lg}}{r}\right) \tag{6.48}$$

Assuming the gas to be ideal ($V^g = RT/p^g$) and noting that $V^g - V^l \approx V^g$ we obtain

$$\frac{RT}{V^l}\frac{dp^g}{p^g} = d\left(\frac{2\sigma^{lg}}{r}\right) \tag{6.49}$$

If the pressure dependence of the molar volume of the liquid is neglected, integration from a flat interface ($r = \infty$) yields

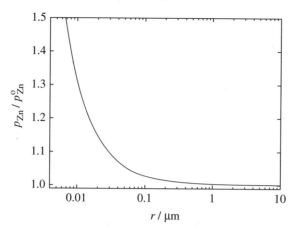

Figure 6.14 The vapour pressure of Zn over a spherical droplet of molten Zn at the melting temperature as a function of the droplet radius. $p_{Zn}^o = 2 \cdot 10^{-4}$ bar, $\sigma^{lg} = 0.78$ J m^{-2} and $\rho = 6.58$ g cm^{-3} [8].

$$\ln\frac{p^g}{p^g_{r=\infty}} = \frac{V^l}{RT}\frac{2\sigma^{lg}}{r} \tag{6.50}$$

Equation (6.50) is often referred to as the **Thomson's** (or Kelvin's) **equation**. As an example of the effect of this equation, the vapour pressure of a spherical droplet of molten Zn at the melting temperature is shown as a function of the droplet radius in Figure 6.14.

A consequence of the decreasing vapour pressure with increasing size of the droplet is that in a distribution of droplets the larger droplets will grow at the expense of the smaller ones; a fact that will be discussed more thoroughly below.

Effect of bubble size on the boiling temperature of pure substances

We now return to the equilibrium condition (eq. 6.44), and assume that the liquid is subjected to a constant pressure, p^l. For reversible changes eq. (6.44) becomes

$$-S^g dT + V^g dp^g = -S^l dT \tag{6.51}$$

which may be rearranged to

$$V^g dp^g = (S^g - S^l)dT = (H^g - H^l)\frac{dT}{T} \tag{6.52}$$

Let us consider a spherical bubble of vapour inside its coexisting liquid. Again the gas phase is assumed to be ideal, and eq. (6.52) becomes

$$\frac{\mathrm{d}T}{T^2} = \frac{R}{\Delta_{vap}H_m} \frac{\mathrm{d}p^g}{p^g} \tag{6.53}$$

The pressure can be substituted by the mean curvature through the Laplace equation, for which

$$\frac{\mathrm{d}T}{T^2} = \frac{R}{\Delta_{vap}H_m} \frac{\mathrm{d}(2\sigma^{lg}/r)}{(p^l + [2\sigma^{lg}/r])} \tag{6.54}$$

If the enthalpy of vaporization is assumed to be independent of the curvature, integration of eq. (6.54) from a flat surface ($r = \infty$) yields (T constant)

$$\left(\frac{1}{T}\right)_{r=\infty} - \left(\frac{1}{T}\right)_r = \frac{R}{\Delta_{vap}H_m} \ln\left(1 + \frac{2\sigma^{lg}/r}{p^l}\right) \tag{6.55}$$

The boiling temperature of molten Na is plotted versus the radius of the vapour bubble in Figure 6.15. The boiling temperature is increased by several hundred degrees for a gas bubble with radius 1 μm relative to a flat gas–liquid interface. If the liquid is free of impurities and heterogeneous interfaces, substantial superheating of the liquid above its bulk boiling temperature is possible, as also discussed in Chapter 5.

The effect of curvature is much more pronounced for the thermodynamics of a gas bubble than for the liquid droplet. The curvature is a pressure effect, which is much larger for gases than for condensed phases, reflecting the much larger molar volume of the gas.

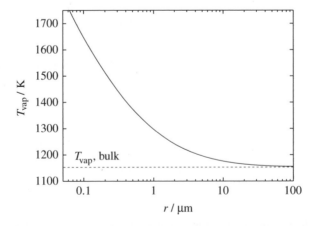

Figure 6.15 The boiling temperature of Na as a function of the radius of a vapour bubble surrounded by molten Na. $\Delta_{vap}H_m = 101.3 \text{ kJ mol}^{-1}$ and $\sigma^{lg} = 0.19 \text{ J m}^{-2}$ [8].

Solubility and nucleation

We will now consider a case where a spherical crystal with radius r of a single component solid phase is surrounded by a liquid with more than one component. The differential of the Laplace equation (6.27) is

$$d(p^s - p^l) = d\left(\frac{2\gamma^{sl}}{r}\right) \tag{6.56}$$

We will consider a reversible change in the system at constant temperature and pressure of the liquid phase p^l. Furthermore, we will assume that the equilibrium concentration of all components in the liquid, except for the single component, i, of the solid phase, is fixed. The equilibrium condition yields

$$d\mu_i^l = d\mu_i^s = V_i^s dp^s = V_i^s d\left(\frac{2\gamma^{sl}}{r}\right) \tag{6.57}$$

Assuming that V_i^s is independent of pressure we obtain

$$(\mu_i^l)_r - (\mu_i^l)_{r=\infty} = V_i^s \frac{2\gamma^{sl}}{r} \tag{6.58}$$

where $(\mu_i^l)_r$ is the chemical potential of i in the liquid in equilibrium with a solid phase of radius r. By expressing the chemical potential in terms of activity (eq. 3.11), eq. (6.58) can be rewritten as

$$\ln\frac{(a_i^l)_r}{(a_i^l)_{r=\infty}} = \frac{V_i^s}{RT}\frac{2\gamma^{sl}}{r} \tag{6.59}$$

Furthermore, if the liquid is assumed to be ideal the activity of a component is equal to the mole fraction of the component. Now the mole fraction of i in the liquid phase can be derived as a function of the radius of the solid phase:

$$\ln\frac{(x_i^l)_r}{(x_i^l)_{r=\infty}} = \frac{V_i^s}{RT}\frac{2\gamma^{sl}}{r} \tag{6.60}$$

The important consequence of eq. (6.60) is that the solubility of the solid increases with decreasing radius of crystal. Although the effect is small this illustrates the need for super-saturation on homogeneous nucleation in a liquid. Super-saturation is necessary in order to obtain nucleation since the solubility of the nuclei is higher than for the bulk. In addition, the interface energy is also important for the kinetics of nucleation, as illustrated below by classical nucleation theory.

In classical nucleation theory the Gibbs energy of a nucleus is considered as the sum of contributions from the bulk and the surface. Let us consider nucleation of a spherical crystal from its liquid below its melting temperature at 1 bar. The difference in Gibbs energy between a nucleus with radius r and its liquid is

$$\Delta_{1-s}G = -\frac{4}{3}\pi r^3 \left(\frac{\rho}{M}\right)\Delta_{fus}G_m + 4\pi r^2 \gamma^{sl} \tag{6.61}$$

where $\Delta_{fus}G_m$ is the molar Gibbs energy of melting, (M/ρ) is molar volume (M is molar mass and ρ is density), $\frac{4}{3}\pi r^3$ is the volume of the spherical nuclei and γ^{sl} is the surface energy of the solid–liquid interface. Here the effect of curvature has been neglected and the surface of the nucleus is assumed to be planar. Since the two terms in eq. (6.61) have opposite sign, the Gibbs energy goes through a maximum as a function of r, the surface term dominating for small r and the bulk term at large r. This maximum corresponds to the **thermodynamic barrier** to nucleation. The **critical radius** corresponding to a maximum in Gibbs energy is determined by differentiating eq. (6.61) with respect to r:

$$\frac{d\Delta_{1-s}G}{dr} = -4\pi r^2 \left(\frac{\rho}{M}\right)\Delta_{fus}G_m + 8\pi r\gamma^{sl} \tag{6.62}$$

At the maximum $d\Delta_{1-s}G/dr = 0$ and the critical radius is $r^* = [2(M/\rho)\gamma^{sl}]/\Delta_{fus}G_m$. Finally, by substituting $r = r^*$ into eq. (6.61) the thermodynamic barrier for nucleation is obtained:

$$\Delta_{1-s}G^* = \frac{16\pi(\gamma^{sl})^3 M^2}{3\rho^2 \Delta_{fus}G_m^2} \tag{6.63}$$

Since $\Delta_{fus}G_m$ increases with decreasing temperature, the critical radius and the thermodynamic barrier decrease with decreasing temperature. At the same time the thermodynamic driving force for nucleation is increasing. This is illustrated for crystallization of aluminium in Figure 6.16(a). The Gibbs energy of a nucleus of aluminium is given as a function of the radius of the nucleus in Figure 6.16(b) (at $T/T_{fus} = 0.95$). For small nuclei the surface term dominates. Above the critical radius the bulk contribution will stabilize the nuclei.

Ostwald ripening

During sintering of granular solids (for example ceramics or hard metals) grain growth may occur by a dissolution–precipitation mechanism if a secondary liquid phase is present. The chemical driving force for grain growth by this mechanism is

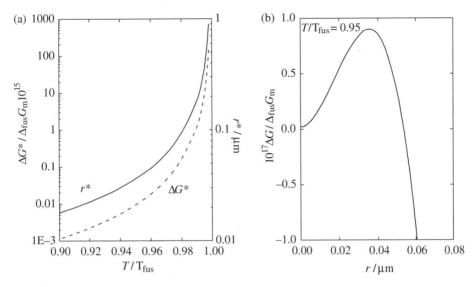

Figure 6.16 (a) The critical radius (r^*) and thermodynamic barrier for nucleation of Al (ΔG^*) versus degree of supercooling T/T_{fus}. (b) The Gibbs energy of a spherical Al crystal relative to the supercooled Al(l) as a function of its radius. $\Delta_{fus}H_m = 10.794$ kJ mol^{-1}, T_{fus} = 933.47 K and $\rho = 2.55$ g cm^{-3} [8].

derived by looking at the system we considered for our analysis of solubility and nucleation. A spherical crystal with radius r of the single-component solid is surrounded by a multi-component liquid phase. Equation (6.58) can be used to find an expression for the difference in chemical potential between to particles of different radii:

$$(\mu_i^s)_{r'} - (\mu_i^s)_{r''} = 2V_i^s \gamma^{sl}\left(\frac{1}{r'} - \frac{1}{r''}\right) \tag{6.64}$$

Equation (6.64) shows that the chemical potential of i in the smaller grain (r') is higher than in the larger (r'') and consequently larger grains will grow at the expense of the smaller ones. This phenomenon is known as **Ostwald ripening**. Ostwald ripening plays an important role in materials science, for example in relation to grain growth and elimination of pores (pore ripening) during the final stage of sintering of refractory metals or ceramics. The difference in the chemical potential between a spherical Au particle with radius 10 μm and a smaller spherical Au particle of radius r is shown in Figure 6.17. The Gibbs energy difference becomes significant below 0.1 μm, i.e. when the ratio between the radius of the two particles is larger than 100.

Effect of particle size on melting temperature

It is becoming increasingly more popular to prepare sub-micrometre sized artificially engineered structures with new properties. In particular, the tailoring of

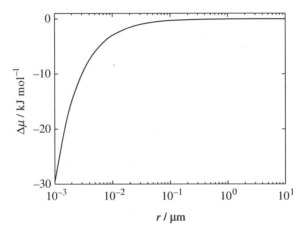

Figure 6.17 The difference in the chemical potential of Au(s) between a spherical particle with radius 10 μm and a smaller particle with radius r. $\rho = 18.4$ g cm^{-3} and $\gamma^{sl} = 1.38$ J m^{-2} [21].

electronic properties of semiconductors by reducing the particle size is important. The thermodynamics is also affected. It has been known for a relatively long time that particle size affects the melting temperature of pure substances. Two simple expressions for freezing temperature depression with origin in size have been derived by Hansen [20] and Buffat and Borel [21]. In the first approach, the equilibrium between a solid and a liquid droplet with the same mass is considered. The pressure and temperature contribution to the chemical potential can be expressed as a power series:

$$\mu(T, p) = \mu^* + \left(\frac{\partial \mu}{\partial T}\right)_p (T - T^*) + \left(\frac{\partial \mu}{\partial p}\right)_T (p - p^*) + \ldots \tag{6.65}$$

where μ^* is the chemical potential at a chosen reference state at T^* and p^*. The partial derivatives of the chemical potential with respect to temperature and pressure are $-S$ and V respectively, and the following expression for the chemical potential is obtained:

$$\mu(T, p) = \mu^*(T^*, p^*) - S(T - T^*) + \frac{M}{\rho}(p - p^*) + \ldots \tag{6.66}$$

Here M/ρ is the molar volume of the substance. At equilibrium, the chemical potential of the liquid and solid droplet are equal:

$$\mu^s(T, p^s) = \mu^l(T, p^l) \tag{6.67}$$

The equilibrium conditions require equal temperatures but different pressures due to the curvature. Using eq. (6.66) for both phases, the equilibrium condition, eq. (6.67), yields

$$\mu_1^*(T^*, p^*) - \mu_s^*(T^*, p^*) + (S^l - S^s)(T - T^*)$$
$$+ \frac{M}{\rho_1}(p^l - p^*) - \frac{M}{\rho_s}(p^s - p^*) = 0 \tag{6.68}$$

Here ρ_s and ρ_1 are the density of the solid and liquid and p^s and p^l the pressures inside the solid and liquid droplets, respectively. Rearrangement of eq. (6.68) yields

$$-\frac{\Delta_{fus}H_m}{T^*}(T - T^*) + \frac{M}{\rho_1}(p^l - p^*) - \frac{M}{\rho_s}(p^s - p^*) = 0 \tag{6.69}$$

where $\Delta_{fus}H_m$ is the enthalpy of fusion of the substance. The pressures inside the particles are given by the Laplace equation as

$$p^s = p^g + \frac{2\gamma^{sg}}{r^s} \tag{6.70}$$

$$p^l = p^g + \frac{2\sigma^{lg}}{r^l} \tag{6.71}$$

where p^g, the external pressure in the surrounding vapour, is set to p^*. Substituting eqs. (6.70) and (6.71) into (6.69) gives

$$-\frac{\Delta_{fus}H_m}{T^*}(T - T^*) + \frac{M}{\rho_1}\frac{2\sigma^{lg}}{r^l} - \frac{M}{\rho_s}\frac{2\gamma^{sg}}{r^s} = 0 \tag{6.72}$$

The difference in curvature between the liquid (r^l) and the solid (r^s) particles can be found since they are related by $\rho_1 V^l = \rho_s V^s$ and $V^i = \frac{4}{3}\pi(r^i)^3$ which gives

$$r^l = \left(\frac{\rho_s}{\rho_1}\right)^{1/3} r^s \tag{6.73}$$

Here, we have used the fact that the masses of the liquid and solid droplet are equal. Substituting eq. (6.73) in eq. (6.72) gives the following equation for the melting temperature as a function of the radius r of the solid phase:

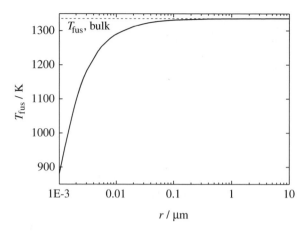

Figure 6.18 The suppression of the melting temperature of a spherical gold particle as a function of its radius. $\Delta_{fus}H_m = 12.35$ kJ mol^{-1}, $T_{fus} = 1336$ K, $\rho_l = 17.28$ g cm^{-3}, $\rho_s = 18.4$ g cm^{-3}, $\sigma^{lg} = 1.135$ J m^{-2}, $\gamma^{sg} = 1.38$ J m^{-2} [21].

$$1 - \frac{(T_{fus})_r}{(T_{fus})_{r=\infty}} = \frac{2M}{r^s \rho_s \Delta_{fus} H_m} \left[\gamma^{sg} - \sigma^{lg} \left(\frac{\rho_s}{\rho_l} \right)^{2/3} \right] \tag{6.74}$$

where $(T_{fus})_{r=\infty} = T^*$ and $(T_{fus})_r$ are the melting temperature of a bulk material and a particle with radius r^s, respectively.

An alternative expression can be derived by an approach in which the solid particle is considered to be embedded in a thin liquid overlayer [20, 21]. Both models give the same qualitative relationship between r and T_{fus}, but the physical interpretation is somewhat different. The calculated melting point of gold versus the radius of the gold particle is shown in Figure 6.18. The estimated temperature of fusion versus particle size is in relatively good agreement with experimental data [21].

The considerations presented above could in principle be extended to binary systems and phase diagrams. In this case one might imagine that the solid particle is embedded in the coexisting liquid, and that the pressure gradient across the solid–liquid interface is determined by the solid–liquid interfacial energy. Like the bulk thermodynamics, the interfacial thermodynamics also depend on composition [22]. An approach to estimating the effect of particle size on phase diagrams has recently been reported [23, 24]. The estimates consider a hypothetical equilibrium between spherical solid and liquid particles at constant pressure. The difference in the Gibbs energy of the liquid and solid spheres is described by one bulk and one surface term. Since the surface tension of liquids is usually lower than the surface energy of solids, the liquidus and solidus lines are suppressed to lower temperatures as the radius is reduced [23, 24].

Particle size-induced phase transitions

It has been shown for several materials that synthesis of fine-grained powders has resulted in the formation of a polymorph other than the stable bulk polymorph. A typical example is that nanocrystalline alumina is usually found as γ-Al$_2$O$_3$, while the stable polymorph for bulk alumina is α-Al$_2$O$_3$. It is commonly assumed that these metastable (relative to the bulk) structures are adopted in order to lower the total Gibbs energy of the material through a decrease in surface energy. The enthalpy difference between γ-Al$_2$O$_3$ and α-Al$_2$O$_3$ is shown as a function of the surface area in Figure 6.19. The average surface energy of γ-Al$_2$O$_3$ is lower than for α-Al$_2$O$_3$, and γ-Al$_2$O$_3$ becomes energetically more stable than α-Al$_2$O$_3$ when the surface area exceeds ~135 m^2 g^{-1} (smaller than 12 nm grain size) [25].

A difference in surface energy will also affect the equilibrium transition temperature between two polymorphs when measured as a function of the particle size. An expression for the variation of the phase transition temperature between a high-surface energy, low-temperature polymorph α and a low-surface energy, high-temperature polymorph β can be described using the phenomenology of the solid–liquid phase transition described above. The transition temperature versus the size of a grain is now expressed as

$$1 - \frac{(T_{trs})_r}{(T_{trs})_{r=\infty}} = \frac{2M}{r\rho_\alpha \Delta_{trs}H_m}\left[\gamma^{\alpha g} - \gamma^{\beta g}\left(\frac{\rho_\alpha}{\rho_\beta}\right)^{2/3}\right] \tag{6.75}$$

where $(T_{trs})_{r=\infty}$ and $(T_{trs})_r$ are the bulk transition temperature and the transition temperature for a particle with radius r respectively. Equation (6.75) establishes that depression of the phase transition temperature is only possible when $\gamma^{\alpha g} > \gamma^{\beta g}(\rho_\alpha/\rho_\beta)^{2/3}$. In Figure 6.20 the monoclinic to tetragonal phase transition

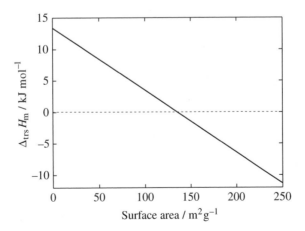

Figure 6.19 The enthalpy of transition of γ-Al$_2$O$_3$ to α-Al$_2$O$_3$ as a function of the surface area of the nanocrystalline particles [25].

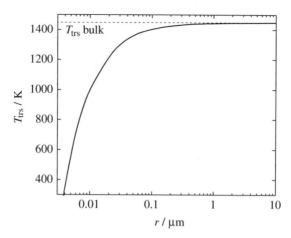

Figure 6.20 The transition temperature between monoclinic and tetragonal zirconia versus grain size. $\Delta_{trs}H_m = 5.94$ kJ mol^{-1}, $T_{trs}^{eq} = 1447$ K, $\rho_m = 5.45$ g cm^{-3}, $\rho_t = 5.77$ g cm^{-3}, $\gamma^{mg} = 1.46$ J m^{-2}, $\gamma^{tg} = 1.1$ J m^{-2} [26].

of ZrO$_2$ is shown as a function of the grain size. Monoclinic zirconia becomes stable below 20 nm grain size at room temperature, in reasonable agreement with experimental observations [26].

Due to recent developments in synthesis, the preparation of nanocrystalline polymorphs, which are usually unstable as bulk phases, has been achieved for several materials such as ZrO$_2$, TiO$_2$ and various perovskites. The appearance of these exotic materials does not necessarily mean that they are thermodynamically stable, since the kinetics (templates and surfactants) are probably more important for the processes than the thermodynamics. Adsorption of water may also play an important role as in the case of alumina, but in the data given in Figure 6.19 the effect of water has been accounted for [25].

6.3 Adsorption and segregation

The adsorption of gases on solids can be classified into **physical** and **chemical adsorption**. Physical adsorption is accompanied by a low enthalpy of adsorption, and the adsorption is reversible. The adsorption/desorption characteristics are in these cases often described by adsorption isotherms. On the other hand, chemical adsorption or segregation involves significantly larger enthalpies and is generally irreversible at low temperatures. It is also often accompanied by reconstruction of the surface due to the formation of strong ionic or covalent bonds.

Gibbs adsorption equation

The adsorption of an **adsorbent** of a solution β on an **adsorbate** α is formally described below. The adsorbent is atoms or molecules and the solution a liquid or a gas, while the adsorbate is a solid or liquid phase. Both the solution β and the phase α are in general multi-component systems. The concentration of the adsorbent at the interface between the two phases is described in terms of the adsorption, Γ, defined by equation (6.3). The following section will relate adsorption to surface tension and in this section we will not use superscripts on σ.

The internal energy of the system described in Figure 6.2 is

$$U = TS - p^\alpha V^\alpha - p^\beta V^\beta + \sum_{i=1}^{C} n_i \mu_i + \sigma A_s \qquad (6.76)$$

By subtracting the internal energy of the two homogeneous phases adjacent to the dividing surface from equation (6.76) the internal energy of the dividing surface is obtained:

$$U^\sigma = TS^\sigma + \sum_{i=1}^{C} n_i^\sigma \mu_i + \sigma A_s \qquad (6.77)$$

Differentiation of eq. (6.77), when combined with eq. (6.14), gives the Gibbs–Duhem equation in internal energy for a system where the surface energy is not negligible:

$$S^\sigma \, dT + \sum_{i=1}^{C} n_i^\sigma \, d\mu_i + A_s \, d\sigma = 0 \qquad (6.78)$$

Reorganization of eq. (6.78) and using eq. (6.3) leads to the **Gibbs adsorption equation**:

$$d\sigma = -\frac{S^\sigma}{A_s} \, dT - \sum_{i=1}^{C} \Gamma_i \, d\mu_i \qquad (6.79)$$

For solid–gas and liquid–gas interfaces, where β is a gas phase (eq. 6.79) can be further simplified if the adsorbate contains only two components, A and B, since changes in the chemical potential of the two components of the adsorbate due to a change in mole fraction are related by eq. (1.92) as

$$d\mu_A = -\frac{x_B}{x_A} \, d\mu_B \qquad (6.80)$$

By substituting eq. (6.80) into eq. (6.79) we obtain at constant temperature

$$-\left(\frac{d\sigma}{d\mu_B}\right)_T = \left[\Gamma_B - \left(\frac{x_A}{x_B}\right)\Gamma_A\right] \qquad (6.81)$$

where x_A and x_B are the mole fraction of A and B in the adsorbate α. This equation can be further developed.

The adsorption Γ depends on the position of the Gibbs dividing surface and it is therefore convenient to define a new function, the relative adsorption, that is not dependent on the dividing surface. The absorption of component i at the interface is defined by eq. (6.3) as

$$\Gamma_i = \frac{n_i^\sigma}{A_s} = \frac{1}{A_s}(n_i - n_i^\alpha - n_i^\beta) = \frac{1}{A_s}(n_i - c_i^\alpha V^\alpha - c_i^\beta V^\beta) \qquad (6.82)$$

where $c_i^\alpha = n_i^\alpha/V^\alpha$ and $c_i^\beta = n_i^\beta/V^\beta$ are the concentrations of i in the two phases adjacent to the interface and V^α and V^β are the volumes of these phases. Since eq. (6.82) can be rewritten in terms of the total volume as

$$\Gamma_i = \frac{1}{A_s}[n_i - c_i^\alpha V - (c_i^\beta - c_i^\alpha)V^\beta] \qquad (6.83)$$

the adsorption of component A can be expressed as

$$\Gamma_A = \frac{1}{A_s}[n_A - c_A^\alpha V - (c_A^\beta - c_A^\alpha)V^\beta] \qquad (6.84)$$

V^β can be eliminated by combining eqs. (6.83) and (6.84). We thus obtain the following important expression:

$$\Gamma_i - \Gamma_A\left(\frac{c_i^\alpha - c_i^\beta}{c_A^\alpha - c_A^\beta}\right) = \frac{1}{A_s}\left[(n_i - c_i^\alpha V) - (n_A - c_A^\alpha V)\left(\frac{c_i^\alpha - c_i^\beta}{c_A^\alpha - c_A^\beta}\right)\right] \qquad (6.85)$$

The advantage of this expression is that although the adsorption of each component depends on the Gibbs dividing surface, the right-hand side is independent of its position. We can thus define the **relative adsorption** of component B with respect to component A:

$$\Gamma_B^{(A)} = \Gamma_B - \Gamma_A\left(\frac{c_B^\alpha - c_B^\beta}{c_A^\alpha - c_A^\beta}\right) \qquad (6.86)$$

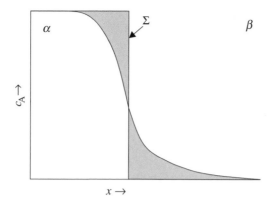

Figure 6.21 Schematic illustration of the concentration of component A across an interface. The Gibbs dividing surface is positioned such that it gives zero adsorption of component A since the algebraic sum of the two shaded areas with opposite sign is zero.

Since $\Gamma_B^{(A)}$ is independent of the position of Σ, we can choose the position of Σ to correspond to $\Gamma_A = 0$, as illustrated for a schematic two-dimensional interface in Figure 6.21. The two shaded areas, above and below the interface, are equal and give zero adsorption of A. Recall that it is only for planar surfaces that the position of the Gibbs dividing surface is arbitrary, and in the following we will restrict our treatment to planar surfaces only.

Relative adsorption and surface segregation

We can now use the relative adsorption to describe a two-component system at constant temperature. The relative adsorption of B with respect to A for $\Gamma_A = 0$ is given by eq. (6.81) as

$$\Gamma_B = \Gamma_B^{(A)} = -\left(\frac{\partial \sigma}{\partial \mu_B}\right)_T \tag{6.87}$$

Thus the adsorption of B at the interface is given by the variation of the surface tension with chemical potential.

Equation (6.87) is the basis for most adsorption measurements associated with liquid solutions. When $(\partial \sigma / \partial \mu_B)_T$ is negative, Γ_B is positive and there is an excess of the solute at the interface. For $(\partial \sigma / \partial \mu_B)_T > 0$, Γ_B is negative and there is a deficiency of the solute at the interface. In other words, solutes that reduce the surface tension are enriched at the surface.

The relative adsorption of component B is clearly affected by its partial pressure. Let us consider a binary system A–B where $\Gamma_A = 0$ and B is an ideal gas with partial pressure p_B. The chemical potential of B can be expressed in terms of the partial pressure of B and this is then also the case for the relative adsorption of B

$$\Gamma_B = \Gamma_B^{(A)} = -\frac{1}{RT}\left(\frac{\partial \sigma}{\partial \ln p_B}\right)_T \tag{6.88}$$

The effect of an impurity on the surface energy is often discussed in terms of the **surface activity** of the impurity B, j_B, defined as the slope of the surface tension or energy versus composition at infinite dilution:

$$j_B = -\left(\frac{\partial \sigma}{\partial x_B}\right)_{x_A \to 1} \tag{6.89}$$

Using the Gibbs absorption equation and assuming Henry's law (eq. 3.42)

$$\begin{aligned} j_B &= -\frac{1}{RT}\left(\frac{\partial \sigma}{\partial x_B}\right)_{x_A \to 1} \\ &= -\frac{1}{RT}\left(\frac{\partial \sigma}{\partial \mu_B}\right)_{x_A \to 1}\left(\frac{\partial \mu_B}{\partial x_B}\right)_{x_A \to 1} \\ &= +\frac{1}{RT}\Gamma_B^{(A)}\left(\frac{\partial \mu_B}{\partial x_B}\right)_{x_A \to 1} \\ &= \Gamma_B^{(A)}\left(\frac{\partial \ln a_B}{\partial x_B}\right)_{x_A \to 1} \end{aligned} \tag{6.90}$$

Experimentally j_B is found to be finite. The slope of the relative adsorption versus composition, which is also finite, is referred to as **Henry's law for surfaces**. For electronegative elements on metallic surfaces the surface activity becomes very high, often of the order of 10^3. This means that very small amounts of these elements have a large effect on the surface energy, and that the experimental determination of reliable surface energies needs systems of extreme purity.

The effect of composition on σ with a focus on the dilute limit, is shown for selected systems in Figure 6.22. A considerable degree of segregation must be expected for these example systems. Physically this situation corresponds to solute atoms that have large positive size misfits, and/or large positive enthalpies of mixing and these solutes are thus expected to segregate readily to the surface. Substances that have a large effect on the interfacial energy even at small concentrations are called surface-active species. For most metals, oxygen, sulfur and other elements of group 16 are generally strong surface-active species. Similarly, the surface tension of liquid oxides or halides is strongly influenced by the addition of small amounts of other components that have the opposite acid–base properties. For example the surface tension of silicates is easily modified by addition of basic oxides such as alkali or alkali earth metal oxides.

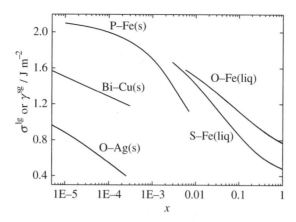

Figure 6.22 Surface tension as a function of the concentration for some surface-active species in solid Fe, Cu and Ag near their melting temperatures [27] and liquid Fe at 1550–1600 °C [28].

In most systems the variation of the surface tension with composition is much lower. Data for some binary liquids are shown in Figure 6.23. A limited degree of segregation is expected in these systems.

Adsorption isotherms

Here we are considering the dynamic equilibrium between molecular species in the gas phase and the adsorbed gas species on a surface. Let us consider the following quasi-chemical equilibrium between the species B in the gas, B_g, and the available sites at the surface of the adsorbate:

$$B_g + V_{MON} \underset{k_d}{\overset{k_a}{\rightleftarrows}} B_{MON} \tag{6.91}$$

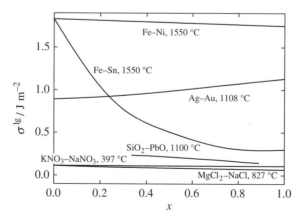

Figure 6.23 Isothermal surface tension versus composition of some binary metals [28], oxide [29] and salt systems [14, 15].

Here V_{MON} and B_{MON} represent the available vacant sites and surface sites occupied by B, respectively, of the first monolayer on a solid absorbate. The equilibrium constant K_L for the reaction is given by the ratio of the rate constant for k_a for adsorption and k_d for desorption

$$K_L = \frac{k_a}{k_d} = \frac{\Gamma_B}{a_B^g(\Gamma_B^{sat} - \Gamma_B)} \tag{6.92}$$

where Γ_B is the adsorption of B at the surface, $\Gamma_B^{sat} - \Gamma_B$ the concentration of vacant sites in the monolayer and a_B^g the activity of B in the gas phase. By introducing the **fractional coverage** of the adsorbate surface θ, eq. (6.92) can be transformed into the **Langmuir adsorption isotherm** [30] given as

$$\frac{\Gamma_B / \Gamma_B^{max}}{1 - \Gamma_B / \Gamma_B^{max}} = \frac{\theta}{1-\theta} = K_L a_B \tag{6.93}$$

where Γ_B^{max} is the maximum adsorption. In this simplest physically realistic adsorption isotherm it is assumed that adsorption cannot precede beyond a monolayer coverage, all sites are equivalent and the ability of a molecule to adsorb is independent of the occupation of the neighbouring sites, i.e. there are no mutual interactions. Typical adsorption isotherms are shown in Figure 6.24 for different values of K_L.

The Langmuir model was extended to include interaction between the adsorbed atoms/molecules by Fowler and Guggenheim [31]. The model now becomes

$$K_L a_B^\beta = \frac{\theta}{1-\theta} \exp\left[-2\left(\frac{z\omega}{k_B T}\right)\theta\right] \tag{6.94}$$

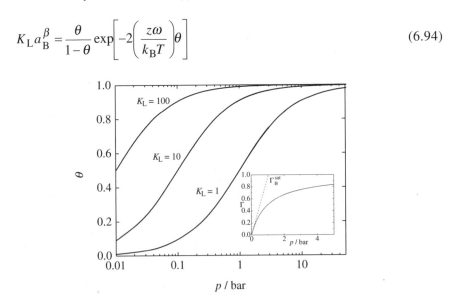

Figure 6.24 Adsorption isotherms following Langmuir adsorption isotherm.

where z is the number of nearest neighbours within the surface layer, ω is the regular solution parameter and a_B^β is the activity of B in β. For $\omega = 0$, the Fowler–Guggenheim adsorption reduces to the Langmuir isotherm.

A major advance in adsorption theory generalized the treatment of monolayer adsorption and incorporated the concept of multilayer adsorption. This is known as the **BET theory** after Brunauer, Emmett and Teller [32]. The adsorption of a gas on a solid surface can be described by

$$\frac{p}{m_A(p_0 - p)} = \frac{1}{m_{MON}C} + \frac{C-1}{m_{MON}C}\left(\frac{p}{p_0}\right)$$

where m_A is the total amount of gas adsorbed and m_{MON} is the quantity of gas adsorbed corresponding to a monolayer. p and p_0 are the pressure and saturation pressure of the absorbent and C is a constant. A plot of $p/[m_A(p_0 - p)]$ versus p/p_0 gives a straight line with intercept $1/m_{MON}C$ and slope $(C-1)/m_{MON}C$. The values of C and m_{MON} may then be obtained from a linear regression of experimental points of m_A versus p/p_0. By using the mean area per molecule adsorbent the surface area of the solid can be calculated from m_{MON}. This method is frequently used to determine specific surface area of porous materials.

Experimentally the enthalpy of adsorption is observed to be a function of the fractional surface coverage θ since it depends on interactions between the adsorbent molecules/atoms and rearrangements of the surface due to the formation of new chemical bonds. In some cases, for example for CO adsorption on single crystal surfaces of metals, the enthalpy of adsorption can change abruptly when the structure of the absorbed layer changes [33]. Although the determination of accurate enthalpies of adsorption is difficult and values reported often vary from one laboratory to another, clear trends in the enthalpy of adsorption are often observed. It is clear that the enthalpy of adsorption of gases like, CO, H_2, N_2 and NH_3 on transition metals decreases when going from left to right along a period [33]. For adsorption of metals on metal oxides, as exemplified by Cu on the (100) surface of MgO, the low-coverage heats of adsorption when the metals are mainly in two-dimensional islands correlate with the bulk sublimation enthalpy of the adsorbent. This suggests that covalent metal–Mg bonding dominates the interaction at low coverage [34]. On adsorption beyond the first monolayer the enthalpy of adsorption approaches the enthalpy of sublimation of Cu. Accurate data on the enthalpy of adsorption are known in only a few systems, and a recent report [35], which shows that the particle size has a much larger effect on the energetics than predicted by equations of the type considered in Section 6.2, suggests that our knowledge on this complex topic is still limited.

References

[1]   J. W. Gibbs, *The Collected Works, Volume I Thermodynamics*. Harvard: Yale University Press, 1948.

[2]   A. I. Rusanov, *Surf. Sci. Reports* 1996, **23**, 247.
[3]   C. H. P. Lupis, *Chemical Thermodynamics of Materials*. Amsterdam: Elsevier Science, 1983.
[4]   G. Wulff, *Z. Krist.* 1901, **23**, 449.
[5]   C. Herring, in *Structure and Properties of Solid Surfaces* (R. Gomer and C. S. Smith, eds.). Chicago: University of Chicago Press, 1953.
[6]   C. Herring, *Phys. Rev.* 1951, **82**, 87.
[7]   W. W. Mullins, *Metal Surfaces: Structure, Energetics and Kinetics*. Metals Park, OH: Am. Soc. Metals, 1963, Ch. 2.
[8]   J. H. Alonso and N. H. March, *Electrons in Metals and Alloys*. London: Academic Press, 1989.
[9]   A. S. Shapski, *J. Chem. Phys.* 1948, **16**, 386.
[10] A. A. V. Grosse, *J. Inorg. Nucl. Chem.* 1964, **6**, 1349.
[11] J. M. Howe, *Interfaces in Materials*. New York: John Wiley & Sons, 1997.
[12] G. A. Somorjai, *Principles of Surface Chemistry*. Englewood Cliffs, NJ: Prentice Hall, 1972.
[13] F. D. Richardson, *Physical Chemistry of Melts in Metallurgy*. London: Academic Press, 1974.
[14] G. J. Jans, U. Krebs, H. F. Siegenthaler and R. P. T. Tomkins, *J. Phys. Chem. Ref. Data*, 1972, **1**, 680.
[15] G. J. Jans, R. P. T. Tomkins , C. B. Allen, J. R. Downey Jr, G. L. Gardner and U. Krebs, *J. Phys. Chem. Ref. Data*, 1975, **4**, 1117.
[16] D. Turnbull, *J. Appl. Phys.* 1950, **21**, 1022.
[17] J. D. van der Waals, *Z. Chem. Phys.* 1894, **13**, 657.
[18] E. A. Guggenheim, *J. Chem. Phys.* 1945, **13**, 253.
[19] C. Herring, in *Physics of Powder Metallurgy* (W. E. Kingston ed.). New York: McGraw-Hill, 1951.
[20] K.-J. Hansen, *Z. Phys.* 1960, **157**, 523.
[21] Ph. Buffat and J.-P. Borel, *Phys. Rev. A* 1976, **13**, 2287.
[22] R. T. DeHoff, *Thermodynamics in Materials Science*. New York: McGraw-Hill, 1993.
[23] T. Tanaka and S. Hara, *Z. Metallkd.* 2001, **92**, 467.
[24] T. Tanaka and S. Hara, *Z. Metallkd.* 2001, **92**, 1236.
[25] J. M. McHale, A. Auroux, A. Perrotta and A. Navrotsky, *Science* 1997, **277**, 788.
[26] R. C. Garvie, *J. Phys. Chem.* 1978, **82**, 218.
[27] E. D. Hondros, in *Precipitation in Solids* (K.C. Russel and H. I. Aaronson eds.). Warrendale, PA: The Metallurgical Society of AIME, 1978.
[28] L. E. Murr, *Interfacial Phenomena in Metals and Alloys*. Reading, MA: Addison-Wesley, 1975.
[29] L. Shartsis, S. Spinner and A. W. Smack, *J. Am. Ceram. Soc.* 1948, **31**, 23.
[30] I. Langmuir, *J. Am. Chem. Soc.* 1918, **40**, 1361.
[31] R. Fowler and E. A. Guggenheim, *Statistical Thermodynamics*. Cambridge: Cambridge University Press, 1939, p. 430.
[32] S. Brunauer, P. H. Emmett and E. Teller, *J. Am. Chem. Soc.* 1938, **60**, 309.
[33] G. A. Somorjai, *Chemistry in Two Dimensions: Surfaces*. Ithaca, NY: Cornell University Press, 1981.
[34] C. T. Campbell and D. E. Starr, *J. Am. Chem. Soc.* 2002, **124**, 9212.
[35] C. T. Campbell, S. C. Parker and D. E. Starr, *Science* 2002, **298**, 811.

Further reading

J. M. Howe, *Interfaces in Materials*. New York: John Wiley & Sons, 1997.

M. J. Jaycock and G. D. Parfitt, *Chemistry of Interfaces*. John Wiley & Sons, New York, 1981.

J. Lyklema, *Fundamentals of Interface and Colloid Science, Vol. 1: Fundamentals*. London: Academic Press, 1991.

L. E. Murr, *Interfacial Phenomena in Metals and Alloys*. Reading, MA: Addison-Wesley, 1975.

G. A. Somorjai, *Chemistry in Two Dimensions: Surfaces*. Ithaca, NY: Cornell University Press, 1981.

G. A. Somorjai, *Introduction to Surface Chemistry and Catalysis*. New York: John Wiley & Sons, 1994.

Trends in enthalpy of formation

The standard enthalpy of formation, $\Delta_f H_m^o$, of a compound at 0 K reflects the strength of the chemical bonds in the compound relative to those in the constituent elements in their standard state. The standard enthalpy of formation of a binary oxide such as CaO is thus the enthalpy change of the reaction

$$Ca(s) + \tfrac{1}{2}O_2\,(g) = CaO(s) \tag{7.1}$$

at $p = 1$ bar.

A number of theoretical approaches can account for the fact that an enthalpy of formation of such a binary oxide or a ternary oxide is large and negative. The stability of a ternary oxide relative to the binary constituent oxides is, however, often small, as demonstrated in Table 7.1 using Mg_2SiO_4 as an example [1]. The enthalpy differences between the three different polymorphs of Mg_2SiO_4 – olivine, β-phase and spinel – are less than 2% of the enthalpy of formation of the polymorphs. These enthalpy differences are comparable in magnitude to the enthalpy

Table 7.1 Magnitudes of enthalpies of various reactions of a ternary oxide using Mg_2SiO_4 as an example (after Navrotsky [1]).

$2Mg(s) + Si(s) + 2O_2(g) = Mg_2SiO_4$ (olivine)	-2170.41 kJ mol^{-1}
$2MgO + SiO_2(s) = Mg_2SiO_4$ (olivine)	-56.61 kJ mol^{-1}
Mg_2SiO_4 (olivine) $= Mg_2SiO_4$ (liq.)	114 kJ mol^{-1}
Mg_2SiO_4 (olivine) $= Mg_2SiO_4$ (β)	29.9 kJ mol^{-1}
Mg_2SiO_4 (β) $= Mg_2SiO_4$ (spinel)	9.1 kJ mol^{-1}

Chemical Thermodynamics of Materials by Svein Stølen and Tor Grande
© 2004 John Wiley & Sons, Ltd ISBN 0 471 492320 2

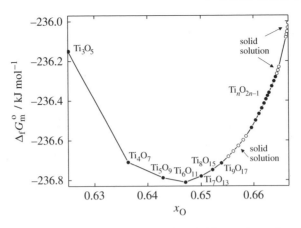

Figure 7.1 The Gibbs energies of formation of stoichiometric and non-stoichiometric compounds in the system $Ti_3O_5-TiO_2$ [4]. Composition given as mole fraction O.

of formation of olivine from the binary oxides, and to the enthalpy of fusion of olivine. The fact that different structural modifications of a given ternary oxide may be close in enthalpy and thereby in Gibbs energy at low temperatures makes predictions of phase diagrams far from trivial. Still, quantum mechanical approaches have in recent years been increasingly able to derive the relative enthalpies of different structural modifications of ordered compounds with a given composition, see e.g. [2, 3] and also Chapter 11.

Although the existence of polymorphism is challenging to theory, the prediction of the compositions of the phases that exist in a given binary or ternary system is in general even more challenging. The Gibbs energy of formation of stoichiometric and non-stoichiometric compounds in the system $Ti_3O_5-TiO_2$, presented graphically in Figure 7.1 [4], constitutes an excellent example. A large number of phases of the homologous series Ti_nO_{2n-1} are close in Gibbs energy and the Gibbs energy differences between different phases or phase mixtures are tiny. In general, a phase with a given stoichiometry must be stable relative to other phases of the same composition, but also relative to the neighbouring phases. A large number of marginally stable and marginally metastable phases exist in many material systems. It is in general much more difficult to predict the compositions of the compounds that exist in a given binary or multi-component system than the relative stability of different polymorphs of a given composition.

The composition and crystal structure of the materials that are formed are complex functions of a large number of factors. Solid compounds may be qualitatively assigned to have ionic, covalent or metallic bonds, and although a number of compounds exist in which one of these bonding schemes dominates, most compounds do not belong to clear-cut categories and even for largely ionic compounds covalent contributions must be taken into consideration. While the valence electrons are localized in ionic and covalent compounds, they are highly delocalized in metals. Schematically the character of bonding in any compound may be indicated

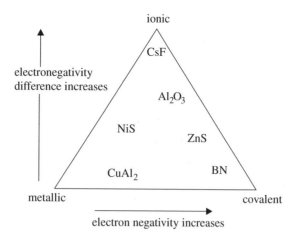

Figure 7.2 Schematic categorization of bonding in solids.

in a triangular diagram, such as Figure 7.2, whose corners are the three extremes of pure covalent, ionic and metallic bonding [5]. The electronegativity difference between the components is a major factor and a number of rationalization schemes have been proposed in which the energetics are characterized in terms of two factors, one related to size and one related to electronegativity.

Most energetic contributions are, as we have discussed, difficult to predict and large experimental efforts have for that reason been devoted to derive systematic trends in the energetics of classes of materials. In this chapter we will try to convey an overview of periodic trends in the thermodynamic properties of inorganic compounds and we will also present selected examples illustrating some of the more usual rationalization schemes. Finally, trends in enthalpy of mixing are treated. Also here we aim to look at trends and rationalization schemes. The chapter is by no means exhaustive – only selected classes of compounds and selected rationalization schemes are discussed.

7.1 Compound energetics: trends

Prelude on the energetics of compound formation

Let us consider the enthalpy of formation of an ionic compound like NaCl, or in general terms MX:

$$M(s) + \tfrac{1}{2}X_2(g) = MX(s) \tag{7.2}$$

This reaction may be analyzed through the thermodynamic cycle given in Figure 7.3 where the following five reactions and associated enthalpy changes are involved:

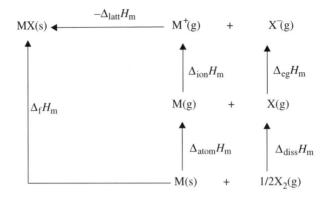

Figure 7.3 Thermodynamic cycle for the formation of MX(s).

$$M^+(g) + X^-(g) = MX(s) \qquad -\Delta_{latt}H_m \qquad (7.3)$$

$$M(s) = M(g) \qquad \Delta_{atom}H_m \qquad (7.4)$$

$$M(g) = M^+(g) + e^-(g) \qquad \Delta_{ion}H_m \qquad (7.5)$$

$$\tfrac{1}{2}X_2(g) = X(g) \qquad \Delta_{diss}H_m \qquad (7.6)$$

$$X(g) + e^- = X^-(g) \qquad \Delta_{eg}H_m \qquad (7.7)$$

The **lattice enthalpy**, $\Delta_{latt}H_m$, is the molar enthalpy change accompanying the formation of a gas of ions from the solid. Since the reaction involves lattice disruption the lattice enthalpy is always large and positive. $\Delta_{atom}H_m$ and $\Delta_{diss}H_m$ are the **enthalpies of atomization** (or sublimation) of the solid, $M(s)$, and the enthalpy of dissociation (or atomization) of the gaseous element, $X_2(g)$. The **enthalpy of ionization** is termed **electron gain enthalpy**, $\Delta_{eg}H_m$, for the anion and **ionization enthalpy**, $\Delta_{ion}H_m$, for the cation.

The enthalpy of formation of the compound MX is now the sum of these five contributions:

$$\Delta_fH_m(MX) = -\Delta_{latt}H_m + \Delta_{atom}H_m + \Delta_{ion}H_m + \Delta_{diss}H_m + \Delta_{eg}H_m \qquad (7.8)$$

While the enthalpy of formation is the property of interest in chemical thermodynamics of materials, many books focus on the lattice enthalpy when considering trends in stability. The static non-vibrational part of the lattice enthalpy can be deconvoluted into contributions of electrostatic nature, due to electron–electron repulsion, dispersion or van der Waals attraction, polarization and crystal field effects. The lattice enthalpy is in the 0 K approximation given as a sum of the potential energies of the different contributions:

$$\Delta_{latt}H_m = \Phi_{electrostatic} + \Phi_{repulsion} + \Phi_{dispersion} + \Phi_{polarization}$$
$$+ \Phi_{crystal\ field} \tag{7.9}$$

and trends in this property become evident only when the relative importance of these different contributions are considered.

The largest contribution, the electrostatic interaction, is due to attraction between ions with opposite charge and repulsion between ions of the same charge. Using the NaCl-type crystal structure as an example, the electrostatic potential energy is evaluated taking one particular M^+ at the body centre of the unit cell as a starting point and calculating the interaction between this particular ion and its neighbours. The central M^+ is surrounded by an octahedron of six X^- ions with each X^--ion at distance r_{MX}. The attractive energy ignoring $X^- - X^-$ interactions is then

$$6\frac{e^2 q_M q_X}{r_{MX}} \tag{7.10}$$

The next nearest neighbours to the central M^+ are 12 M^+ at distance $\sqrt{2}r$. The repulsive cation–cation interaction term is given as

$$-12\frac{e^2 q_M q_X}{\sqrt{2}r_{MX}} \tag{7.11}$$

Correspondingly, the contribution from the third nearest neighbours, 8 X^- at $\sqrt{3}r$ is

$$8\frac{e^2 q_M q_X}{\sqrt{3}r_{MX}} \tag{7.12}$$

and so on. The net attractive energy between the central M^+ and all other ions in the crystals is thus given by the infinite series

$$\Phi = \frac{e^2 q_M q_X}{r_{MX}}\left(6 - \frac{12}{\sqrt{2}} + \frac{8}{\sqrt{3}} - \frac{6}{\sqrt{4}} + \cdots\right) \tag{7.13}$$

The crystal arrangement is hence important for the lattice enthalpy and $V_{electrostatic}$ can be extracted for a specific crystal structure as

$$\Phi_{electrostatic} = \frac{NMe^2 q_M q_X}{r_{MX}} \tag{7.14}$$

where M is the Madelung constant for a given structure and r_{MX} is the shortest MX distance. The Madelung constant depends only on the geometrical arrangement of the point charges defined by the crystal structure.

A second and repulsive energy term must be introduced to take account of the electron–electron repulsion that arises at very short interatomic distances. Several models are used to describe this repulsive term. Often used is the Buckingham potential, which, however, includes both attractive and repulsive components:

$$V_{vdw}(r_{ij}) = A \exp\left(-\frac{r_{ij}}{\rho}\right) - \frac{C}{r_{ij}^{-6}} \tag{7.15}$$

Here the first term reproduces the effective two-body repulsion and the second the effective two-body van der Waals attraction. r_{ij} is the interatomic distance, whereas A, ρ and C are constants that are usually determined empirically. The repulsive term is steeply increasing at low interatomic distances but quickly becomes negligible beyond the nearest neighbour distance.

In general, the electrostatic terms contribute 75 to 90% to the total lattice enthalpy, while the repulsive contribution is about 10 to 20%. It follows that the lattice enthalpy depends largely on the charge and on the relative size of the cation and anion, since these factors dominate the electrostatic term in eq. (7.14). This is reflected in the lattice enthalpies presented in Figures 7.4(a) and (b). The lattice enthalpy of the Na halides becomes less positive from NaF to NaI. Similarly, the lattice enthalpy of the alkali iodides becomes less positive from LiI to CsI. While this effect is clearly important when considering enthalpies of formation, this is not the only factor to be taken into account [6, 7]. As indicated by the thermodynamic cycle in Figure 7.3, the variation in the atomization and ionization enthalpies of the metal atom and in the dissociation and electron gain enthalpies of the non-metal atom must also be considered. The relevant data are given in Figures 7.4(c) and (d).

Periodic trends in the enthalpy of formation of binary compounds

Let us now consider trends in enthalpy of formation and first trends for the alkali halides with a given halide ion, i.e. the alkali fluorides and the alkali iodides. The ease with which a free gaseous ion is formed from the solid metal increases down group 1 of the periodic system. Similarly, the ionization enthalpy becomes less positive as seen in Figure 7.4(d). The enthalpy of formation analyzed in terms of eq. (7.8) involves the formation of the compound from the gaseous ions and thus involves $-\Delta_{latt}H_m$. The effect of the lattice enthalpy is thus opposite to the trend in the atomization and ionization enthalpies and as a result of the balance of these two opposing trends there is often little change in the enthalpy of formation down these groups, as is evident in Figure 7.5(a). For the alkali fluorides the size of the anion implies that the most negative enthalpy of formation is that of LiF. The less endothermic enthalpy of atomization and ionization of Cs gives CsI the higher stability among the alkali iodides.

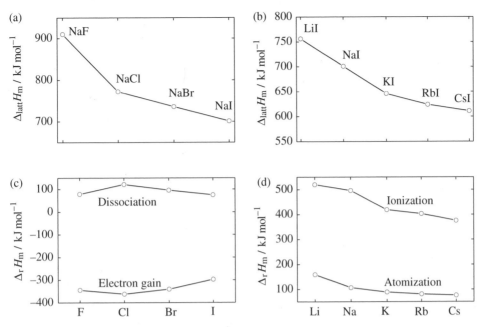

Figure 7.4 Thermodynamic data needed in evaluation of the enthalpy of formation of MX(s). (a) Lattice enthalpy of sodium halides; (b) lattice enthalpy of alkali iodides; (c) electron gain and dissociation enthalpies of halides; (d) ionization and atomization enthalpies of alkali metals.

Larger differences are observed when comparing the enthalpy of formation of the different halides of a given alkali metal. The enthalpy of formation of gaseous halide ions is exothermic since the exothermic electron gain enthalpy in absolute value is larger than the endothermic dissociation enthalpy. Furthermore, the enthalpy of formation of gaseous halide ions becomes less favourable with

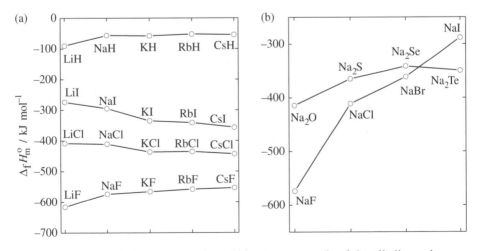

Figure 7.5 Enthalpy of formation of binary compounds of the alkali metals.

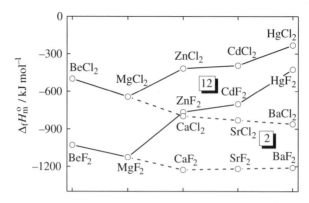

Figure 7.6 Enthalpy of formation of group 2 and 12 dichlorides and difluorides.

increasing size of the halogen atom. The trend in the lattice enthalpy amplifies this effect and the enthalpy of formation of the Na halides varies much more with the halide ion than the alkali iodides did with the alkali metal; see Figure 7.5(b).

The enthalpies of formation of the group 2 halides given in Figure 7.6 show many of the same effects. The fluorides are more stable than the chlorides and the stability increases with increasing size of the cation. The ease of atomization of the metal here increases more on a relative scale than it did for the alkali metals. The group 12 halides are also included in Figure 7.6. Their enthalpies of formation are distinctly less exothermic than those of group 2. The reason for this is the increase by +10 in nuclear charge between Ca and Zn and between Sr and Cd, and of +24 between Ba and Hg. These extra electrons are not screening the nuclear charge effectively, resulting in a much higher ionization enthalpy for the group 12 elements and a less negative enthalpy of formation for the group 12 compounds compared with the group 2 compounds. This destabilization effect is visible also in the enthalpy of formation of the metal oxides of groups 2 and 12, 3 and 13, 4 and 14 and 5 and 15; see Figure 7.7. Here, the enthalpies of formation of the metal oxides of group 13 and periods 4, 5 and 6 are much less exothermic than the for corresponding metal oxides of group 3. The same argument is valid for metal oxides of group 4 versus 14 and for group 5 versus 15.

While we have interpreted the trends observed using a simple thermodynamic cycle where the electrostatic enthalpy is given in terms of formal charges of the ions and their interatomic distance (eq. 7.14), most solid compounds cannot be described properly without taking polarization into consideration. The larger the electronegativity difference between the two elements of a compound, the more polar the compound is. Polarization is more extensive for the lower coordination numbers and depends on the polarizing power and the polarizability of the ions involved in the bonding. The polarizing power increases the smaller the ion is and the higher charge it has, while polarizability is usually larger for large ions – anions with loosely bond electrons. Polarization always leads to a decrease in interatomic distance and thus to an increase in the lattice enthalpy. The difference between

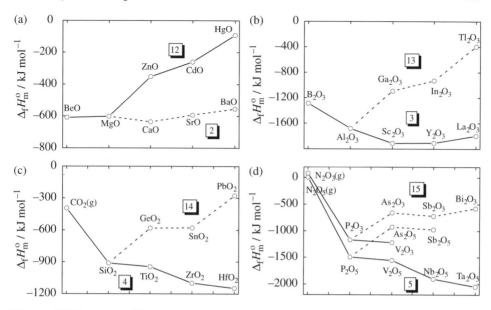

Figure 7.7 Enthalpy of formation of binary oxides of (a) group 2 and 12, (b) 3 and 13, (c) 4 and 14 and (d) 5 and 15 metals.

lattice enthalpies derived using experimental data and the thermodynamic cycle given in Figure 7.3 and calculated lattice enthalpies is often ascribed to polarization effects. However, the use of both pairwise potentials and models that neglect anion–anion short-range interactions is in general questionable for compounds where there is a covalent contribution and directionality of the bonds starts to play a role.

In general, overlap of incompletely filled p orbitals results in large deviations from pure ionic bonding, and covalent interactions result. Incompletely filled f orbitals are usually well shielded from the crystal field and behave as essentially spherical orbitals. Incompletely filled d orbitals, on the other hand, have a large effect on the energetics of transition metal compounds and here the so-called crystal field effects become important.

All the d orbitals are equal in energy for an isolated atom. In an electric field of lower than spherical symmetry caused by the surrounding ligands, ligand field or crystal field splitting is observed. In an octahedral environment the five d orbitals of a transition metal are no longer degenerate but split into two groups; the lower energy t_{2g} and the higher energy e_g groups. This splitting is, according to the crystal field theory, due to the interaction between the negative charges on the ligands and the electrons in the d orbitals. Electrons in the two d orbitals that are pointing directly along the Cartesian axes, $d_{x^2-y^2}$ and d_{z^2} are repelled more strongly by the negative charges on the ligands (which in the octahedral case are placed along the Cartesian axes) than the electrons in the three d orbitals that are pointing between the ligands: d_{xy}, d_{xz}, d_{yz}. For a tetrahedral local environment (ligands placed between the Cartesian axes) similar arguments give stabilization of the

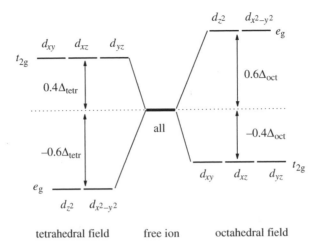

Figure 7.8 *d* level splitting for octahedral and tetrahedral crystal fields.

e_g orbitals and destabilization of the t_{2g} orbitals. The splitting is schematically shown in Figure 7.8 for octahedral and tetrahedral symmetry. The extent of the splitting is often called crystal field (ionic picture) or ligand field (covalent picture) splitting, and the accompanying energy is often given the symbol Δ. This splitting energy measures the interaction between the cation and the surrounding anions, and depends strongly on the interatomic distance that again depends on the nature of the ligand. It is also affected by temperature, pressure, crystal structure and composition of the compound. For an octahedral cation the crystal field stabilization energy for one electron in the d_{xy}, d_{xz} or d_{yz} orbital is $-0.4\Delta_{oct}$ while an electron in a d_{z^2} or $d_{x^2-y^2}$ are destabilized by $0.6\Delta_{oct}$; see Figure 7.8.

The energy effect of the crystal field is, although considerable, not necessarily directly reflected in the enthalpy of formation of transition metal compounds; other effects may dominate. Even so, the relative stability of the binary compounds of the *d* elements varies in a characteristic way. The enthalpy of formation of transition metal dichlorides, difluorides and monoxides for the first series transition metals are shown in Figure 7.9(a). The stabilization of the manganese (II) compounds relative to their nearest neighbours in terms of *d* electrons, Cr and Fe, is mainly due to a low atomization enthalpy for Mn (Figure 7.9(b)). The sum of the first and second ionization enthalpies varies more regularly with the number of *d* electrons (Figure 7.9(c)). By using the experimental enthalpies of formation, the lattice enthalpies of the compounds can be derived. The data for the dichlorides given in Figure 7.9(d) show a characteristic variation with the number of *d* electrons that reflects crystal field stabilization. Ions that do not show crystal field stabilization are the d^0(Ca), d^5(Mn) and d^{10}(Zn) ions. For these three ions, the distribution of *d* electrons around the core is spherically symmetric since the *d* orbitals are either empty (Ca), singly (Mn) or doubly (Zn) occupied, and their lattice enthalpies thus fall on the lower, dotted curve. Although the crystal field effect is clearly present in the lattice enthalpies derived by the thermodynamic cycle

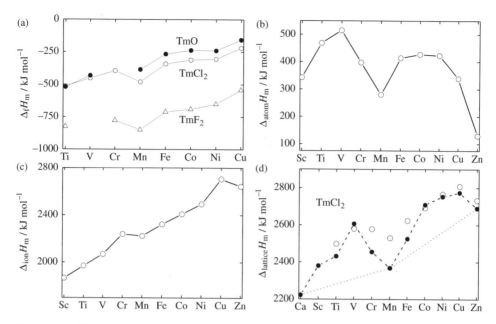

Figure 7.9 Thermodynamic data (b)–(d) needed in analysis of the enthalpy of formation of the binary transition metal compounds given in (a). (b) Atomization enthalpy of first series transition metals; (c) sum of first and second ionization enthalpies of first series transition metals; (d) derived lattice enthalpy of transition metal dihalides.

using experimental enthalpies of formation, the effect is seen even more clearly in the calculated lattice enthalpies given by filled symbols and a dashed line in Figure 7.9(d). The stabilization due to the crystal field is considerable and for the halides the stabilization increases in size from F to I in accordance with the spectrochemical series. Water, oxide ions, hydroxide ions and fluoride ions have comparable crystal field strengths and thus comparable stabilization energies. For typical oxides and hydrates of divalent first row transition elements Δ_{oct} is of the order 8–15 kJ mol^{-1} and typically twice as large for the corresponding trivalent ions. Values for second and third row transition elements are substantially higher than for the first row elements.

Regularities corresponding to those observed for the transition metal compounds are seen for the oxides, nitrates and chlorides of the lanthanide metals in Figure 7.10(a). While the variation in the enthalpies of formation for the different types of lanthanide compounds shows a large degree of similarity, small deviations from a close to linear variation with the number of f electrons are observed for the europium and ytterbium compounds in Figure 7.10(a). These 'anomalous' effects are largely due to high ionization enthalpies for Eu and Yb (Figure 7.10(b)). The relatively low atomization enthalpies (Figure 7.10(c)) of Eu and Yb counteract the large ionization energies to a limited degree only. The calculated lattice enthalpies for the lanthanide trichlorides are given in Figure 7.10(d).

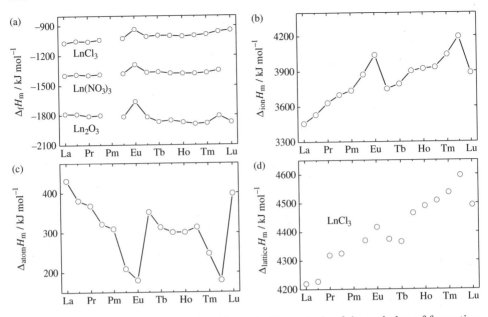

Figure 7.10 Thermodynamic data (b)–(d) needed in analysis of the enthalpy of formation of the binary lanthanide metal compounds given in (a). (b) Sum of first, second and third ionization enthalpies of lanthanide metals; (c) atomization enthalpy of lanthanide metals; (d) derived lattice enthalpy of lanthanide trichlorides.

Transition metals are characterized by their ability to form compounds in several oxidation states. We will here use the $3d$ transition metal oxides to illustrate trends in stability with oxidation state. In general, there is a gradual decrease in stability of the oxides in any given oxidation state relative to the metal across the transition metal series, as shown in Figure 7.11(a). The decrease in stability of oxides of a given oxidation state within a period is most marked in the higher oxidation states. It follows that the ease of oxidation of metals, or of metals in a lower oxidation state oxide to a higher oxidation state one, decreases going to the right within a period. An analogous behaviour is observed for the metals in acidic aqueous solutions where the complexes of the later transition metals are powerful oxidizing agents. While the stability of an oxide is given in terms the Gibbs energies of formation, the stability of the different species of an element M in aqueous solutions can be represented in what is termed Frost diagrams. Here the Gibbs energy is given in terms of the standard potential E^o as $-NFE^o$, where N is the oxidation number of the metal. In the Frost diagram NE^o for the reaction

$$M(N) + Ne^- \rightarrow M(0) \tag{7.16}$$

is plotted versus the oxidation number N.

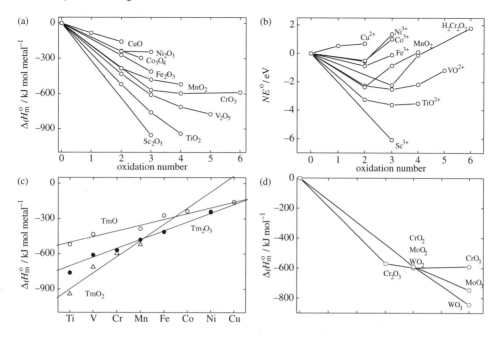

Figure 7.11 Enthalpy of formation of binary oxides of the $3d$ transition metals (a) and (c), Frost diagram for the same $3d$ metals (b) Enthalpy of formation of binary oxides of the group 6 transition metals.

The Frost diagrams for the first series of the d block elements in acidic solution, pH = 0, given in Figure 7.11(b) show many similarities with the variation of the enthalpy of formation of the oxides. Only the oxidation states observed for solid oxides are included.

The relative stabilities of the dioxides, sesquioxides and monoxides for first period transition metals are given in Figure 7.11(c). The stability of the higher oxidation state oxides decreases across the period. As we will discuss later, higher oxidation states can be stabilized in a ternary oxide if the second metal is a basic oxide like an alkaline earth metal. The lines in Figure 7.11(c) can in such cases be used to estimate enthalpies of formation for unstable oxidation states in order to determine the enthalpy stabilization in the acid–base reactions; see below. Finally, it should be noted that the relative stability of the oxides in the higher oxidation states increases from the $3d$ via $4d$ to the $5d$ elements, as illustrated for the Cr, Mo and W oxides in Figure 7.11(d).

We have in the preceding treatment largely confined our discussions to oxides and halides. Similar arguments could also be used on sulfides or nitrides, for example. The variation of the enthalpy of formation of selected binary compounds of group 15 and 16 anions with a common cation are shown in Figure 7.12. The enthalpy of formation becomes more negative the larger the electronegative difference, and thus with increasing group number and decreasing period number for the anion.

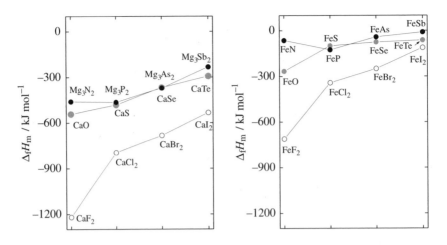

Figure 7.12 Enthalpy of formation of selected binary compounds of group 15, 16 and 17 anions.

Intermetallic compounds and alloys

In metals the electrons lose their association with individual atoms and the number of valence electrons is often used in rationalization schemes. Estimated enthalpies of formation for equi-atomic alloys, MM′, of two elements of the first transition metal series are given as a function of the difference in number of valence electrons in Figure 7.13 [8]. Compounds of a given common metal are given a specific symbol. For example, the scandium compounds ScM′ where M′ = Ti, V, Cr, Mn, Fe, Co, Ni and Zn, are given by open circles. The metal M′ of the compound MM′ is

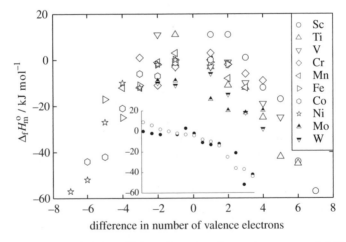

Figure 7.13 Estimated enthalpies of formation of equi-atomic compounds MM′ of the first transition metal series. Data for MoM′ and WM′ are also given [8]. The insert shows agreement between experimental and calculated values for cases where experimental values are available.

given by the difference in number of valence electrons between the specific metal M (in our case Sc) and the other element M'. This difference in number of valence electrons between M and M' is given as the ordinate in Figure 7.13. The data used in the figure is, as mentioned above, estimated since this enables better visualization of the trends in enthalpy of formation. Values are taken from the semi-empirical two-parameter model by Miedema *et al.* [8] that have proven to give good estimates of the enthalpy of formation of a large number of intermetallic compounds and alloys. The insert to the figure shows the agreement between the estimated and experimental values for selected intermetallic compounds as a function of the enthalpy of formation of the compound.

The enthalpy of formation of MM' varies in a systematic way for all the elements of the first transition metal series. Small differences in the number of valence electrons correspond to small differences in electronegativity and thereby a low degree of stabilization due to a small degree of electron transfer between the two elements. The size mismatch between the two elements is in this case the more important enthalpy contribution, giving a destabilization effect. The strain energy due to size mismatch is further discussed in Section 7.3. Larger differences in the number of valence electrons and thus in electronegativity give more negative enthalpies of formation.

Data for Mo and W compounds (MoM' and WM') are included in Figure 7.13 to show the effect of going from one period to the next. The variation in enthalpy of formation with the difference in number of valence electrons is similar; however, the enthalpies of formation are more exothermic for the MoM' and WM' compounds compared with the corresponding first transition metal series element compounds CrM'. Finally, it should be added that the enthalpies of formation of equiatomic alloys of elements of the same group are close to zero.

7.2 Compound energetics: rationalization schemes

Acid–base rationalization

Although periodic trends in enthalpies of formation are often striking, these trends can in general not be used to estimate accurate data for compounds where experimental data are not available. Other schemes are frequently used and these estimates are often based on atomic size and electronegativity-related arguments. As an example, the enthalpy of formation of a ternary oxide from the binary constituent oxides, i.e. the enthalpy of a reaction like

$$AO + B_xO_y = AB_xO_{1+y} \tag{7.17}$$

is often interpreted in terms of factors related to electron transfer between the two oxides. The electron donor/acceptor quality of an oxide is considered in terms of acidity/basicity. A basic oxide, e.g. Na_2O, is one that easily transfers its oxygens to the coordination sphere of an acid oxide like SiO_2 forming a complex, well-defined

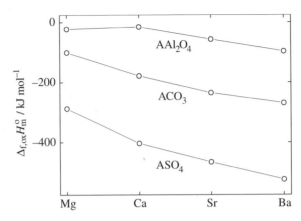

Figure 7.14 Enthalpy of formation of selected ternary oxides from their binary constituents.

covalently bonded anion SiO_4^{2-}. The larger the difference in acidity of AO and B_xO_y, the more exothermic is the enthalpy of reaction (7.17). This concept, developed by Flood and Førland [9, 10], following the initial idea of Lux [11], is frequently used in modified form, not only for oxides (see e.g. Navrotsky [1]), but even for alloys [8, 12]. In many ways this approach is analogous to what we observe in aqueous solutions. Still, in oxides the complex ions formed are often not easily defined and the oxygen atoms are bonded to several types of cations. Although the arguments are mostly used qualitatively, a quantitative optical basicity scale has been developed based on UV–visible spectroscopy and redox properties in general [13].

While the s block oxides are usually basic, the p block oxides are acidic. The basicity increases when going down a specific group e.g. MgO < CaO < SrO < BaO. The same trend is observed for the p block oxides and the oxides in general become more acidic across a specific period, e.g. Al < Si < P < S. The acidity/basicity of transition metal oxides depends on several factors. The acidity increases with increasing number of d electrons and with the oxidation state. It follows that oxides of the early transition elements in their lower oxidation states are the more basic ones. Using these simple rules the relative stability of the ternary 'oxides' presented in Fig. 7.14 is rationalized. BaO is the most basic oxide among those considered and the ternary Ba oxides have the most negative enthalpy of formation in the three cases considered. Similarly, the sulfates are more stable than the carbonates and aluminates relative to the binary constituent oxides, since the acidity of SO_3 is larger than for CO_2 and Al_2O_3. For compounds consisting of binary oxides of similar basicity, the enthalpy of formation is much smaller in magnitude. Here the enthalpy difference between the ternary oxide and its binary constituent oxides is small and the entropy of formation may become decisive for the stability. A ternary oxide like mullite $3Al_2O_3 \cdot 2SiO_2$, where the enthalpy of formation from the oxides is positive, is stabilized by entropy [14].

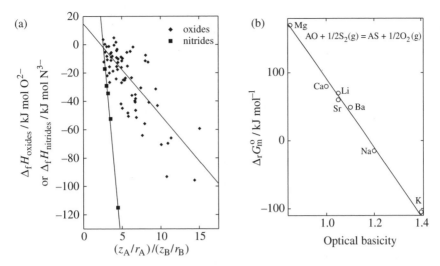

Figure 7.15 (a) Enthalpy of formation of ternary oxides and nitrides from their binary constituent compounds as a function of the ratio of ionic potential [16]. Reprinted with permission from [16] Copyright (1997) American Chemical Society. (b) Gibbs energy of the oxide–sulfide equilibrium for group 1 and 2 metals at 1773 K as a function of the optical basicity of the metal.

Two other examples of acid–base rationalizations are given in Figure 7.15. Figure 7.15(a) shows a large amount of data on the enthalpy of formation of ternary oxides and nitrides from the binary constituents. For both types of materials the enthalpy of formation scales with the acid–base ratio. The difference in acidity is here represented in terms of ionic potentials. The **ionic potential** is defined as the formal charge divided by the ionic radius. A rough grouping relating acidity and ionic potential can be made as follows: $q/r < 2$, strongly basic; $2 < q/r < 4$, basic; $4 < q/r < 7$, amphoteric; $q/r > 7$, acidic (here r is given in Å) [15]. The stabilization observed for a given ratio of ionic potential appears to be larger for nitrides than for oxides. The slope obtained using data for over 80 ternary oxides is smaller in magnitude than the slope observed for the ternary nitrides considered; see Figure 7.15(a) [16]. The difference may result from the greater polarizability of the N^{3-} anion and the related higher degree of covalency in the bonding in the ternary nitrides.

Figure 7.15(b) show the Gibbs energy of the oxide–sulfide equilibrium

$$A_yO\,(s) + \tfrac{1}{2}S_2(g) = A_yS(s) + \tfrac{1}{2}O_2(g) \qquad (7.18)$$

for group 1 and 2 metals at 1773 K as a function of the optical basicity of the metal. The linear relation observed shows the applicability of these concepts for such reactions and acid–base arguments are often successfully used in considerations of metal–slag systems [17].

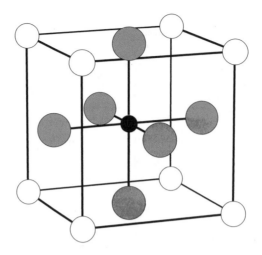

Figure 7.16 The perovskite-type structure. Small black circles represent the B atom, large grey circles represent O atoms and open circles represent the A atom.

Atomic size considerations

Systematization of thermodynamic data in terms of the relative atomic size of the different ions in a compound or in terms of the ionic radius of a given cation of an iso-structural series of compounds are frequently seen. The perovskite-type structure ABO_3 named after the mineral with ideal composition $CaTiO_3$ is a well-known example. The crystal structure of the perovskite is shown in Figure 7.16. A is in the perovskite-type structure 12-coordinated, while B is octahedrally surrounded by oxygen atoms. The BO_6 octahedra are linked via corners and form a network. The bond lengths and angles of the BO_6 octahedra can be distorted to allow a wide range of cation sizes to be accommodated into the structure compared to what would have been possible for a purely cubic arrangement. For a cubic perovskite, the **tolerance factor**, t, introduced already in 1926 by Goldschmidt [18], is defined as

$$t = \frac{r_{AO}}{\sqrt{2}r_{BO}} = \frac{r_A + r_O}{\sqrt{2}(r_B + r_O)} \tag{7.19}$$

where r_i is the ionic radius of species i. Although $t = 1$ correspond to the optimum cation–anion bond length, values in the range $0.8 < t < 1.1$ are common.

The enthalpies of formation of selected perovskite-type oxides are given as a function of the tolerance factor in Figure 7.17. Perovskites where the A atom is a Group 2 element and B is a d or f element that readily takes a tetravalent state [19, 20] show a regular variation with the tolerance factor. Empirically, it is suggested that the cations that give t close to 1 have the most exothermic enthalpies of formation. When t is reduced, the crystal structure becomes distorted from cubic symmetry and this also appears to reduce the thermodynamic stability of the

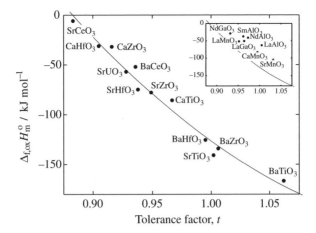

Figure 7.17 Enthalpy of formation of selected perovskite-type oxides as a function of the tolerance factor. Main figure show data for perovskites where the A atom is a Group 2 element and B is a d or f element that readily takes a tetravalent state [19, 20]. The insert shows enthalpies of formation of perovskite-type oxides where the A atom is a trivalent lanthanide metal [21] or a divalent alkaline earth metal [22] whereas the B atom is a late transition metal atom or Ga/Al.

compound. Although this rationalization may be useful for chemically related compounds, compounds that are significantly different in chemical nature should not be expected to follow the same correlation. This is evident from the insert, which shows enthalpies of formation of selected perovskite-type oxides that are chemically different from those considered in the main graph. Here, the A atom is a trivalent lanthanide metal [21] or a divalent alkaline earth metal [22], whereas the B atom is a late transition metal atom or Ga/Al. However, even here we could draw a curve roughly parallel to the one for the oxides in the main figure. Hence this rationalization scheme works well for chemically related compounds, but should in general be used with care. Similar thermodynamic regularities have been reported for the structurally closely related K_2NiF_4-type oxides [23].

This and related schemes require that we have well-defined ionic radii for the different elements. There have been many reports on approaches for deriving internally consistent radii from bond lengths and atomic, ionic, covalent and metallic radii are found in literature. The radii reported by Shannon and Prewitt [24] are commonly used for oxides and fluorides. Cation radii vary much more than anion radii and increase markedly when we move down a group in the periodic table. It decreases with increasing cation charge for a given element and the radii of ions of the same charge decreases across a period. Finally, the cation radius increases with increasing coordination number.

Electron count rationalization

Experimental studies have shown that in many alloy systems the number of valence electrons is the dominant factor. Electron density is usually taken to denote the

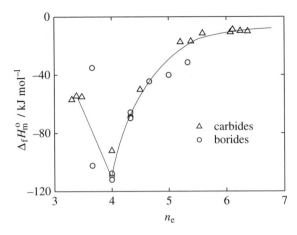

Figure 7.18 Enthalpy of formation of binary carbides and diborides [25, 26] as a function of the average number of valence electrons per atom.

number of valence electrons per unit cell (provided all atomic sites are occupied), or alternatively it may be taken as the ratio of all valence electrons to the number of atoms. The trends in the enthalpy of formation of intermetallic compounds clearly relates to this factor. Importantly, the same type of arguments may also be used for less metallic compounds. The enthalpy of formation of complex binary carbides and diborides [25, 26] are shown in Figure 7.18. A good correlation with the average number of valence electrons per atom, n_e, is observed. A similar variation with the average number of valence electrons is also observed for nitrides. Zhukov *et al.* [27] showed that an observed linear decrease in the enthalpy of formation of selected NiAs-type oxides and carbides for $4 < n_e < 5.5$ could be understood from simple electron band structure considerations. TiC with $n_e = 4$ has a high enthalpy of formation since the Fermi level falls in a pronounced gap in density of electronic states separating bonding and anti-bonding electron bands. Further increase in the number of valence electrons reduces the enthalpy of formation since anti-bonding states are filled. Correspondingly, the decrease in enthalpy of formation at lower electron concentrations shown by Fernandez-Guillermet and Grimvall [28] is due to a reduced number of electrons in the bonding states. This general approach can be used also for more complex structures.

Volume effects in microporous materials

Microporous materials is an important class of compounds with many applications and also interesting since the energetic factors affecting the enthalpy of formation must be expected to be different from those considered in the previous sections. Although most zeolites are kinetically stabilized through the use of templates during synthesis, there are several aspects of the energetics that are worth noting.

Pure silica zeolites or molecular sieves are metastable with regards to the thermodynamic stable polymorph at ambient conditions, α-quartz. However, they are

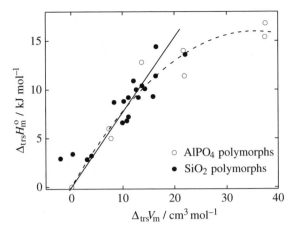

Figure 7.19 Enthalpy of transition from the stable polymorph versus volume correlations for micro- and mesoporous SiO_2 and $AlPO_4$ materials [29, 34].

only 7–14 kJ mol^{-1} less stable than α-quartz and only 0.5–7 kJ mol^{-1} less stable than dense SiO_2 glass [29]. The low barrier between different local energy minima explains the number of structural polymorphs. There exist around 30 different pure SiO_2 structures. The enthalpy of formation correlates well with framework density and with molar volume [29]; see Figure 7.19. The packing of the tetrahedra is thus the main factor for structural stability and this relates directly to the internal surface area of the zeolites. It has been shown that the enthalpy differences between the stable modification, α-quartz and the metastable silica zeolites can be described based on an average internal surface enthalpy of 0.093 ± 0.0009 J m^{-2} [30]. This surface energy is similar to that reported for various amorphous silicas. The correlation between framework density and enthalpy of formation is reproduced in lattice enthalpy calculations on 26 structures with widely different framework densities [31]. While the enthalpy is largely dependent on the framework density, the standard entropies of the zeolites at 298 K are not. They are only slightly higher (3–4 J K^{-1} mol^{-1}) than for α-quartz despite the much larger molar volume [32] since the local structural elements are the same as in the denser compounds; the structures are built as rigid frameworks of SiO_4 tetrahedra that is expected to have similar vibrational characteristics that does not vary largely with connectivity. From the magnitudes of the enthalpy and entropy contributions it can be concluded that the Gibbs energies of the zeolites relative to α-quartz are in total within twice the thermal energy at 298 K.

The importance of framework density and molar volume is evident also for large pore, mesoporous silica [33] and for $AlPO_4$ polymorphs [34]. Data for the latter are included in Figure 7.19. For mesoporous silica a transition from a regime where cages and pores affects the energetics to one in which the large pores act as inert diluent is reported. A further increase in pore diameter does not appear to increase the enthalpy of the compound [33]. The similarity in enthalpy of many different structures shows that the synthesis of metastable microporous framework structures is not limited by energetic constraints. Similar effects have been shown, e.g. for layered manganese dioxides [35] and iron oxides [36].

7.3 Solution energetics: trends and rationalization schemes

Solid solutions: strain versus electron transfer

The factors that affect the energetics of solid solutions and indirectly solid solubility are to a large extent the same as those that control the enthalpy of formation of compounds. Most often the differences between the atomic radii of the participating elements, in electronegativity and in valence electron density are considered for solutions of elements. For solid solutions of binary compounds, similar factors are used, but some measure of the volume of the compounds is often used instead of atomic radii.

Two elements or compounds that do not adopt the same crystal structure cannot exhibit complete solid solubility except when one of the space groups is a subgroup of the other. The energetics of solid solutions of compounds with different structures are obviously difficult to treat systematically and trends may be impossible to obtain, since the energetics is largely related to structural short-range order. We will thus confine our discussion of solid solutions to systems where the two end-members take the same crystal structure.

Let us first consider metallic alloys. The enthalpies of formation of intermetallic compounds, and also the enthalpy of mixing of solid intermetallic solutions, can largely be interpreted in terms of the relative atomic size of the elements being mixed and the difference in electronegativity. The difference in size generates a local misfit or distortion and thus results in a strain energy that increases with the size difference. The empirically derived rule of Hume-Rothery states that restricted solid solubility is expected if the difference between the atomic sizes of the component elements forming the alloy exceeds 15% [37]. A second and negative term is related to the electronegativity of the elements. The more electronegative elements tend to attract the electrons from the less electronegative elements on compound or solution formation and the excess enthalpy of mixing varies from large negative values in systems with large electron transfer between the two elements, e.g. Pd–Zr, to about zero for elements of similar size and similar electronegativity, e.g. Ti–Zr. A large negative enthalpy of formation implies that different atoms attract each other, and a consequence of this is that compound formation becomes more likely. Generally, the number of intermetallic phases found in a binary system correlates with the enthalpy of mixing. An empirically relationship is shown in Figure 7.20 [8].

For ionic solutions the strain energy seem to be relatively more important than for the metallic alloy systems [38–40] and the size difference between the two components being mixed dominates the energetics, although other factors are also of importance. In cases where the the covalency or ionicity of the components being mixed are largely different a limited solid solubility also must be expected, even

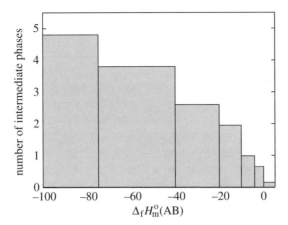

Figure 7.20 Empirical relationship between the number of intermetallic phases in binary systems and the enthalpy of formation of AB [8].

when the ions being mixed are of similar size. It follows that both the NaCl–AgCl and NaBr–AgBr systems show large positive enthalpies of mixing, although atomic size considerations would indicate the opposite. The oxidation state of the components of the ionic solution is also important, higher valence difference leading to lower solubility. Finally, the electron configuration is of particular importance for transition metal systems, where a similar electron configuration is necessary for a large degree of solid solubility.

Still, the strain enthalpy is of particular importance. An elastic continuum model for this size mismatch enthalpy shows that, within the limitations of the model, this enthalpy contribution correlates with the square of the volume difference [41, 42]. The model furthermore predicts what is often observed experimentally: for a given size difference it is easier to put a smaller atom in a larger host than vice versa. Both the excess enthalpy of mixing and the solubility limits are often asymmetric with regard to composition. This elastic contribution to the enthalpy of mixing scales with the two-parameter sub-regular solution model described in Chapter 3 (see eq. 3.74):

$$\Delta_{mix} H_m = x_A x_B^2 c_A \frac{\Delta V^2}{3V_A} + x_B x_A^2 c_B \frac{\Delta V^2}{3V_B} \tag{7.20}$$

where V_A and V_B are the molar volumes of A and B and ΔV is volume change on solution formation. The proportionality constants c_A and c_B are related to the shear and bulk moduli of the two components [42]. The suggested proportionality between the enthalpy of mixing and the square of the volume mismatch is also supported by computer simulations [43].

An analysis of a large amount of experimental data by Davies and Navrotsky has also shown that the enthalpy of mixing of ionic solid solutions correlates with the volume mismatch [39]. The volume mismatch was in the simplest case assumed to

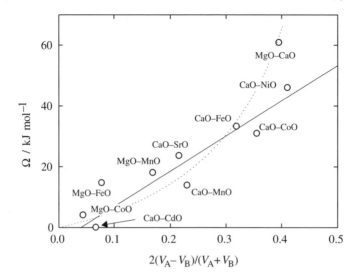

Figure 7.21 Enthalpies of mixing of selected NaCl-type systems involving the alkali earth oxides. The solid and dashed lines show scaling with the volume mismatch and with the square of the volume mismatch, respectively.

correlate with the interaction coefficient Ω of the regular solution model. Within the experimental accuracy of the data, a close to linear relationship was observed for rock salt oxides and chalcogenide solid solutions, and also for alkali halide solutions. The deduced experimental enthalpies of mixing of selected NaCl-type systems involving the alkali earth oxides are compared with theoretical expectations in Figure 7.21. The experimental data do not allow discrimination between variation with volume mismatch or with the square of the volume mismatch. In addition, other contributions to the enthalpy of mixing than the strain enthalpy will affect the experimental data.

The importance of the size of the solute relative to that of the solvent mentioned above is evident also from experimental determinations of the extent of solid solubility in complex oxides and from theoretical evaluations of the enthalpy of solution of large ranges of solutes in a given solvent (e.g. a mineral). The enthalpy of solution for mono-, di- and trivalent trace elements in pyrope and similar systems shows an approximately parabolic variation with ionic radius [44]. For the pure mineral, the calculated solution energies always show a minimum at a radius close to that of the host cation.

Solubility of gases in metals

For interstitial solid solutions, too, the criteria used historically for the degree of solid solubility relates to elastic and electronic interactions. Experimentally observed maximum interstitial solubilities of H, B, C and N in Pd are inversely proportional to the sum of the s and p electrons, and hence are controlled by the valence electron concentration. Thus the electronic interactions dominate the energetics; the electronic interaction between solute and matrix is more important than the elastic contribution [45].

The solubility of gases in metals is of particular importance. In these systems, the concentration of a solute like hydrogen can be varied by controlling the temperature and partial pressure of the solute in the gas phase. Hydrogen dissolves interstitially in many metals, and at low concentrations the solubility of a diatomic gas is proportional to the square root of its partial pressure. At higher concentrations of hydrogen, large repulsive interactions between the hydrogen atoms give rise to immiscibility at low temperatures. The resulting two-phase regions are reflected in the corresponding partial pressure–composition isotherms.

Focusing on the low concentrations situation, all isotherms are characterized by

$$x \propto \sqrt{p} \tag{7.21}$$

The slope is, for a specific system, given by the enthalpy and entropy of solution of hydrogen in the metal; in other words by the energetics of the following reaction

$$M + \tfrac{x}{2} H_2(g) = MH_x \tag{7.22}$$

The relationship between the concentration of hydrogen in the metal and the partial pressure of hydrogen is now

$$x \propto \sqrt{\frac{p}{p_0}} \exp\left(\frac{\Delta_{7.22} S}{R}\right) \exp\left(\frac{\Delta_{7.22} H}{RT}\right) \tag{7.23}$$

This relation is termed **Sievert's law**.

Values for the enthalpy of solution of hydrogen in transition metals at infinite dilution shown in Figure 7.22 are more negative for the early transition metals. It should be noted that the enthalpies of solution in general are functions of the concentration of the solute. Still, the values at infinite dilution are useful when looking for systematic variations, particularly since changes with composition are often limited.

Non-stoichiometry and redox energetics

Hydrides of variable composition are not only formed with pure metals as solvents. A large number of the binary metal hydrides are non-stoichiometric compounds. Non-stoichiometric compounds are in general common for d, f and some p block metals in combination with soft anions such as sulfur, selenium and hydrogen, and also for somewhat harder anions like oxygen. Hard anions such as the halides, sulfates and nitrides form few non-stoichiometric compounds. Two factors are important: the crystal structures must allow changes in composition, and the transition metal must have accessible oxidation states. These factors are partly related. FeO,

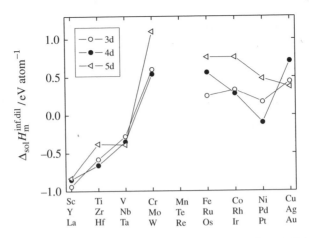

Figure 7.22 Enthalpy of solution of hydrogen in transition metals at infinite dilution [45].

CoO and NiO all take the NaCl-type structure and the difference in non-stoichiometry relates to the relative stability of the formal di- and trivalent oxidation states. The stability of the trivalent state and the degree of non-stoichiometry decreases from Fe^{3+} to Ni^{2+}. Hence the non-stoichiometric nature of $Fe_{1-y}O$ is made possible by the relatively high stability of Fe^{3+} that is reflected in the fact that Fe_2O_3 is a stable compound in the Fe–O system, whereas Ni_2O_3 is not in the Ni–O system. This relative stability of the different oxidation states is also reflected in Figure 7.11(c).

As indicated above, the crystal structure is also important and the difference between hexagonal and cubic $SrMnO_{3-\delta}$ may serve as example. Acid–base factors affect the relative stability of the different oxidation states of a given metal, and in general the redox energetics of ternary oxides must be expected to be quite different from that of the binary ones. As an example Fe(IV) is stabilized in $SrFeO_3$, and while iron dioxide is a non-existing binary compound, the enthalpy of oxidation of Fe(III) to Fe(IV) in $SrFeO_{3-\delta}$ is large and negative, -120 kJ per mol O_2 at 800 K [46]. Similarly, the enthalpy of oxidation of $Mn(III)_2O_3$ to $Mn(IV)O_2$ is -158 kJ per mol O_2, while the corresponding enthalpies of oxidation for $CaMnO_{3-\delta}$ and $SrMnO_{3-\delta}$ are -356 and -293 kJ per mol O_2 [47]. Hence Mn(IV) is stabilized by the basic alkaline earth oxide relative to Mn(III).

Even though the difference in enthalpy of formation between the cubic and hexagonal modification of $SrMnO_3$ is only about 6 kJ mol^{-1}, the temperature of initial reduction of the hexagonal modification in air takes place 600 K above the temperature where the initial reduction appears for cubic $SrMnO_3$. This difference is due to a large difference in the relative Gibbs energy of the oxidized and reduced limiting compounds of the two solid solutions. While cubic $SrMnO_{2.50}$ is relatively stable (can be prepared in the laboratory), hexagonal $SrMnO_{2.50}$ is unstable. The Gibbs energy difference between the oxidized and reduced compounds is hence much larger for the hexagonal case than for the cubic case. The reason for this can

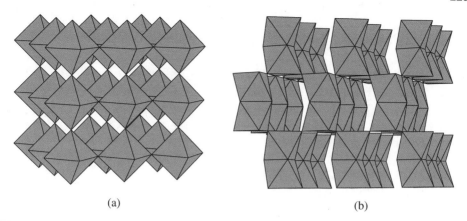

(a)   (b)

Figure 7.23 Connectivity of transition metal BO_6 octahedra in (a) cubic and (b) hexagonal perovskite $SrMnO_3$.

be understood by taking the structure of hexagonal $SrMnO_3$ into consideration. While the $Mn-O_6$ octahedra share corners in the usual cubic perovskite-type structure, they share faces in the hexagonal structure: see Figure 7.23. Reduction of the cubic structure gives rise to square pyramidal coordinated manganese while reduction of the hexagonal structure in the end would lead to face-shared octahedra with high vacancy concentrations in localized areas of the crystal structure. The latter structure must be expected to be energetically unfavourable. In conclusion, the redox energetics of a phase depends strongly on the crystal structure, a fact that should be taken into account when looking for trends in redox properties. In terms of defect chemistry the defect–defect interaction energy is much larger for hexagonal $SrMnO_3$ than for cubic $SrMnO_3$. The enthalpy of oxidation of Mn(III) to Mn(IV) for hexagonal $SrMnO_{3-d}$ is estimated to be -590 kJ per mol O_2, i.e. twice the value for the cubic structure [47].

Liquid solutions

In the crystalline state the solution is restricted by the crystal structures taken by the solid solution. The coordination numbers of the atoms or ions are not allowed to change unless the solid solution involve occupation of interstitial lattice sites. Liquid solutions, on the other hand, have in principle no such structural restrictions and an endothermic enthalpy of mixing is overcome by the entropy of mixing, provided the temperature is high enough. However, in cases where the positive enthalpy of mixing is large, the solubility of one liquid in the other is limited. Liquid immiscibility occurs typically when the two liquids have significantly different chemical and physical properties. One type of system that shows limited solubility is mixtures of liquid metals and molten salts.

Another classical example of demixing in the liquid state occurs in the system $CaO-SiO_2$ and other binary silicate systems where SiO_2 is mixed with basic oxides. While the solubility of $SiO_2(l)$ in the basic oxide is high, there is a

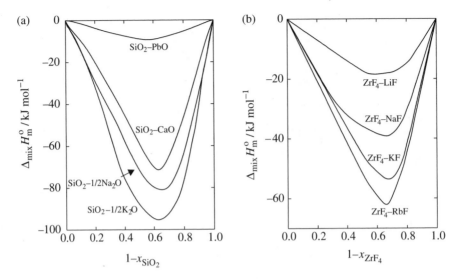

Figure 7.24 Enthalpy of mixing of (a) binary silicate [48] (reprinted by permission of A. Navrotsky) and (b) fluoride systems [49].

immiscibility gap at high SiO_2 concentrations. In some of these systems the immiscibility occurs in the metastable supercooled region, as discussed for the system $Na_2B_8O_{13}$–SiO_2 in Chapter 5. The solubility of CaO(l) in SiO_2(l) is strongly limited, since Si^{4+} in SiO_2 is 4-coordinated, while Ca^{2+} would prefer a near 6-coordinated environment, as in CaO(l). Thus, Ca^{2+} is energetically not favoured to replace Si^{4+} in the covalently bonded liquid SiO_2, where a high degree of short-range order is present. On the other hand, the solubility of SiO_2(l) in CaO(l) or other basic oxides is energetically strongly favourable. This reflects the different acid–base properties of CaO and SiO_2 [9–11].

The enthalpy of mixing of several binary silicate systems is shown in Figure 7.24(a) [48]. The enthalpies of mixing of SiO_2 with alkali and alkali earth oxides are exothermic and become increasingly more exothermic with increasing basicity of the alkali and alkali earth oxide. The shape of the curves reflects the local structure of the liquid. A relatively sharp minimum is evident near the ortho-silicate composition, for example Ca_2SiO_4. Here a high degree of local order exists in the liquid that consists of Ca^{2+} cations and SiO_4^{4-} anions. At higher concentration of the basic oxide, SiO_4^{4-} is mixed with O^{2-} on an anion quasi-lattice, and the enthalpy of mixing becomes less exothermic. Moreover, the SiO_4^{4-} complex anions start to polymerize to form, for example, dimer $Si_2O_7^{6-}$ or linear chains of $(SiO_3^{2-})_n$. Even though the enthalpy of mixing increases with increasing concentration of SiO_2, melts with compositions corresponding to meta-silicates, e.g. $CaSiO_3$, also show large negative enthalpies of mixing; see Figure 7.24(a). At even higher concentrations of SiO_2 the enthalpy of mixing increases rapidly and the immiscibility at high SiO_2 concentration is reflected by a change in sign of the second derivative (inflection point) of the enthalpy of mixing that is not, however, that easily seen at the resolution of the figure. The strong local order found in silicate systems due to the energetically preferred structural SiO_4 entity is also a characteristic of other types of system, such as phosphates and borates where PO_4^{3-} and BO_4^{5-} are the energetically preferred entities. In borate melts, local order due to formation of planar trigonal BO_3 entities is also known.

A crude analysis of the enthalpy of mixing based on the acidity/basicity of the end members involved in the solution can also be applied for fluoride melts. One example is the alkali fluoride–zirconium fluoride mixtures in Figure 7.24(b). Here the enthalpies of mixing of several systems are displayed [49]. ZrF_4 can be regarded as a strong Lewis acid, while the basicity of the alkali fluoride increases with increasing size of the alkali metal. The enthalpy of mixing becomes increasingly more exothermic with increasing basicity of the alkali metal, and for all the systems a minimum is evident near the composition corresponding to the complex anion ZrF_6^{2-}. The minima become increasingly more pronounced with increasing basicity of the alkali metal. Other complexes may also be present in the liquid [49]. The similarity to the binary oxide mixtures shown in Figure 7.24(a) is evident.

The enthalpies of mixing of other binary halide or oxide systems that have an asymmetrical charge distribution display similar behaviour. The enthalpy of mixing becomes more exothermic with increasing difference in acidity, and a sharp minimum in the enthalpy of mixing for a given composition points to a high degree of local order corresponding to a particular complex anion. Most halide systems are ionic in nature. Still, halides like $AlCl_3$ and BeF_2 are characterized by formation of liquids with a high degree of local order due to a preference for fourfold coordination. While, BeF_2 has a similar local structure to SiO_2, $AlCl_3$, is a molecular liquid consisting of Al_2Cl_6 molecules. $AlCl_3$–XCl melts, where X is an alkali metal, are dominated by the formation of $AlCl_4^-$ complex anions.

Large sets of experimental data exist also for simpler ionic solutions such as mixtures of simple molten salt of alkali or alkali earth cations and halogen anions or complex anions like NO_3^-, CO_3^{2-} and SO_4^{2-}. Enthalpies of mixing of such systems are reviewed by Kleppa [50]. Generally, the enthalpy of mixing is small in the range of a few kJ mol^{-1} at $x = 0.5$. For charge symmetrical systems, where the cations and anions have equal charge, the enthalpy of mixing can be understood in terms of Coulombic interactions and polarization of the ions. These interactions are described using the size parameter $\delta = (d_A - d_B)/(d_A + d_B)$, where d_A and d_B are the sums of the cation and anion radii of the two salts A and B [50]. For most systems with a common anion, the enthalpy of mixing is exothermic, while with common cations, anion–anion repulsion may dominate, leading to a positive enthalpy of mixing.

The energetics of liquid metal alloys mimics the energetics of the solid state, and the semi-empirical approach by Miedema and co-workers seems to be in reasonable agreement with experimental observations [8]. However, for mixtures containing d block metals, a third interaction due to p–d hybridization must be added [51]. Generally, the enthalpy of mixing is strongly exothermic in systems, in which intermediate compounds are stable, while the lack of intermediate phases reflects a less exothermic enthalpy of mixing in accordance with Figure 7.20. One example of a system where no intermetallic phases are formed is the binary system Au–Ag, where the enthalpy of mixing at $x_{Ag} = 0.50$ is about 4.0 kJ mol^{-1} [51]. The formation of intermediate compounds also implies that the liquid phase will be energetically stabilized relative to the end members.

References

[1]   A. Navrotsky, *Am. Mineral.* 1994, **79**, 589.

[2]   D. G. Pettifor, *Bonding and Structure of Molecules and Solids*. Oxford: Clarendon Press, 1995.

[3]   T. B. Massalski, *Met. Trans.* 1989, **20A**, 1295.

[4]   R. R. Merritt and B. G. Hyde, *Phil. Trans. Roy. Soc.* 1973, **274A**, 627.

[5]   W. B. Jensen, *J. Chem. Educ.* 1998, **8**, 817.

[6]   C. S. G. Phillips and R. J. P. Williams, *Inorganic Chemistry*. Oxford: Clarendon Press, 1965.

[7]   D. A. Johnson, *Some Thermodynamic Aspects of Inorganic Chemistry*. Cambridge: Cambridge University Press, 1982.

[8]   F. R. de Boer, R. Boom, W. C. M. Mattens, A. R. Miedema and A. K. Niessen, *Cohesion in Metals*. Amsterdam: Elsevier Science Publishers, 1988.

[9]   H. Flood and T. Førland, *Acta Chem. Scand.* 1947, **1**, 592.

[10] H. Flood, T. Førland and B. Roald, *Acta Chem. Scand.* 1947, **1**, 790.

[11] H. Lux, *Z. Electrochem.* 1939, **45**, 303.

[12] H.-F. Wu and L. Brewer, *J. Alloys Comp.* 1997, **247**, 1.

[13] J. A. Duffy, *J. Non-Cryst. Solids.* 1989, **109**, 35.

[14] O. J. Kleppa and J. L. Holm, *Amer. Mineral.* 1966, **51**, 1608.

[15] A. Navrotsky, *Physics and Chemistry of Earth Materials*. Cambridge: Cambridge University Press, 1994.

[16] J. M. McHale, A. Navrotsky, G. R. Kowach, V. E. Balbarin and F. J. DiSalvo, *Chem. Mater.* 1997, **9**, 1538.

[17] C. Wagner, *Metal. Trans.* 1975, **6B**, 405.

[18] V. M. Goldschmidt, *Skr. Nors. Vidensk.-Akad. Oslo* 1926, **1**, 1.

[19] J. Goudiakas, R. G. Haire and J. Fuger, *J. Chem. Thermodyn.* 1990, **22**, 577.

[20] L. R. Morss, *J. Less-Common Metals.* 1983, **93**, 301.

[21] Y. Kanke and A. Navrotsky, *J. Solid State Chem.* 1998, **141**, 424.

[22] L. Rørmark, S. Stølen, K. Wiik and T. Grande, *J. Solid State Chem.* 2002, **163**, 186.

[23] H. Yokokawa, T. Kawada and M. Dokiya, *J. Am. Ceram. Soc.* 1989, **72**, 152.

[24] R. D. Shannon, *Acta Cryst.* 1976, **A32**, 751.

[25] A. Fernandez Guillermet and G. Grimvall, *Phys. Rev. B* 1989, **40**, 10582.

[26] P. Vajeeston, P. Ravindran, C. Ravi and R. Asokamani, *Phys. Rev. B* 2001, **63**, 045115.

[27] P. Zhukov, V. A. Gubanov, O. Jepsen, N. E. Christensen and O. K. Andersen, *J. Phys. Chem. Solids.* 1988, **49**, 841.

[28] A. Fernandez Guillermet and G. Grimvall, *J. Phys. Chem. Solids.* 1992, **53**, 105.

[29] P. M. Piccione, C. Laberty, S. Yang, M. A. Camblor, A. Navrotsky and M. E. Davis, *J. Phys. Chem. B.* 2000, **104**, 10001.

[30] E. C. Moloy, L. P. Davila, J. F. Shackelford and A. Navrotsky, *Micropor. Mesopor. Mater.* 2002, **54**, 1.

[31] N. J. Henson, A. K. Cheetham and J. D. Gale, *Chem. Mater.* 1994, **6**, 1647.

[32] P. M. Piccione, B. F. Woodfield, J. Boerio-Goates, A. Navrotsky and M. E. Davis, *J. Phys. Chem. B* 2001, **105**, 6025.

[33] Y. Hu, A. Navrotsky, C.-Y. Chen and M. E. Davis, *Chem. Mater.* 1995, **7**, 1816.

[34] A. Navrotsky, I. Petrovic, Y. Hu, C.-Y. Chen and M. E. Davis, *Microporous Mater.* 1995, **4**, 95.

[35] S. Fritsch, J. E. Post and A. Navrotsky, *Geochim. Cosmochim. Acta.* 1997, **61**, 2613.

[36] C. Laberty and A. Navrotsky, *Geochim. Cosmochim. Acta.* 1998, **62**, 2905.

[37] W. Hume-Rothery, *Electrons, Atoms, Metals and Alloys.* New York: Dover Publications, 1963.

[38] D. M. Kerrick and L. S. Darken, *Geochim. Cosmochim. Acta.* 1975, **39**, 1431.

[39] P. K. Davies and A. Navrotsky, *J. Solid State Chem.* 1983, **46**, 1.

[40] C. A. Geiger, *Amer. Mineral.* 2000, **85**, 893.

[41] T. B. Massalski and H. W. King, *Progr. Mater. Sci.* 1961, **10**, 1.

[42] H. J. Greenwood, *Geochim. Cosmochim. Acta.* 1979, **43**, 1873.

[43] A. Bosenick, M. T. Dove, V. Heine and C. A. Geiger, *Phys. Chem. Minerals.* 2001, **28**, 177.

[44] W. van Westrenen, N. L. Allan, J. D. Blundy, J. A. Purton and B. J. Wood, *Geochim. Cosmochim. Acta.* 2000, **64**, 1629.

[45] Y. Fukai, *The Metal–Hydrogen System.* Berlin: Springer-Verlag, 1993.

[46] C. Haavik, T. Atake and S. Stølen, *Phys. Chem. Chem. Phys.* 2002, **4**, 1082.

[47] L. Rørmark, A. B. Mørch, K. Wiik, T. Grande and S. Stølen, *Chem. Mater.* 2001, **13**, 4005.

[48] A. Navrotsky in *Silicate Melts, Short Course Handbook* (C. M. Scarfe ed.). Toronto: Mineralogical Association of Canada, 1986, vol. 12, p. 130.

[49] T. Grande, S. Aasland and S. Julsrud, *J. Am. Ceram. Soc.* 1997, **80** 1405.

[50] O. J. Kleppa in *Molten Salt Chemistry – An Introduction and Selected Applications* (G. Mamantov and R. Marassi eds.). NATO ASI Series C: Mathematical and Physical Sciences vol. 202, Holland: Reidel Publishing Company, 1986.

[51] O. J. Kleppa, *J. Non-Cryst. Solids.* 1984, **61&62**, 101.

Further reading

A. Navrotsky, *Physics and Chemistry of Earth Materials.* Cambridge: Cambridge University Press, 1994.

C. S. G. Phillips and R. J. P. Williams, *Inorganic Chemistry.* Oxford: Clarendon Press, 1965.

D. A. Johnson, *Some Thermodynamic Aspects of Inorganic Chemistry.* Cambridge: Cambridge University Press, 1982.

8

Heat capacity and entropy

The Gibbs energy of formation of a compound can be expressed as

$$\Delta_f G_m^o = \Delta_f H_m^o - T\Delta_f S_m^o + p\Delta_f V_m^o \tag{8.1}$$

At low temperatures the enthalpy of formation usually constitutes the largest energetic contribution. And while high pressures favour the formation of dense compounds, the entropy become increasingly more important at high temperatures. We saw in the previous chapter that the enthalpy of formation of a ternary oxide from the binary constituent oxides is small compared to the enthalpy of formation from the elements and comparable to the enthalpy of melting and to the enthalpy difference between different modifications of the compound. Similarly, small differences in the enthalpy of formation are observed between different structure types of metallic compounds. These small differences in enthalpy of formation explain the occurrence of several polymorphs of many substances, which again points to the importance of entropy and volume contributions to the total Gibbs energy. Even for simple elements complex phase relations are observed. Using bismuth as an example, five different polymorphs are known, as shown in the p,T phase diagram in Figure 8.1.

The absolute value of the entropy of a compound is obtained directly by integration of the heat capacity from 0 K. The main contributions to the heat capacity and thus to the entropy are discussed in this chapter. Microscopic descriptions of the heat capacity of solids, liquids and gases range from simple classical approaches to complex lattice dynamical treatments. The relatively simple models that have been around for some time will be described in some detail. These models are, because of their simplicity, very useful for estimating heat capacities and for relating the heat capacity to the physical and chemical

Chemical Thermodynamics of Materials by Svein Stølen and Tor Grande
© 2004 John Wiley & Sons, Ltd ISBN 0 471 492320 2

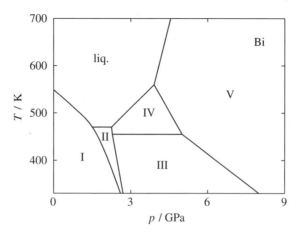

Figure 8.1 p–T phase diagram of Bi.

characteristics of the phases. Gases are briefly treated, followed by a more thorough discussion of the heat capacity of solids. Here we focus on physically based models, some of which may be used to deduce trends in the vibrational contribution to the entropy. Finally, heat capacity contributions of electronic origin and treatments of disordered systems are considered.

8.1 Simple models for molecules and crystals

Let us first look at a monoatomic perfect gas consisting of N atoms. The internal energy of the gas is

$$U = \sum_{i=1}^{N} \frac{1}{2} m_i c_i^2 + \Phi(r_1, r_2, r_3, ..., r_N) \tag{8.2}$$

where $c(c_x, c_y, c_z)$ and $r(r_x, r_y, r_z)$ are the velocity and position of atom i. For a perfect monoatomic gas there are no atomic interactions and the potential energy, $\Phi_p(r_1, r_2, r_3, ..., r_N)$, is negligible. The thermal energy of the system is thereby the mean of the kinetic energy of the atoms, and is given by the first term in eq. (8.2). A monoatomic gas has three translational degrees of freedom represented by three independent quadratic variables in the internal energy. Each of these contributes $\frac{1}{2} k_B T$ to the total internal energy, and the mean internal energy per atom follows:

$$\frac{1}{2} m < c^2 > = \frac{3 k_B T}{2} \tag{8.3}$$

The molar internal energy of a monoatomic ideal gas is therefore

$$U_m = L \frac{3}{2} k_B T = \frac{3}{2} RT \tag{8.4}$$

and the heat capacity at constant volume, obtained by differentiation with respect to temperature, is

$$C_{V,m} = \left(\frac{\partial U_m}{\partial T}\right)_V = \frac{3}{2}R \tag{8.5}$$

The internal energy is, as indicated above, connected to the number of degrees of freedom of the molecule: that is the number of squared terms in the Hamiltonian function or the number of independent coordinates needed to describe the motion of the system. Each degree of freedom contributes $\frac{1}{2}RT$ to the molar internal energy in the classical limit, e.g. at sufficiently high temperatures. A monoatomic gas has three translational degrees of freedom and hence, as shown above, $U_m = 3/2RT$ and $C_{V,m} = 3/2R$.

A linear gas molecule can in addition rotate about any pair of directions perpendicular to each other and perpendicular to the axis of the linear molecule. A diatomic molecule therefore has two additional rotational degrees of freedom. Non-linear polyatomic molecules, which can rotate about the three principal axes, have three rotational degrees of freedom and a total of six degrees of freedom of translational and rotational nature. In addition, the molecular vibrations must be taken into account. The energy of each mode of vibration has associated with it two terms: one kinetic and one potential energy term. Each mode therefore contributes R to the molar heat capacity. In general, a polyatomic molecule has $(3n - 6)$ vibrational modes, where n is equal to the number of atoms in the molecule. If the molecule is linear the number of modes is $(3n - 5)$. The number of translational, rotational and vibrational modes and the resulting limiting molar heat capacities of gases at constant volume and at constant pressure are given in Table 8.1. The difference between the heat capacity at constant pressure and constant volume, the **dilational heat capacity**, is, for an ideal gas:

$$C_{p,m} - C_{V,m} = \frac{\alpha^2 TV}{\kappa_T} = R \tag{8.6}$$

where α and κ_T are the isobaric expansivity and isothermal compressibility respectively. The molar heat capacities at constant pressure of H(g), H_2(g) and H_2O(g) are given in Figure 8.2. The classical heat capacity is in each case marked with open symbols at $T = 5000$ K. Monoatomic H(g) with only translational degrees of freedom is already fully excited at low temperatures. The vibrational frequencies (ν) of H_2(g) and H_2O(g) are much higher, in the range of 100 THz, and the associated energy levels are significantly excited only at temperatures above 1000 K. At room temperature only a few molecules will have enough energy to excite the vibrational modes, and the heat capacity is much lower than the classical value. The rotational frequencies are of the order 100 times smaller, so they are fully excited above ~10 K.

Table 8.1 Number of modes and heat capacity of gases in the classical limit.

	Number of modes			Classical	
	Translational	Rotational	Vibrational	$C_{V,m}/R$	$C_{p,m}/R$
A(g)	3			3/2	5/2
AB(g)	3	2	1	7/2	9/2
AB_2(g) non-linear	3	3	3	6	7
AB_2(g) linear	3	2	4	13/2	15/2
AB_{n-1}(g) non-linear	3	3	$(3n-6)$	$3+(3n-6)$	$4+(3n-6)$
AB_{n-1}(g) linear	3	2	$(3n-5)$	$7/2+(3n-6)$	$9/2+(3n-6)$

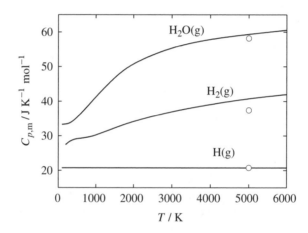

Figure 8.2 Molar heat capacity at constant pressure of H(g), H_2(g) and H_2O(g). The open symbols at 5000 K represent the limiting classical heat capacity.

The statistical treatment of the vibrational degrees of freedom of crystals is far more difficult compared to gases. Let us initially consider a monoatomic crystal. An atom in a crystal vibrates about its equilibrium lattice position. In the simplest approach, three non-interacting superimposed linear harmonic oscillators represent the vibrations of each atom. The total energy, given by the sum of the kinetic and potential energies for the harmonic oscillators, is

$$U = \tfrac{1}{2}mc^2 + \tfrac{1}{2}Kx^2 = \tfrac{1}{2}mA^2\omega^2\cos^2\omega t + \tfrac{1}{2}KA^2\sin^2\omega t \qquad (8.7)$$

where m is the mass of the atom, ω is the angular frequency of the harmonic oscillator, A is the amplitude and $x = A\sin\omega t$ is the time-dependent displacement of the atom from its equilibrium lattice position. The angular frequency of the harmonic oscillator is given by the force constant K and the mass m of the atom:

$$\omega = 2\pi v = \sqrt{\frac{K}{m}} \tag{8.8}$$

Since $\omega^2 = K/m$, the mean potential and kinetic energy terms are equal and the total energy of the linear oscillator is twice its mean kinetic energy. Since there are three oscillators per atom, for a monoatomic crystal $U_m = 3RT$ and $C_{V,m} = 3R = 24.94\,\text{J K}^{-1}\,\text{mol}^{-1}$. This first useful model for the heat capacity of crystals (solids), proposed by Dulong and Petit in 1819, states that the molar heat capacity has a universal value for all chemical elements independent of the atomic mass and crystal structure and furthermore independent of temperature. **Dulong–Petit's law** works well at high temperatures, but fails at lower temperatures where the heat capacity decreases and approaches zero at 0 K. More thorough models are thus needed for the lattice heat capacity of crystals.

8.2 Lattice heat capacity

The Einstein model

The decrease in the heat capacity at low temperatures was not explained until 1907, when Einstein demonstrated that the temperature dependence of the heat capacity arose from quantum mechanical effects [1]. Einstein also assumed that all atoms in a solid vibrate independently of each other and that they behave like harmonic oscillators. The motion of a single atom is again seen as the sum of three linear oscillators along three perpendicular axes, and one mole of atoms is treated by looking at $3L$ identical linear harmonic oscillators. Whereas the harmonic oscillator can take any energy in the classical limit, quantum theory allows the energy of the harmonic oscillator (ε_n) to have only certain discrete values (n):

$$\varepsilon_n = (n + \tfrac{1}{2})\hbar\omega \tag{8.9}$$

The probability that an oscillator at a given temperature occupies a given energy state, ε_n, is given by Bose–Einstein statistics (see e.g. C. Kittel and H. Kroemer, Further reading) and the mean value of n at a given temperature is given by

$$\bar{n} = \frac{1}{\exp(\hbar\omega/k_BT) - 1} \tag{8.10}$$

In the Einstein model, all the independent oscillators have the same angular frequency, ω_E, and the average total internal energy is

$$\bar{U} = 3N(\tfrac{1}{2} + \bar{n})\hbar\omega_E = 3N\left(\frac{\hbar\omega_E}{2} + \frac{\hbar\omega_E}{\exp(\hbar\omega_E / k_BT) - 1} \right) \tag{8.11}$$

The heat capacity is derived by differentiation with respect to temperature:

$$C_{V,m} = \left(\frac{d\bar{U}}{dT}\right)_V = 3R\left(\frac{\Theta_E}{T}\right)^2 \frac{\exp(\Theta_E/T)}{[\exp(\Theta_E/T)-1]^2} \tag{8.12}$$

where Θ_E, the **Einstein temperature**, is defined by

$$\Theta_E = \frac{\hbar\omega_E}{k_B} \tag{8.13}$$

This model, the **Einstein model** for heat capacity, predicts that the heat capacity is reduced on cooling and that the heat capacity becomes zero at 0 K. At high temperatures the constant-volume heat capacity approaches the classical value $3R$. The Einstein model represented a substantial improvement compared with the classical models. The experimental heat capacity of copper at constant pressure is compared in Figure 8.3 to $C_{V,m}$ calculated using the Einstein model with $\Theta_E = 244$ K. The insert to the figure shows the Einstein frequency of Cu. All $3L$ vibrational modes have the same frequency, $v = 32$ THz. However, whereas $C_{V,m}$ is observed experimentally to vary proportionally with T^3 at low temperatures, the Einstein heat capacity decreases more rapidly; it is proportional to $\exp(\Theta_E/T)$ at low temperatures. In order to reproduce the observed low temperature behaviour qualitatively, one more essential factor must be taken into account; the lattice vibrations of each individual atom are not independent of each other – collective lattice vibrations must be considered.

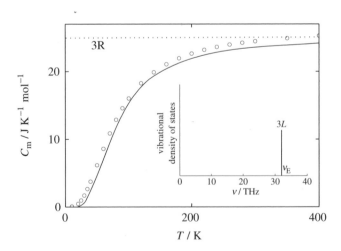

Figure 8.3 Experimental heat capacity of Cu at constant pressure compared with $C_{V,m}$ calculated by the Einstein model using $\Theta_E = 244$ K. The vibrational frequency used in the Einstein model is shown in the insert.

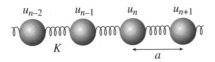

Figure 8.4 One-dimensional chain of atoms with interatomic distance a and force constant K.

Collective modes of vibration

A first impression of collective lattice vibrations in a crystal is obtained by considering one-dimensional chains of atoms. Let us first consider a chain with only one type of atom. The interaction between the atoms is represented by a harmonic force with force constant K. A schematic representation is displayed in Figure 8.4. The average interatomic distance at equilibrium is a, and the equilibrium rest position of atom n is thus $u_n = na$. The motion of the chain of atoms is described by the time-dependent displacement of the atoms, $u_n(t)$, relative to their rest positions. We assume that each atom only feels the force from its two neighbours. The resultant restoring force (F) acting on the nth atom of the one dimensional chain is now in the harmonic approximation

$$F = -K(u_n - u_{n+1} + u_n - u_{n-1})$$ (8.14)

The equation of motion of this nth atom is given by

$$m \frac{\partial^2 u_n}{\partial t^2} = -\frac{\partial U}{\partial u_n} = F = -K(2u_n - u_{n+1} - u_{n-1})$$ (8.15)

Let us describe the displacement of the nth atom from its equilibrium rest position by a cosine-wave with amplitude u_0, angular frequency ω and **wave vector** $q = 2\pi/\lambda$:[1]

$$u_n = u_0 \cos(\omega t - qna)$$ (8.16)

Substituting this expression into the equation of motion (eq. 8.15) we obtain the expression for the angular frequency of the vibrational modes as a function of the wave vector:

$$\omega(q) = \sqrt{\frac{4K}{m}} \left| \sin\left(\frac{qa}{2} \right) \right|$$ (8.17)

The frequency versus wave vector graph, shown in Figure 8.5, is known as a **dispersion** curve. We only need to consider waves with wave vectors lying between

---

1 In two and three dimensions q is a vector. Usually q is also referred to as a vector even in one dimension.

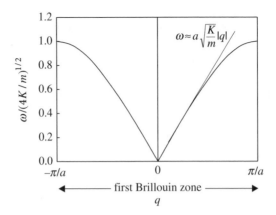

Figure 8.5 The dispersion curve for a one-dimensional monoatomic chain of atoms.

$-\pi/a < q < \pi/a$. This range of wave vectors is within what is termed the **first Brillouin zone** of the one-dimensional lattice. At longer wave vectors or shorter wavelengths, $|q| > \pi/a$, the periodicity of the wave is shorter than the interatomic distance. This has no physical meaning.

For short wave vectors (or long wavelengths) corresponding to waves in the acoustic or ultrasonic range, eq. (8.17) reduces to

$$\omega \approx a \sqrt{\frac{K}{m}} |q| \qquad (8.18)$$

The **group velocity**, $d\omega/dq$, is constant and independent of the wavelength, as shown in Figure 8.5. The discontinuous nature of matter can be neglected, and for these wavelengths the vibrational characteristics of the material can be described by its macroscopic properties (density and elastic constants). The group velocity is here equal to the speed of sound in the solid.

At short wavelengths the frequency is no longer proportional to q and the velocity of propagation decreases as q increases. This is called dispersion and is reflected in Figure 8.5. The wavelength becomes comparable to the interatomic distance and the discrete nature of the atoms becomes important. The vibration frequency has its maximum value at the Brillouin zone boundary where $q = \pi/a$ and the group velocity, $d\omega/dq$, is equal to zero. Thus here the wave vector corresponds to a standing wave. Successive atoms have displacements of alternating sign.

The linear model can be extended to include more distant neighbours and to three dimensions. Let us consider an elastic lattice wave with wave vector q. The collective vibrational modes of the lattice are illustrated in Figure 8.6. The formation of small local deformations (strain) in the direction of the incoming wave gives rise to stresses in the same direction (upper part of Figure 8.6) but also perpendicular (lower part of Figure 8.6) to the incoming wave because of the elasticity of the material. The cohesive forces between the atoms then transport the deformation of the lattice to the

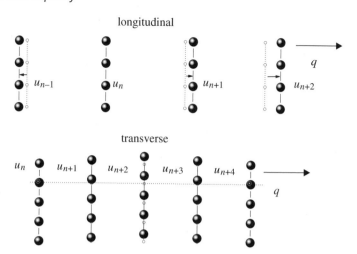

Figure 8.6 Schematic representation of transverse and longitudinal collective vibrational waves.

neighbourhood. It is clear that the waves are vibrating not independently but collectively. When a wave propagates along q, entire planes of atoms move in phase with displacements either parallel or perpendicular to the direction of the wave vector. For each wave vector there are three modes: one of longitudinal polarization (**longitudinal mode**) and two of transverse polarization (**transverse modes**). For an isotropic solid the two transverse waves are degenerate. Also, in crystals of high symmetry, there are three-, four- or six-fold rotation axes along the direction of the wave vector, where the two transverse modes will be degenerate. When the symmetry of the crystal or a specific symmetry operator is lower, the complexity of the vibrational modes will increase. The dispersion relations in the [100], [110] and [111] directions for Pb (fcc structure) at 100 K [2] are given in Figure 8.7. The two transverse modes are degenerate for [100] and [111], which are four- and three-fold rotation axes, but not for [110], which is a two-fold rotation axis.

The method delineated in the preceding sections can readily be extended to the case of two atoms in the chain. An illustration of the diatomic chain model is given in Figure 8.8. The two atoms are characterized by having different masses m_1 and m_2. The two equations of motion obtained (one for each type of atom) must be solved simultaneously, giving two solutions for the angular frequency known as the **acoustic** and **optic branches**

$$\omega_a^2 = K\left[\left(\frac{m_1 + m_2}{m_1 m_2}\right) - \frac{\sqrt{(m_1 - m_2)^2 + 4m_1 m_2 \cos^2 qa}}{m_1 m_2}\right] \tag{8.19}$$

$$\omega_o^2 = K\left[\left(\frac{m_1 + m_2}{m_1 m_2}\right) + \frac{\sqrt{(m_1 - m_2)^2 + 4m_1 m_2 \cos^2 qa}}{m_1 m_2}\right] \tag{8.20}$$

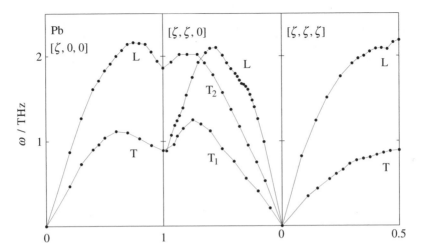

Figure 8.7 Experimental dispersion relations for acoustic modes for lead at 100 K [2]. Reproduced by permission of B. N. Brockhouse and the American Physical Society.

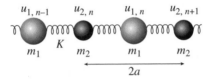

Figure 8.8 One-dimensional diatomic chain with lattice parameter $2a$ and force constant K.

The dispersion curve for a diatomic chain is given in Figure 8.9. The curve consists of two distinct branches: the acoustic and the optic. In the first of these the frequency varies from zero to a maximum ω_1. The second one has a maximum value of ω_0 at $q = 0$ and decreases to ω_2 at $q = q_{max}$. There are no allowed frequencies in the gap between ω_1 and ω_2.

The solution for the acoustic branch approaches zero as q goes to zero, and for small q:

$$\omega_a \approx a \sqrt{\frac{2K}{m_1 + m_2}} |q| \tag{8.21}$$

For $m_1 = m_2$, the expression reduces to that obtained for a monoatomic chain (eq. 8.18). When q approaches zero, the amplitudes of the two types of atom become equal and the two types of atom vibrate in phase, as depicted in the upper part of Figure 8.10. Two neighbouring atoms vibrate together without an appreciable variation in their interatomic distance. The waves are termed acoustic vibrations, acoustic vibrational modes or **acoustic phonons**. When q is increased, the unit cell, which consists of one atom of each type, becomes increasingly deformed. At q_{max} the heavier atoms vibrate in phase while the lighter atoms are stationary.

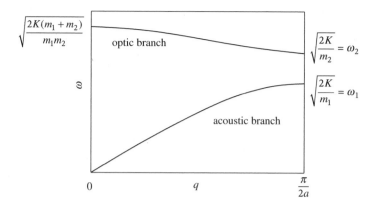

Figure 8.9 The dispersion curve for a one-dimensional diatomic chain of atoms. $m_2 < m_1$.

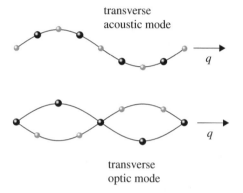

Figure 8.10 Transverse acoustic and optic modes of motion in a one-dimensional diatomic chain at the same wavelength.

The optical branch takes a non-zero limiting value at $q = 0$:

$$\omega_o(q = 0) = \sqrt{\frac{2K(m_1 + m_2)}{m_1 m_2}} \tag{8.22}$$

Close to this limit the displacements of the two types of atom have opposite sign and the two types of atom vibrate out of phase, as illustrated in the lower part of Figure 8.10. Thus close to $q = 0$, the two atoms in the unit cell vibrate around their centre of mass which remains stationary. Each set of atoms vibrates in phase and the two sets with opposite phases. There is no propagation and no overall displacement of the unit cell, but a periodic deformation. These modes have frequencies corresponding to the optical region in the electromagnetic spectrum and since the atomic motions associated with these modes are similar to those formed as response to an electromagnetic field, they are termed optical modes. The optical branch has frequency maximum at $q = 0$. As q increases ω slowly decreases and

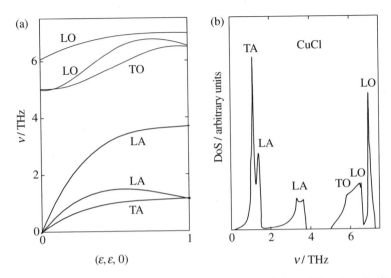

Figure 8.11 (a) Dispersion curve for CuCl(s) along [110] of the cubic unit cell. (b) Density of vibrational modes [3]. Here L, T, A and O denote longitudinal, transverse, acoustic and optic. Reproduced by permission of B. Hennion and The Institute of Physics.

approaches a minimum at q_{max}. As q increases the vibration becomes a mixture of displacement and deformation of the unit cell. For $q = q_{max}$ the heavier atoms are stationary, whereas the lighter atoms vibrate in phase.

Although acoustic and optic mode behaviour can be distinguished at the zone centre, this is not generally so at other values of q. Nevertheless, the convention is to refer to the whole branch as being optic or acoustic.

In three dimensions, transverse and longitudinal optic and acoustic modes result. The dispersion curve for CuCl along [100] of the cubic unit cell [3] is shown in Figure 8.11(a) as an example. The number of discrete modes with frequencies in a defined interval can be displayed as a function of the frequency. This gives what is termed the density of vibrational modes or the vibrational **density of states** (DoS). The vibrational DoS of CuCl is given in Figure 8.11(b).

In general a crystal that contains n atoms per unit cell have a total of $3L·n$ vibrational modes. Of these there are $3L$ acoustic modes in which the unit cell vibrates as an entity. The remaining $3L(n-1)$ modes are optic and correspond to different deformations of the unit cell. At high temperatures where classical theory is valid each mode has an energy k_BT and the total heat capacity is $3R$, in line with the Dulong–Petit law.

The collective modes of vibration of the crystal introduced in the previous paragraph involve all the atoms, and there is no longer a single vibrational frequency, as was the case in the Einstein model. Different modes of vibration have different frequencies, and in general the number of vibrational modes with frequency between v and $v + dv$ are given by

$$3N_A g(v)dv \quad \text{where} \quad \int_0^\infty g(v)dv = 1 \tag{8.23}$$

and $g(v)$ is the density of vibrational modes at a particular frequency. When the density of vibrational modes is known as a function of v, the total internal energy can be calculated and the heat capacity obtained through differentiation with respect to temperature.

The Debye model

The quantization of vibrational energy implies that at low temperatures only the low-frequency modes of lattice vibrations will be appreciably excited. The usual very low-frequency vibrations of a solid are the acoustic modes with wavelengths much longer than the atomic dimensions. **Debye** calculated the distribution of frequencies that result from the propagation of acoustic waves of permitted wavelengths in a continuous isotropic solid and assumed the same distribution to hold in a crystal [4]. The distribution of frequencies is then given by (see e.g. Grimvall, Further reading)

$$g(\omega) = \frac{3\omega^2}{\omega_D^3} \quad \text{for} \quad \omega_D \geq \omega \tag{8.24}$$

$$g(\omega) = 0 \quad \text{for} \quad \omega > \omega_D \tag{8.25}$$

The limiting angular vibrational frequency, ω_D, that exists defines the **Debye temperature**, Θ_D, as

$$\Theta_D = \frac{\hbar\omega_D}{k_B} = 2\pi \frac{hv_D}{k_B} \tag{8.26}$$

and it can be shown that the heat capacity corresponding to the Debye model is given by

$$C_{V,m} = 9R\left(\frac{T}{\Theta_D}\right)^3 \int_0^{\Theta_D/T} \frac{e^x}{(e^x - 1)^2} x^4 dx \tag{8.27}$$

where $x = \hbar\omega/k_B T$.

Whereas the Debye heat capacity is equal to $3R$ at high temperatures, it reduces to

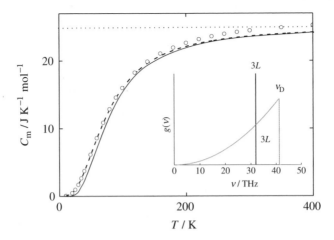

Figure 8.12 Experimental heat capacity of Cu at constant pressure compared with the Debye and Einstein $C_{V,m}$ calculated by using $\Theta_E = 244$ K and $\Theta_D = 314$ K. The vibrational density of states according to the two models is shown in the insert.

$$ C_V = \frac{12\pi^4}{5} R \left(\frac{T}{\Theta_D} \right)^3 \qquad\qquad (8.28) $$

at low temperatures. When the temperature approaches 0 K, the heat capacity tends to zero as T^3, in agreement with experiment.

The experimental constant-pressure heat capacity of copper is given together with the Einstein and Debye constant volume heat capacities in Figure 8.12 (recall that the difference between the heat capacity at constant pressure and constant volume is small at low temperatures). The Einstein and Debye temperatures that give the best representation of the experimental heat capacity are $\Theta_E = 244$ K and $\Theta_D = 315$ K and schematic representations of the resulting density of vibrational modes in the Einstein and Debye approximations are given in the insert to Figure 8.12. The Debye model clearly represents the low-temperature behaviour better than the Einstein model.

Both the Einstein and the Debye temperatures reflect the bonding strength in a particular compound. A higher characteristic temperature represents stronger bonding between the atoms and the classical limit will be reached at higher temperatures. The lattice heat capacity of different polymorphs of carbon is displayed in Figure 8.13. The soft molecule C_{60} with weak intermolecular bonds has a low Debye temperature ($\Theta_D = 46$ K) [5]. Graphite with weak van der Waals bonds between the graphite layers has considerably higher $\Theta_D = 760$ K, while diamond with only strong covalent bonds between all the C atoms displays a very high Debye temperature (2050 K). Hence, while C_{60} approaches the classical value $3R$ (per carbon atom) far below 100 K, the heat capacity of diamond is only 20% of this value at 300 K.

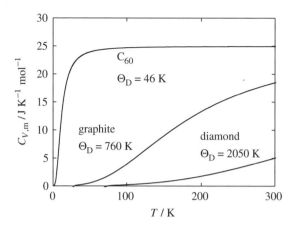

Figure 8.13 Lattice heat capacity of three different polymorphs of carbon; C_{60} [5], graphite and diamond.

Since the vibrational contribution is generally the largest contribution to the heat capacity, trends in the Debye temperature largely defines trends in entropy. Periodic variations of the Debye temperature of the elements, and for Li, Na, K and Rb halides, are given in Table 8.2 and Figure 8.14, respectively. The Debye temperature for the halides decreases with the mass of both the cation and anion. The charge of the ions is constant and the variation in Θ_D reflects the interatomic distance and the bonding strength. The fluorides, which in Chapter 7 were shown to have the more negative enthalpy of formation among the halides, have the highest Debye temperatures and thus the lowest entropies.

Although the Debye model reproduces the essential features of the low- and high-temperature behaviour of crystals, the model has its limitations. A temperature-dependent Debye temperature, $\Theta_D(T)$, can be calculated by reproducing the heat capacity at each single temperature using the equation

$$C_{V,\text{Debye}}(T, \Theta_D) = C_{V,\text{exper}}(T) \tag{8.29}$$

A dip in Θ_D versus temperature is typically observed for $\Theta_D/50 < T/K < \Theta_D/2$. It follows that a constant Debye temperature is not able to reproduce the experimental observations over large temperature ranges.

More importantly, the Debye model and also the Einstein model work best for materials that take crystal structures of high symmetry, for pure elements as well as for binary compounds where the two elements have similar masses and bonding. The approach is less applicable to complex materials that contain elements that differ largely in chemical bonding, coordination and/or mass. Furthermore, groups like carbonate anions and the SiO_4 tetrahedra in silicates will retain their intrinsic properties to a large degree even in complex compounds. An alternative is to represent the massive vibrational modes of the unit cell as a whole by a Debye function and the internal vibrational modes of covalent groups by one or more Einstein functions. This approach will be further discussed below.

Table 8.2. Debye temperature (Θ_D in K) and electronic heat capacity coefficient (see Section 8.4) (γ in mJ K^{-1} mol^{-1}) of the elements.

Li 344 18	Be 1440 2						A Θ_D γ						B	C 2050 0	N	O	F	Ne 75
Na 158 14	Mg 400 14												Al 428 14	Si 645	P	S	Cl	Ar 92
K 91 21	Ca 230 77	Sc 360	Ti 420 36	V 380 92	Cr 630 16	Mn 410 180	Fe 470 50	Co 445 48	Ni 450 73	Cu 315 7	Zn 327 6	Ga 320 6	Ge 374	As 282	Se 90	Br	Kr 72
Rb 56 24	Sr 147 37	Y 280 30	Zr 291 88	Nb 275 21	Mo 450	Tc	Ru 600 34	Rh 480 49	Pd 274 100	Ag 225 6	Cd 209 7	In 108 18	Sn 200 18	Sb 211	Te 153	I	Xe 64
Cs 33 32	Ba 110 27	La 142	Hf 252 26	Ta 240 59	W 400 12	Re 430 25	Os 500 24	Ir 420 31	Pt 240 66	Au 165 7	Hg 72 19	Tl 79 15	Pb 105 34	Bi 119	Po	At	Rn

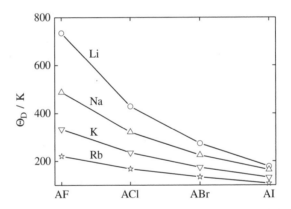

Figure 8.14 Debye temperature of the alkali halides.

The relationship between elastic properties and heat capacity

Both the Einstein and Debye theories show a clear relationship between apparently unrelated properties: heat capacity and elastic properties. The Einstein temperature for copper is 244 K and corresponds to a vibrational frequency of 32 THz. Assuming that the elastic properties are due to the sum of the forces acting between two atoms this frequency can be calculated from the Young's modulus of copper, $E = 13 \times 10^{10}$ N m^{-2}. The force constant K is obtained by dividing E by the number of atoms in a plane per m^2 and by the distance between two neighbouring planes of atoms. K thus obtained is 14.4 N m^{-1} and the Einstein frequency, obtained using the mass of a copper atom into account, 18 THz, is in reasonable agreement with that deduced from the calorimetric Einstein temperature.

Table 8.3 Comparison of Debye temperatures derived from heat capacity data and from elastic properties.

	Ag	Cu	Al	NaCl	KBr	LiF
Θ_D(Elastic)	226.4	344.4	428.2	321.9	182.8	834.1
Θ_D(CV)	226.2	345.1	426	320	184	838

Even better agreement is observed between calorimetric and elastic Debye temperatures. The Debye temperature is based on a continuum model for long wavelengths, and hence the discrete nature of the atoms is neglected. The wave velocity is constant and the Debye temperature can be expressed through the average speed of sound in longitudinal and transverse directions (parallel and normal to the wave vector). Calorimetric and elastic Debye temperatures are compared in Table 8.3 for some selected elements and compounds.

Dilational contributions to the heat capacity

In general, C_V is obtained from C_p by taking the isobaric expansivity and isothermal compressibility of the crystal into consideration:

$$C_{p,\mathrm{m}} - C_{V,\mathrm{m}} = \frac{\alpha^2 VT}{\kappa_T} \tag{8.30}$$

The thermal expansivity of a solid is in general low at low temperatures and the anharmonic contribution to the heat capacity is therefore small in this temperature region and $C_{V,\mathrm{m}} \approx C_{p,\mathrm{m}}$. At high temperatures the difference between the heat capacity at constant pressure and at constant volume must be taken into consideration.

The heat capacity models described so far were all based on a harmonic oscillator approximation. This implies that the volume of the simple crystals considered does not vary with temperature and $C_{V,\mathrm{m}}$ is derived as a function of temperature for a crystal having a fixed volume. Anharmonic lattice vibrations give rise to a finite isobaric thermal expansivity. These vibrations contribute both directly and indirectly to the total heat capacity; directly since the anharmonic vibrations themselves contribute, and indirectly since the volume of a real crystal increases with increasing temperature, changing all frequencies. The constant volume heat capacity derived from experimental heat capacity data is different from that for a fixed volume. The difference in heat capacity at constant volume for a crystal that is allowed to relax at each temperature and the heat capacity at constant volume for a crystal where the volume is fixed to correspond to that at the Debye temperature represents a considerable part of $C_{p,\mathrm{m}} - C_{V,\mathrm{m}}$. This is shown for Mo and W [6] in Figure 8.15.

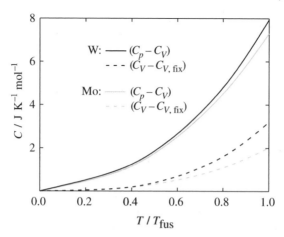

Figure 8.15 $C_{p,\mathrm{m}} - C_{V,\mathrm{m}}$ and $C_{V,\mathrm{m}} - C_{V,\mathrm{m,fix}}$ for Mo and W [6].

In order to calculate the dilational contribution exactly a considerable quantity of data is needed. The temperature dependence of the volume, the isobaric expansivity and the isothermal compressibility is seldom available from 0 K to elevated temperatures and approximate equations are needed. The **Nernst–Lindeman relationship** [7] is one alternative. In this approximation $C_{p,\mathrm{m}} - C_{V,\mathrm{m}}$ is given by

$$C_{p,\mathrm{m}} - C_{V,\mathrm{m}} = \frac{V\alpha^2}{\kappa_T C_{p,\mathrm{m}}^2} C_{p,\mathrm{m}}^2 T = A C_{p,\mathrm{m}}^2 T \tag{8.31}$$

The parameter A is nearly constant over a wide range of temperatures and A can be calculated at any temperature if the data needed to derive its value are available at one temperature. The equation can be used over a wide temperature range without introducing large errors.

In an alternative approach [7], applicable if the isobaric thermal expansivity is known as a function of temperature, $C_{p,\mathrm{m}} - C_{V,\mathrm{m}}$ is given by

$$C_{p,\mathrm{m}} - C_{V,\mathrm{m}} = \gamma_G \alpha C_{V,\mathrm{m}} T \tag{8.32}$$

where γ_G is the **Grüneisen parameter**, defined as

$$\gamma_G = \left[\frac{\partial p}{\partial (U/V)} \right]_V = \frac{\alpha V}{\kappa_T C_{V,\mathrm{m}}} \tag{8.33}$$

γ_G is largely independent of temperature at high temperatures.

Estimates of heat capacity from crystallographic, elastic and vibrational characteristics

Most thermodynamic data are limited to a certain range in temperature and pressure and cannot be extrapolated to other p,T conditions. The data tabulated in major data compilations and thermodynamic tables are often obtained by fitting non-physical models to experimental data and the fits are only valid over the temperature and pressure ranges where measurements have been carried out. Extrapolations to other T and p conditions are frequently done using simplified models of the vibrational density of states. These estimation schemes have been shown often to be accurate at temperatures above about 200 K. A model is typically developed by taking available crystallographic, vibrational and thermoelastic data into consideration and can be used over larger pressure and temperature ranges.

Kieffer has estimated the heat capacity of a large number of minerals from readily available data [8]. The model, which may be used for many kinds of materials, consists of three parts. There are three acoustic branches whose maximum cut-off frequencies are determined from speed of sound data or from elastic constants. The corresponding heat capacity contributions are calculated using a modified Debye model where dispersion is taken into account. High-frequency optic modes are determined from specific localized internal vibrations (Si–O, C–O and O–H stretches in different groups of atoms) as observed by IR and Raman spectroscopy. The heat capacity contributions are here calculated using the Einstein model. The remaining modes are ascribed to an optic continuum, where the density of states is constant in an interval from v_L to v_H and where the frequency limits v_L and v_H are estimated from Raman and IR spectra.

This approach can be briefly illustrated through an analysis of the heat capacity of calcite. The primitive unit cell contains two $CaCO_3$ formula units (10 atoms) and calcite thus has 30 degrees of freedom. Among these modes, 3 are acoustic and 12 are internal modes related to the vibrations of the two CO_3^{2-} groups (with $3n - 6$ vibrational modes), whereas the remaining 15 are optic modes that will be taken into account by the optic continuum. Directly measured acoustic velocities are available. The frequency of the internal optic modes represented by Einstein oscillators and the limits of the optic continuum are derived from the IR and Raman spectra given in Figures 8.16(a) and (b). The resulting approximate phonon density of states is shown in Figure 8.16(c), while the heat capacity and the relative contribution from the acoustic and internal optic modes and the modes of the optic continuum are given in Figure 8.16(d).

The Kieffer approach uses a harmonic description of the lattice dynamics in which the phonon frequencies are independent of temperature and pressure. A further improvement of the accuracy of the model is achieved by taking the effect of temperature and pressure on the vibrational frequencies explicitly into account. This gives better agreement with experimental heat capacity data that usually are collected at constant pressure [9].

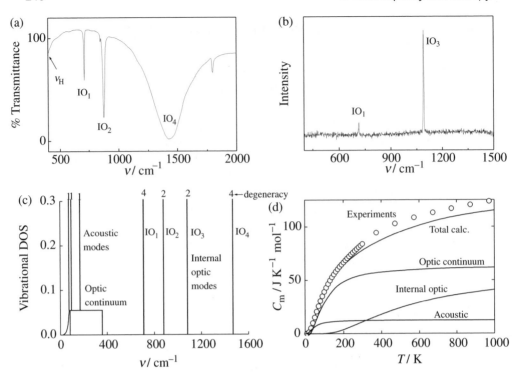

Figure 8.16 (a) IR and (b) Raman spectra for the mineral calcite, $CaCO_3$. The estimated density of vibrational states is given in (c) while the deconvolution of the total heat capacity into contributions from the acoustic and internal optic modes as well as from the optic continuum is given in (d).

8.3 Vibrational entropy

The Einstein and Debye models revisited

The vibrational heat capacity is the largest contribution to the total heat capacity and determines to a large extent the entropy. Analytical expressions for the entropy of the models described in the previous section can be derived. The entropy corresponding to the Einstein heat capacity is

$$S_E = 3R \left[\frac{\Theta_E / T}{[\exp(\Theta_E / T) - 1]} - \ln[1 - \exp(-\Theta_E / T)] \right] \qquad (8.34)$$

where $\theta_E = \hbar \omega_E / k_B$.

Similarly, the expression for the entropy of the Debye model is

$$S_D = 3R \left[\frac{4T^3}{\Theta_D^3} \int_0^{\Theta_D / T} \frac{x^3 dx}{[\exp(x) - 1]} - \ln[1 - \exp(-\Theta_D / T)] \right] \qquad (8.35)$$

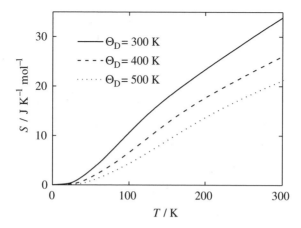

Figure 8.17 Entropy of a monoatomic solid for different values of the Debye temperature, Θ_D.

and here $\theta_D = \hbar\omega_D/k_B$ and $x = \hbar\omega/k_B T$.

Whereas the latter expression must be solved numerically for low temperatures, the entropy at high temperatures can be derived by a series expansion [4]. For the Debye or Einstein models the entropy is essentially given in terms of a single parameter at high temperature:

$$S = 3R\left[\frac{4}{3} + \ln\left(\frac{T}{\theta_i}\right) + \frac{1}{40}\left(\frac{\theta_i}{T}\right)^2 - \frac{1}{2240}\left(\frac{\theta_i}{T}\right)^4 + \ldots \right] \qquad (8.36)$$

The entropy of a monoatomic solid is given as a function of the Debye temperature and the thermodynamic temperature in Figure 8.17.

An alternative to deriving the Debye temperature from experimental heat capacities is to derive an entropy-based Debye temperature by calculation of the Θ_S that reproduces the observed entropy for each single temperature using

$$S_{Debye}(T, \Theta_S) = S_{exper}(T) \qquad (8.37)$$

This relation yields $\Theta_S(T)$, in an analogous manner to the heat capacity-based Debye temperature (eq. 8.29). While the derivation of Debye temperatures from heat capacities is done using low-temperature data, the entropy-based Debye temperature can be derived from high-temperature data since the vibrational part of the standard entropy usually dominates at high temperatures. As evident from eq. (8.36), at high temperatures the entropy depends on the logarithmic average of the phonon frequency. These phonon frequencies, as we have seen, depend on the interatomic forces and the atomic masses. The logarithmic average can be deconvoluted so that the effect of atomic mass is separated from the effect of the interatomic force constant. The entropy contribution related to the force constant

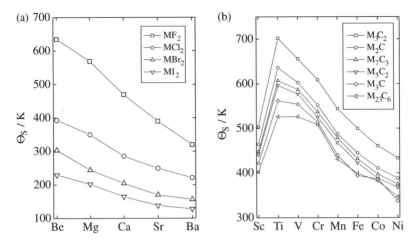

Figure 8.18 Entropy Debye temperature, Θ_S, for (a) alkali earth dihalides [10] and (b) first series transition metal carbides [11].

depends on the electronic structure of the compounds and strong regularities are observed when chemically related materials are compared. The method was developed for estimation of the entropy of materials (stable and metastable) for which experimental entropy data are not available. A large number of compounds have been analyzed. Trends observed in the entropy Debye temperature are given for the alkaline earth dihalides [10] and transition metal carbides [11] in Figure 8.18. A close resemblance is observed between the trends for the normal Debye temperature of the alkali halides in Figure 8.14 and the entropy Debye temperatures for the alkaline earth dihalides in Figure 8.18(a).

Effect of volume and coordination

First-order estimates of entropy are often based on the observation that heat capacities and thereby entropies of complex compounds often are well represented by summing in stoichiometric proportions the heat capacities or entropies of simpler chemical entities. Latimer [12] used entropies of elements and molecular groups to estimate the entropy of more complex compounds: see Spencer for revised tabulated values [13]. Fyfe *et al.* [14] pointed out a correlation between entropy and molar volume and introduced a simple volume correction factor in their scheme for estimation of the entropy of complex oxides based on the entropy of binary oxides. The latter approach was further developed by Holland [15], who looked into the effect of volume on the vibrational entropy derived from the Einstein and Debye models.

Even though the Einstein and Debye models are not exact, these simple one-parameter models illustrate the properties of crystals and should give reliable estimates of the volume dependence of the vibrational entropy [15]. The entropy is given by the characteristic vibrational frequency and is thus related to some kind of mean interatomic distance or simpler, the volume of a compound. If the unit cell volume is expanded, the average interatomic distance becomes larger and the bonding strength is usually reduced. Weaker bonding strength obviously results in a reduced vibrational frequency for the longitudinal phonons.

The effect of a change in volume on the entropy is given by the ratio of the isobaric expansivity and isothermal compressibility of a compound:

$$\left(\frac{\partial S}{\partial V}\right)_T = \frac{\alpha}{\kappa_T} \tag{8.38}$$

The volume dependence of the entropy can alternatively be expressed as a function of the Einstein or Debye temperature as

$$\left(\frac{\partial S}{\partial V}\right)_T = \left(\frac{\partial \Theta}{\partial V}\right)_T \left(\frac{\partial S}{\partial \Theta}\right)_T \tag{8.39}$$

$(\partial S/\partial \Theta)_T$ is given by differentiation of eqs. (8.34) or (8.35), while the variation of the characteristic temperature with volume is given by [15]

$$\Theta = \Theta_0 \left(\frac{V_0}{V}\right)^{1/3} \tag{8.40}$$

Here Θ_0 is the characteristic temperature at volume V_0. An average value for the volume dependence of the standard entropy at 298 K for around 60 oxides based on the Einstein model is 1.1 ± 0.1 J K^{-1} cm^{-3} [15]. A corresponding analysis using the Debye model gives approximately the same numeric value.

In the preceding treatment we have assumed a volume change without taking the structure of the material into consideration; for a phase transition where the cation coordination is preserved $\Delta_{trs}S$ generally has the same sign as $\Delta_{trs}V$ and the transition temperature increases with pressure since $dp/dT = \Delta_{trs}S/\Delta_{trs}V$. A different conclusion can be derived for phase transitions which involves a change in the nearest coordination number. The vibrational properties of an atom will to a large degree depend on the local environment and a change in coordination necessarily implies a change in the vibrational density of states. For a transition that increases the coordination, the interatomic distances in the first coordination sphere become longer. Although the number of bonds increases, the vibrational density of states is generally shifted to lower frequencies. The denser compound thus has higher entropy than otherwise expected and may even in some cases have a higher entropy than the less dense modification, i.e. the Clapeyron slope becomes negative since $\Delta_{trs}S > 0$ for $\Delta_{trs}V < 0$.

The effect of coordination may be illustrated by the vibrational properties of the four modifications of $MgSiO_3$ [16]. Above 1000 K, the entropy of these modifications of $MgSiO_3$ decreases in the order

pyroxene > perovskite > garnet > ilmenite

The density increases in the order

pyroxene < garnet < ilmenite < perovskite

Thus garnet, ileminite and perovskite are modifications stabilized by pressure and they are all stable at some specific pressure–temperature conditions.

Using the volume argument, pyroxene is the high-entropy modification, in line with experiments. On the other hand, the perovskite should have the lowest entropy. This is not observed and this reflects the opposing effect of a coordination change. All the Si atoms are tetrahedrally coordinated in pyroxene, while 50% are tetrahedrally coordinated and 50% octahedrally coordinated in garnet. In the ilmenite and perovskite modifications all Si atoms are octahedrally coordinated. When Si transforms from a tetrahedral to an octahedral coordination, the frequency of the Si–O stretching modes decreases from typically 900–1100 cm^{-1} (SiO_4 tetra-hedra) to 600–800 cm^{-1} (SiO_6 octahedra). The bonds are stronger in the tetrahedra than in the octahedra [16]. A lower bond strength gives larger vibrational ampli-tudes, lower frequencies and thereby a larger entropy for the 6-coordinated com-pounds. These factors explain why the perovskite has higher entropy than the garnet at high temperatures and that the slope of the p–T phase boundary between the two phases is negative.

8.4 Heat capacity contributions of electronic origin

Electronic and magnetic heat capacity

Quantization is important also for understanding the electronic contribution to the heat capacity. In classical theory the electrons were considered as small particles moving freely through the crystal, and the three translational degrees of freedom were expected in total to contribute $3/2R$ to the total heat capacity. The heat capacity of a monovalent metal, like copper, was by these considerations expected to be $3R + 3/2R$. The experimental heat capacity is, however, as we have seen, close to $3R$. The small electronic contribution to the heat capacity is due to quantization. Electrons follow Fermi–Dirac statistics (see e.g. C. Kittel and H. Kroemer, Further reading) and only electrons near the Fermi level can be excited at a given tempera-ture. When the temperature rises from 0 K to T the total increase in internal energy is roughly

$$\Delta U = N_1 k_B T \tag{8.41}$$

where N_1 is the number of electrons that are excited by the thermal energy $k_B T$. These electrons occupy electron states in a band of thickness $k_B T$ about the Fermi level, ε_F, as visualized in Figure 8.19, where the energy distribution for a free elec-tron gas is given. It follows that the number of excited electrons, N_1, at a given tem-perature is given by

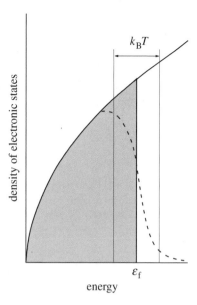

Figure 8.19 Energy distribution for a free electron gas at 0 K (shaded) and an elevated temperature (dashed line), T.

$$N_1 = n(\varepsilon_F)k_BT \tag{8.42}$$

where $n(\varepsilon_F)$ is the density of electronic states at the Fermi level. The internal energy of the electrons follows

$$\Delta U = n(\varepsilon_F)k_B^2 T^2 \tag{8.43}$$

Using this simple argument, the electronic heat capacity, C_E, of a free electron gas is

$$C_E = \frac{\partial \Delta U}{\partial T} = 2n(\varepsilon_F)k_B^2 T \tag{8.44}$$

An exact derivation that takes into account the variation of the Fermi level with temperature gives basically the same result:

$$C_E = \frac{\pi^2}{3} n(\varepsilon_F)k_B^2 T \tag{8.45}$$

The electronic heat capacity thus varies linearly with temperature and is often represented as

$$C_E = \gamma T \tag{8.46}$$

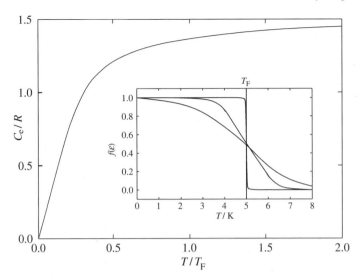

Figure 8.20 Heat capacity of a free electron gas. The population of the electronic states at different temperatures is shown in the insert. T_F is typically of the order of 10^5 K.

where γ is termed the electronic heat capacity coefficient. The variation of γ across the periodic table is given in Table 8.2.

The electronic heat capacity for the free electron model is a linear function of temperature only for $T << T_F = \varepsilon_F / k_B$. Nevertheless, the Fermi temperature T_F is of the order of 10^5 K and eq. (8.46) holds for most practical purposes. The population of the electronic states at different temperatures as well as the variation of the electronic heat capacity with temperature for a free electron gas is shown in Figure 8.20. Complete excitation is only expected at very high temperatures, $T > T_F$. Here the limiting value for a gas of structureless mass points $3/2R$ is approached.

Since the lattice contribution to the heat capacity varies with T^3, the total heat capacity at low temperatures (typically $T < 10$ K or lower) to a first approximation is given by

$$C_V = \beta T^3 + \gamma T \qquad (8.47)$$

A plot of C_V/T against T^2 should therefore give a straight line as illustrated for Cu in Figure 8.21. From this plot the electronic density of states at the Fermi level can be calculated, and calorimetry has frequently been used for this purpose. A striking example is the electronic heat capacity coefficients observed for Rh–Pd–Ag alloys given in Figure 8.22 [17]. In the rigid band approach the addition of Ag to Pd gives an extra electron per atom of silver and these electrons fill the band to a higher energy level. Correspondingly, alloying with Rh gives an electron hole per Rh atom and the Fermi level is moved to a lower energy. The variation of the electronic heat capacity coefficient with composition of the alloy maps approximately the shape of such an electron band. The electronic density of states for Rh–Pd–Ag alloys can be

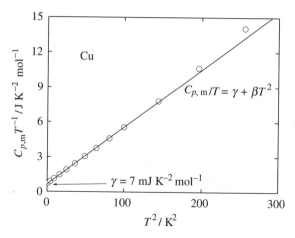

Figure 8.21 Heat capacity of Cu plotted as $C_{p,\mathrm{m}} \cdot T^{-1}$ versus T^2.

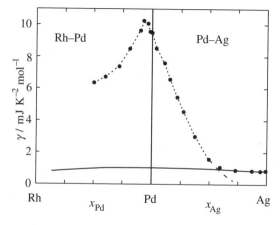

Figure 8.22 Variation of the electronic heat capacity coefficient with composition for the alloys Rh–Pd and Pd–Ag [17]. Solid and dotted lines represent the electronic DoS for the 5s and 4d bands, respectively.

understood by the superposition of a narrow 4d band on a flat 5s band. This is also indicated in Figure 8.22; the solid line represents the 5s band and the dotted line the 4d band.

Whereas the vibrational and electronic contributions to the heat capacity are readily treated in terms of simple models, the treatment of the heat capacity due to excitation of collective **magnetic excitation** modes of ordered magnetic structures, often termed **spin waves** or **magnons**, is more complex. The ideally ordered magnetic states exist only in the absence of thermal agitation. At finite temperatures the spins at some sites are excited. By analogy with collective vibrational modes, spin waves are formed. If the magnon frequency for a particular wave vector is $v(q)$ and $g_{\mathrm{magn}}(v)$ is the density of states of magnon frequencies, the magnetic heat capacity can be expressed by analogy with that for the lattice heat capacity as (see e.g. Grimvall, Further reading)

$$C_{\text{magn}} = R \int \frac{x^2 e^x}{(e^x - 1)^2} g_{\text{magn}}(\omega) d\omega \tag{8.48}$$

where $x = \hbar\omega / k_B T$. The magnon density of states is in general not easily calculated and the functional variation depends on the type of magnetic order. It can be shown that the magnetic frequencies for a ferromagnetic insulator scale with the square of the wave vector. The magnetic heat capacity resulting from this characteristic is proportional to $T^{3/2}$ at low temperature. The heat capacity of an antiferromagnet shows a T^3 variation at low temperatures. At higher temperatures, interactions between the magnons give rise to other terms.

Electronic and magnetic transitions

Electronic transitions like insulator–metal transitions, magnetic order–disorder transitions, spin transitions and Schottky-type transitions (due to crystal field splitting in the ground state in f element-containing compounds) profoundly influence the phase stability of compounds. A short description of the main characteristics of these transitions will be given below, together with references to more thorough treatments.

The electronic heat capacity naturally has a pronounced effect on the energetics of **insulator–metal transitions** and the entropy of a first-order transition between an insulating phase with $\gamma = 0$ and a metallic phase with $\gamma = \gamma_{\text{met}}$ at T_{trs} is in the first approximation $\Delta_{\text{ins–met}} S_m = \gamma_{\text{met}} T_{\text{trs}}$.

Differences in the Debye temperature, or in other words the vibrational character of the two phases, will modify the transitional entropy to some extent. Still, this entropy change is normally not large for transitions where the coordination number is preserved.

Magnetic order–disorder transitions give rise to a heat capacity and entropy contribution of configurational nature. While the magnon heat capacity is complex, a very general expression is available for the maximum total order–disorder entropy. A particle with N_{un} unpaired electrons and total spin quantum number $\mathscr{S} = \Sigma \frac{1}{2} N_{\text{un}}$ has $(2\mathscr{S} + 1)$ quantized orientations. While all spins have the same orientation at low temperatures for an ordered magnetic structure, all orientations are degenerate in the paramagnetic state. The entropy change corresponding to this disordering is thus $k_B \ln(2\mathscr{S} + 1)$ per particle or

$$\Delta S = R \ln(2\mathscr{S} + 1) \tag{8.49}$$

for one mole of atoms. For Fe^{3+} (d^5) in its electronic ground state $\mathscr{S} = 5/2$ and $\Delta\mathscr{S} = R \ln 6 = 14.89$ J K^{-1} mol^{-1}. If the disordering takes place through a second-order transition, the entropy contribution at a given temperature may be estimated

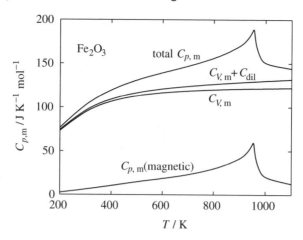

Figure 8.23 Heat capacity of Fe_2O_3 [18]. The heat capacity is deconvoluted to show the relative magnitude of the main contributions. $C_{dil} = C_{p,m} - C_{V,m} = \alpha^2 TV/\kappa_T$.

from Landau theory (see Section 2.3). It should in this context be noted that a magnetic transition that takes place at high temperatures may give significant contributions to the entropy at temperatures far below T_C, as evident from the deconvoluted heat capacity of Fe_2O_3 in Figure 8.23 [18]. Although the spin-only approximation has been shown to give reasonable estimates for iron oxides, for example, deviations from the spin-only entropy must in general be expected as for example shown for MnO [19]. An alternative approximation for the maximum magnetic entropy frequently used in thermodynamic representations of phase diagrams involves the saturation magnetization at 0 K instead of the spin quantum number [20].

Transitions with large entropy changes take place also in the paramagnetic state. Octahedrally coordinated transition metal ions having electronic configurations d^4, d^5, d^6 or d^7 can exist in the low-spin or high-spin ground state depending on the strength of the ligand field. The same is the case for tetrahedrally coordinated transition metal ions with d^3, d^4, d^5 or d^6 configuration. Temperature-induced **spin transitions** are expected in compounds where the ligand field splitting energy is comparable to the thermal energy and larger than the electron spin-pairing energy. The ligand field splitting energy depends on the metal, its valence and the ligand, and for a given metal in a given oxidation state, the ligand determines the spin state. Some trivalent oxides of cobalt (for example Co_2O_3) are found to be in a high-spin state at all temperatures, others are in a low-spin state (for example $ZnCo_2O_4$ and $LiCoO_2$), while others show temperature-induced spin transitions (for example Co_3O_4 and $LaCoO_3$). The ligand field splitting is in general smaller for iron than for cobalt, and although Co^{3+} and Fe^{2+} have the same electronic configuration, all iron oxides are in a high-spin state at ambient pressure. Reviews focusing on metal–organic compounds by Gütlich *et al.* [21] and by Sorai [22] give more detailed accounts of the physics of spin transitions.

Large entropy increments accompany the spin transitions. For Co^{3+} in an octahedral crystal field the high-spin state has a degeneracy of 15 (the product of orbital

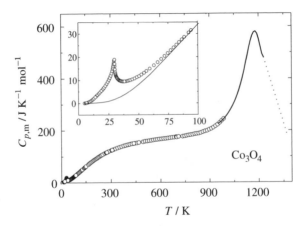

Figure 8.24 Heat capacity of Co_3O_4 [23–25]. The insert shows the magnetic order–disorder transition at around 30 K [24] in detail.

and spin degeneracy), whereas the low-spin state is singly degenerate. A transition from the low-spin to the high-spin ground state should thus give a configurational entropy increment of $R \ln 15 = 22.5$ J·K^{-1} mol^{-1} in the high-temperature limit and also substantial contributions to the heat capacity. The experimental heat capacity for Co_3O_4, which undergoes a temperature-induced spin transition at high temperatures is shown in Figure 8.24 [23–25]. The normal spinel contains Co^{2+} at tetrahedral sites and low-spin Co^{3+} at octahedral sites. The heat capacity effect observed at $T > 900$ K is in part a low- to high-spin transition of the Co^{3+} ions and in part a partial transition from normal toward random distribution of Co^{3+} and Co^{2+} on the tetrahedral and octahedral sites of the spinel structure. The entropy change connected with the spin change is by far the largest contribution to the transitional entropy. The insert to the figure shows the magnetic order–disorder transition of Co_3O_4 at around 30 K.

The magnetic entropies are large and indicate the need for thermodynamic models, atomistic or purely mathematical, that can handle these effects. Although spin transitions in principle may be simple two-level transitions, they may also be affected by spin–orbit coupling (frequently observed for cobalt ions), magnetic ordering, and associated changes in the crystal structure. Since the spin transition affects the bonding in the crystal and thus the vibrational frequencies, an associated change in the Debye temperature of the compound must be expected. The size of this effect is limited for inorganic compounds. For organometallic compounds, on the other hand, the entropy connected with this lattice effect often is larger than the entropy of the spin state transition itself [22].

A low-spin to high-spin transition relates to the crystal field splitting of the d-orbitals in an octahedral or tetrahedral crystal field. However, even in cases where the energy difference between two spin states is much larger, electronic transitions are observed. An atom with total spin quantum number $\mathscr{S}$ has $(2\mathscr{S} + 1)$ orientations. In a magnetic field the atom will have a number of discrete energy levels with energy spacing dependent on the strength of the magnetic field. Although the crystal field experienced by a metal-atom in a crystal is small compared to what can be achieved by external magnets, the crystal-field typically induces splitting of the f-electron levels in actinides, and the effects of this are visible in the heat capacity in large temperature regions at relatively high temperatures [26]. Similar heat capacity effects are observed also in hydrated paramagnetic salts used for magnetic cooling [27]. The spins on transition metals like Ni and Fe are here localized since they are diluted by the presence of non-magnetic ions. This dilution gives well-defined localized energy levels and thus sharp heat capacity effects.

The heat capacity of transitions of this type is given by the energy level splitting in an electrostatic field and by the degeneracy of the energy levels. Let us for simplicity assume that we have a system with two levels with energy spacing ε/k_B. For $T \ll \varepsilon/k_B$ the upper level is occupied to a negligible extent only. For $T \gg \varepsilon/k_B$ both levels will be approximately equally occupied. Using Boltzmann statistics the heat capacity is

$$C_{sch} = R\left(\frac{\varepsilon}{k_B T}\right)^2 \left(\frac{g_0}{g_1}\right)\left[\frac{\exp(\varepsilon/k_B T)}{\{1+(g_0/g_1)\exp(\varepsilon/k_B T)\}^2}\right] \tag{8.50}$$

where g_0 and g_1 are the degeneracies of the ground level and the excited level, respectively. Experimental Schottky-type heat capacities for Nd_2S_3 [28] and $ErFeO_3$ [29] are shown in Figure 8.25. While the transition in the sulfide is spread over a large temperature range, the $ErFeO_3$ transition is at low temperatures and occurs over a narrow range. A large number of experimental studies by Westrum and co-workers have shown that the energies of the different energy levels can be derived from accurate heat capacity data [26]. In these analyses, the number of energy levels and their corresponding degeneracies are derived based on the symmetry of the lanthanide cations in the particular compounds by crystal field theory.

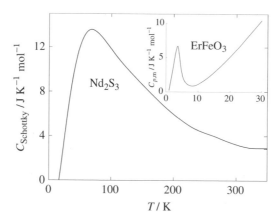

Figure 8.25 The Schottky-type heat capacity of Nd_2S_3 [28]. The insert shows the total heat capacity of $ErFeO_3$ [29].

Table 8.4 Energy levels of Nd_2S_3 [28].

| Level | Degeneracy | E / cm^{-1} | |
		Spectroscopy	Calorimetry
0	2	0	0
1	2	76	80
2	2	140	140
3	2	180	180
4	2	358	340

The derived energy levels are shown to correspond closely to those obtained by spectroscopy [26]. Calorimetrically and spectroscopically obtained energy levels for Nd_2S_3 are given as an example in Table 8.4 [28].

8.5 Heat capacity of disordered systems

Crystal defects

Real crystals contain lattice defects and other extended defects at temperatures above absolute zero (see e.g. Wollenberger, Further reading). The creation of an intrinsic defect is always an endothermic process. However, as the temperature is raised the increased entropy associated with the large number of possible positions for the defect becomes important. Thus the counteracting effects of the entropy and enthalpy of defect formation determine the variation of the Gibbs energy with concentration of defects, which will have a minimum for a given defect concentration at a specific temperature. For a stoichiometric compound the temperature-induced intrinsic disorder may be due to the formation of Frenkel or Schottky defects. This defect formation gives rise to a heat capacity contribution given by the molar entropy and enthalpy of formation of the defects, which for the Schottky defect case is given by

$$C_{V,m} = \frac{1}{RT^2}\left[\Delta_{vac}H_m^2 \exp\left(\frac{\Delta_{vac}S_m}{R}\right)\right]\exp\left(-\frac{\Delta_{vac}H_m}{RT}\right) \qquad (8.51)$$

Here $\Delta_{vac}S$ and $\Delta_{vac}H$ are the molar entropy and enthalpy of formation of the defects. Using a pure metal like aluminium as an example, the fractional number of defects, heat capacity and enthalpy due to defect formation close to the fusion temperature are $5 \cdot 10^{-4}$, 0.3 J K^{-1} mol^{-1} and 30 J mol^{-1} [30].

In other cases more complex disordering mechanisms are observed. Non-stoichiometric oxides in which the oxygen vacancies are ordered at low temperatures illustrate convergent disordering. The oxygen atoms and oxygen vacancies are here distributed at the same sub-lattice at high temperatures. An example is

$Sr_2Fe_2O_5$ in which the entropy of the first-order disordering reaction suggests that the oxygen atoms are far from randomly distributed at high temperatures [31]. Computer simulations described in Section 11.1 suggest that the disorder may be described in terms of a mixture of different structural entities: FeO_4 tetrahedra, FeO_5 square pyramids and FeO_6 octahedra [32]. Cation disorder by what is termed non-convergent disordering takes place in other systems, e.g. in spinels. Here the cations occupy two different sub-lattices, which also remain distinct in the disordered high-temperature state. This will be discussed further in Section 9.2.

Fast ion conductors, liquids and glasses

Fast ion conductors have many physical properties, typical of solids, while others are to some extent similar to those of liquids. For some fast ionic conductors, for example AgI, the high ionic conductivity is associated with a first-order phase transition where the sub-lattice associated with the mobile Ag ion becomes almost completely disordered. The other sub-lattice remains unaffected and constitutes a structural frame, which gives the substance the expected mechanical properties of a solid. In some cases, the entropy of the order–disorder transition forming the fast ion conductor is larger than the entropy of the subsequent melting of the disordered fast ion conductor [33]. The large degree of disorder on the sub-lattice of the mobile species results in a high configurational entropy. This is a characteristic property of liquids as well. Another similarity between liquids and some fast ion conductors is a decreasing heat capacity with increasing temperature. This is observed for many liquids and also for some fast ion conductors like Cu_2S [34] and Ag_2S [35].

In contrast to crystalline solids characterized by translational symmetry, the vibrational properties of liquid or amorphous materials are not easily described. There is no firm theoretical interpretation of the heat capacity of liquids and glasses since these non-crystalline states lack a periodic lattice. While this lack of long-range order distinguishes liquids from solids, short-range order, on the other hand, distinguishes a liquid from a gas. Overall, the vibrational density of state of a liquid or a glass is more diffuse, but is still expected to show the main characteristics of the vibrational density of states of a crystalline compound.

The most characteristic feature of liquids relative to crystalline solids is the configurational entropy associated with the large degree of disorder in the liquid state. The high configurational entropy allows for fast motion by rotation or diffusion, and liquids are usually non-viscous in the stable region above the freezing point. The viscous properties of SiO_2 and some other glass-forming liquids are rare exceptions, but in this case the configurational entropy associated with the melting transition is unusually low [36].

The heat capacity of liquids at constant pressure appears in general to go through a broad minimum as a function of temperature. A classic example is Se, for which the heat capacity is given in Figure 5.1. The heat capacity is expected to decrease at high temperatures, where the intermolecular bonding becomes weaker and vibrational degrees of freedom may be lost when the short-range order is gradually

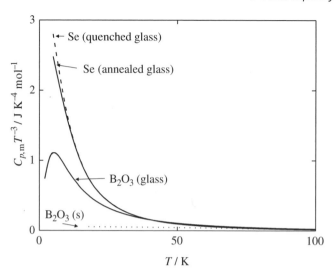

Figure 8.26 Heat capacity of glassy and crystalline B_2O_3 [42–44] and glassy Se [41] plotted as $C_{p,\mathrm{m}} \cdot T^{-3}$ versus T.

destroyed. Finally, it is worth noting that the heat capacity of inorganic and metallic liquids is not that well established experimentally and considerable spread is present even in the data reported for pure metallic elements like Cd [37] and Zn [38].

Two aspects of the thermodynamics of supercooled liquids have generated considerable interest: the non-equilibrium transition to the glassy state at the glass transition [39] as well as the heat capacity of the glass close to the absolute zero of temperature. As already discussed, the vibrational density of states of glasses resembles that of the stable crystalline compound. This is especially true for the vibrational modes associated with short-range order. Still, anomalous heat capacity behaviour is often observed close to absolute zero. This effect is ascribed to excitations in a two-level system associated with highly anharmonic atomic configurations [40]. The 'anomalous' heat capacity effects observed for Se [41] and B_2O_3 [42–44] glasses are shown in Figure 8.26. For a normal Debye-like solid the heat capacity divided by the cube of the temperature should be approximately constant, as indicated for crystalline B_2O_3 by a dotted line. The glasses show heat capacities that largely exceed this Debye-like value.

The glass transition has also received much attention since it is most easily observed experimentally by measurement of the heat capacity and because it is an important characteristic of glasses. The heat capacity displays a jump at the glass transition associated with the onset of the configurational degrees of freedom of the liquid, as exemplified using a range of types of glass in Figure 8.27 [45–49]. The glass transition is a non-equilibrium transition and the glass transition temperature is observed experimentally to depend on the thermal history of the glass. The heat capacities of glassy B_2O_3 observed on heating at different heating rates are shown in Figure 8.28 [50].

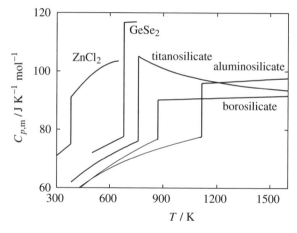

Figure 8.27 Heat capacity of some glass-forming liquids close to their glass transition temperatures: $ZnCl_2$ [45], $GeSe_2$ [46], and a selected titanosilicate [47], aluminosilicate [48] and borosilicate [49] system.

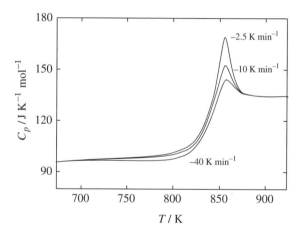

Figure 8.28 Heat capacity of glassy B_2O_3 at different heating rates [50].

This effect of thermal history is also illustrated schematically in Figure 8.29, where the entropy of a liquid during cooling is displayed. Above the glass transition temperature the liquid is ergodic and the entropy is not dependent on the thermal history. The departure from equilibrium behaviour in the glass transition region depends on the cooling rate. At fast cooling rates the supercooled liquid is transformed into a glass at higher temperature relative to a liquid cooled at a slow cooling rate. Slow reheating of a glass prepared by fast cooling may relax, as shown by the short dotted line in Figure 8.29. This is a characteristic of ageing of glasses, which means that the entropy of the glass relaxes towards the entropy of the equilibrium supercooled liquid shown by the dotted line. The relaxation effect is reflected by a peak in the heat capacity at the glass transition temperature, as seen in Figure 8.28

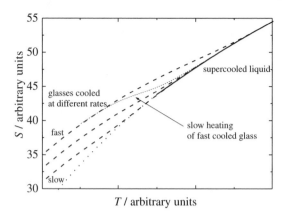

Figure 8.29 Entropy of a supercooled liquid and glasses formed by fast and slow cooling of this liquid (the different dashed lines). The short dashed line represents slow heating of a glass first prepared by fast cooling.

for B_2O_3. This effect is thus a kinetic phenomenon and the effect becomes more pronounced with increasing heating rate. The current understanding of relaxation of glass-forming liquids and amorphous solids is reviewed in [51].

The rate of loss of the configurational entropy of a supercooled liquid on cooling has been used to categorize different glass-forming liquids into fragile and strong [52]. Originally, this was a pure kinetic concept based on the non-Arrhenius behaviour of the viscosity in a T_g-scaled Arrhenius plot. Fragile liquids are strongly non-Arrhenius, while the viscosity of strong liquids follows the Arrhenius law. During heating a fragile liquid becomes non-viscous at a considerably lower temperature than a strong glass with a comparable glass transition temperature. A corresponding thermodynamic fragility is defined by plotting the excess configurational entropy of a supercooled liquid as a function of the reciprocal temperature divided by T_g. The configurational entropy of the supercooled liquid is scaled by the frozen-in configurational entropy at the glass transition. A fragile liquid displays a fast loss of configurational entropy during cooling (a high excess heat capacity relative to the crystalline solid), while strong liquids lose the scaled configurational entropy at a much lower rate (small excess heat capacity of the supercooled liquid relatively to the crystalline solid). The concept of thermodynamic and kinetic fragility therefore seems to open up a qualitative correlation between the thermodynamic and dynamic behaviour of liquids [52].

References

[1]   A. Einstein, *Ann. Physik.* 1907, **22**, 180.
[2]   B. N. Brockhouse, T. Arase, G. Caglioti, K. R. Rao and A. D. B. Woods, *Phys. Rev.* 1962, **128**, 1099.
[3]   B. Prevot, B. Hennion and B. Dorner, *J. Phys. C.: Solid State Phys.* 1977, **10**, 3999.
[4]   P. Debye, *Ann. Physik.* 1912, **39**, 789.

[5]   Y. Miyazaki, M. Sorai, R. Lin, A. Dworkin, H. Szwarc and J. Godard, *Chem. Phys. Lett.* 1999, **305**, 293.

[6]   A. F. Guillermet and G. Grimvall, *Phys. Rev. B*. 1991, **44**, 4332.

[7]   G. Grimvall, *Thermophysical Properties of Materials*. Amsterdam: Elsevier, 1999.

[8]   S. W. Kieffer, *Rev. Geophys. Space Phys.* 1998, **18**, 1; 1989, **18**, 20; 1989, **18**, 35 and 1980, **18**, 862.

[9]   J. Matas, P. Gillet, Y. Ricard and I. Martinez, *Eur. J. Mineral.* 2000, **12**, 703.

[10]   S. Peng and G. Grimvall, *J. Phys. Chem. Solids*. 1994, **55**, 707.

[11]   A. F. Guillermet and G. Grimvall, *J. Phys. Chem. Solids*. 1992, **53**, 105.

[12]   W. M. Latimer, *J. Am. Chem. Soc.* 1951, **73**, 1480.

[13]   W. S. Fyfe, F. J. Turner and J. Verhoogen, *Geol. Soc. Amer. Mem.* 1958, **73**, 259.

[14]   P. J. Spencer, *Thermochim. Acta*. 1998, **314**, 1.

[15]   T. J. B. Holland, *Am. Mineral.* 1989, **74**, 5.

[16]   A. Navrotsky, *Am. Mineral.* 1994, **79**, 589.

[17]   F. E. Hoare in *Electronic Structure and Alloy Chemistry of Transition Elements* (P. A. Beck ed). New York: Interscience, 1963, p. 29.

[18]   F. Grønvold and E. J. Samuelsen, *J. Phys. Chem. Solids*. 1975, **36**, 249.

[19]   B. F. Woodfield, J. L. Shapiro, R. Stevens, J. Boerio-Goates and M. L. Wilson, *Phys. Rev. B*. 1999, **60**, 7335.

[20]   A. P. Miodownik, *CALPHAD*. 1977, **1**, 133.

[21]   P. Gütlich, Y. Garcia and H. A. Goodwin, *Chem. Soc. Rev.* 2000, **26**, 419.

[22]   M. Sorai, *Bull. Chem. Soc. Jpn.* 2001, **74**, 2223.

[23]   A. Navrotsky and D. M. Sherman, *Phys. Chem. Mineral.* 1992, **19**, 88.

[24]   L. M. Khriplovich, E. V. Kholopov and I. E. Paukov, *J. Chem. Thermodyn.* 1982, **14**, 207.

[25]   F. Grønvold, University of Oslo, personal communication.

[26]   E. F. Westrum, Jr, *J. Chem. Thermodyn.* 1983, **15**, 305.

[27]   E. S. R. Gopal, *Specific Heats at Low Temperatures*. London: Heywood Books, 1966.

[28]   E. F. Westrum, Jr, R. Burriel, J. B. Gruber, P. E. Palmer, B. J. Beaudry and W. A. Plautz, *J. Chem. Phys.* 1989, **91**, 4838.

[29]   K. Saito, Y. Yamamura, J. Mayer, H. Kobayashi, Y. Miyazaki, J. Ensling, P. Gütlich, B. Lesniewska and M. Sorai, *J. Magn. Magn. Mater.* 2001, **225**, 381.

[30]   S. Stølen and F. Grønvold, *Thermochim. Acta*. 1999, **327**, 1.

[31]   C. Haavik, E. Bakken, T. Norby, S. Stølen, T. Atake and T. Tojo, *Dalton Trans.* 2003, 361.

[32]   E. Bakken, N. L. Allan, T. H. K. Barron, C. E. Mohn, I. T. Todorov and S. Stølen, *Phys. Chem. Chem. Phys.* 2003, **5**, 2237.

[33]   See e.g. S. Geller (ed.) *Solid Electrolytes*, Topics in Applied Physics, Vol. 21. Berlin: Springer-Verlag, 1977.

[34]   F. Grønvold and E. F. Westrum, Jr, *J. Chem. Thermodyn.* 1987, **19**, 1183.

[35]   F. Grønvold and E. F. Westrum, Jr, *J. Chem. Thermodyn.* 1986, **18**, 381.

[36]   P. Richet, Y. Bottinga, L. Denielou, J. P. Petitet and C. Tequi, *Geochim. Cosmochim. Acta*. 1982, **46**, 2639.

[37]   S. Stølen and F. Grønvold, *Thermochim. Acta*. 2000, **392**, 169.

[38]   F. Grønvold and S. Stølen, *Thermochim. Acta*. 2001, **395**, 127.

[39]   P. G. Debenedetti and F. H. Stillinger, *Nature*. 2001, **410**, 259.

[40]   R. C. Zeller and R. O. Pohl, *Phys. Rev. B*. 1971, **4**, 2029; see also S. Elliott, *The Physics and Chemistry of Solids*. New York: John Wiley & Sons, 1998.

[41] S. S. Chang and A. B. Bestul, *J. Chem. Thermodyn.* 1974, **6**, 325.

[42] E. C. Kerr, H. N. Hersh and H. L. Johnston, *J. Am. Chem. Soc.* 1950, **72**, 4738.

[43] G. K. White, S. J. Collocott and J. S. Cook, *Phys. Rev. B.* 1984, **29**, 4778.

[44] P. Richet, D. de Ligny and E. F. Westrum, Jr, *J. Non-Cryst. Solids.* 2003, **315**, 20.

[45] C. A. Angell, E. Williams, K. J. Rao and J. C. Turner, *J. Phys. Chem.* 1977, **81**, 238.

[46] S. Stølen, T. Grande and H. B. Johnsen, *Phys. Chem. Chem. Phys.* 2002, **4**, 3396.

[47] M. A. Bouhifd, A. Sipp and P. Richet, *Geochim. Cosmochim. Acta.* 1999, **63**, 2429.

[48] M. A. Bouhifd, P. Courtial and P. Richet, *J. Non-Cryst. Solids.* 1998, **231**, 169.

[49] P. Richet, M. A. Bouhifd, P. Courtial and C. Tequi, *J. Non-Cryst. Solids.* 1997, **211**, 271.

[50] M. A. Debolt, A. J. Easteal, P. B. Macedo and C. T. Moynihan, *J. Am. Ceram. Soc.* 1976, **59**, 16.

[51] C. A. Angell, K. L. Ngai, G. B. McKenna, P. F. McMillan and S. W. Martin, *J. Appl. Phys.* 2000, **88**, 3113.

[52] L.-M. Martinez and C. A. Angell, *Nature.* 2001, **410**, 663.

Further reading

T. H. K. Barron and G. K. White, *Heat Capacity and Thermal Expansion at Low Temperatures*. New York: Kluwer Academic, Plenum Publishers, 1999.

M. T. Dove, *Introduction to Lattice Dynamics*. Cambridge: Cambridge University Press, 1993.

M. T. Dove, *Structure and Dynamics, an Atomic View of Materials*. Oxford: Oxford University Press, 2003.

S. Elliott, *The Physics and Chemistry of Solids*. New York: John Wiley & Sons, 1998.

E. S. R. Gopal, *Specific Heats at Low Temperatures*. London: Heywood Books, 1966.

G. Grimvall, *Thermophysical Properties of Materials*. Amsterdam: Elsevier, 1999.

C. Kittel, *Introduction to Solid State Physics*, 7th edn. New York: John Wiley & Sons, 1996.

C. Kittel and H. Kroemer, *Thermal Physics*, 2nd edn. New York: W.H. Freeman and Company, 1980.

H. J. Wollenberger, Point defects, in *Physical Metallurgy*, 4th edn (R. W. Cahn and P. Haasen, eds.). Amsterdam: Elsevier Science, 1996.

Atomistic solution models

The formalism of the thermodynamics of solutions was described in Chapter 3. In this chapter we shall revisit the topic of solutions and apply statistical mechanics to relate the thermodynamic properties of solutions to atomistic models for their structure. Although we will not give a rigorous presentation of the methods of statistical mechanics, we need some elements of the theory as a background for the solution models to be treated. These elements of the theory are presented in Section 9.1.

Statistical mechanical models for solutions were first derived for binary solutions where the two components were distributed and mixed in a single lattice with a fixed number of lattice sites and constant coordination number. We will here confine our discussion to substitutional solutions, although similar models are applicable to interstitial solutions. The models are presented starting with the ideal solution model and continuing with the regular and quasi-chemical models. The latter is interesting in that it introduces some degree of order in the solutions. Finally, the Flory model for polymeric solutions is presented. The thermodynamics of polymers have not been considered in this book before, but the Flory model illustrate a simple approach for the statistical mechanical treatment of solutions of molecules/entities with substantial variation in molecular weight and volume. Here the entropy of mixing is a function of the volume fractions of the components rather than the mole fractions.

In simple solutions such as binary alloys, the components are distributed on a single lattice. More complex solutions may consist of two or more sub-lattices, and in a solution of simple ionic salts like NaCl and NaBr there is one sub-lattice for cations and one for anions. In these cases the interactions considered in the models are between next neighbouring pairs of atoms rather than nearest neighbour atoms, as is the case with a single lattice. Two sub-lattice models can also be applied to

Chemical Thermodynamics of Materials by Svein Stølen and Tor Grande
© 2004 John Wiley & Sons, Ltd ISBN 0 471 492320 2

treat order–disorder in solid solutions. Here the given species are distributed on two different sub-lattices in the ordered structures at low temperatures. These sites may become crystallographically equivalent in the disordered structure at high temperatures and the disordering process in such cases is termed convergent. The treatment of non-convergent disordering, e.g. in spinels, where the crystallo-graphic sites remain distinct even in the disordered high-temperature state, is closely related. For that reason order–disorder in spinels is briefly described.

The final topic of the chapter is materials with a significant concentration of vacant lattice sites. Attention is given to statistical models giving a quantitative link between defect equilibria and chemical thermodynamics. The significance of such a link is seen by the fact that mixed conductor ceramics, for example, a large group of materials that include oxygen permeable membranes, are both electronic and ionic conductors. Both the electrical conductivity and the stability of these materials are governed by the defect chemistry and a link to chemical thermody-namics is thus important.

9.1 Lattice models for solutions

Partition function

The thermodynamics of a system consisting of N interacting particles is in statis-tical mechanics given in terms of the partition function, Z, which is defined as [1]

$$Z = \sum_j \exp\left(-\frac{U_j}{k_B T} \right) \tag{9.1}$$

Here U_j is the energy of the system in the state j, k_B is Boltzmann's constant and T is the temperature. The summation is over all possible energy states of the system. If the summation were instead allowed over all energy levels i, Z would be written

$$Z = \sum_i g_i \exp\left(-\frac{U_i}{k_B T} \right) \tag{9.2}$$

where g_i is the number of states (the degeneracy) with energy i.

All thermodynamic properties can be derived from the partition function. It can be shown that the Helmholtz energy, A, is related to Z by the simple expression

$$A = -k_B T \ln Z \tag{9.3}$$

The contribution from the partition function (right-hand side of eq. (9.3)) should be interpreted as the value of the Helmholtz energy relative to its value in the lowest energy state, or in other words at the absolute zero of temperature, $A(0)$. In the fur-ther discussion we will only consider the values relative to 0 K. Thus

$$A = A(T) - A(0) \tag{9.4}$$

The Gibbs energy, G, is often a more appropriate variable at isobaric conditions. For condensed systems, G can be assumed to be equivalent to A because their difference, the term pV, is usually negligible. The Gibbs energy of condensed phases can therefore in most cases be approximated as

$$G \approx -k_B T \ln Z \tag{9.5}$$

Here again $G = G(T) - G(0)$. Other thermodynamic functions such as enthalpy, entropy and volume may be derived by the thermodynamic relationships discussed in Chapter 1.

Ideal solution model

In Section 3.2 the ideal solution model was introduced. The essential assumption of the ideal model is that there is no energy change associated with rearrangements of the atoms A and B. In other words the energies associated with a random distribution of A and B atoms and a severely non-random distribution, in which the A and B atoms are clustered, are equal.

We are now going to consider such an ideal solution using statistical mechanics. Whereas eq. (9.3) defines the energy relative to the absolute zero, we are now interested in the variation of the energy and the degeneracy with composition at a given temperature. The average energy of an atom A or B is at this temperature $\langle u_A \rangle$ and $\langle u_B \rangle$, respectively. These energies are defined by the partition functions for the two types of atoms taking into consideration all kind of excitations: vibrational, magnetic, electronic and others. The ideal solution is described, taking these energy states as a starting point. It is assumed that the non-configurational part of the entropy is not affected by the mixing and this term is thus not considered in the simplest treatments of solutions.

The ideal solution has only one energy level for a given composition and the average internal energy of this state is simply given in terms of the weighted average internal energies of the A and B atoms before mixing:

$$U = N_A \langle u_A \rangle + N_B \langle u_B \rangle = U_A + U_B \tag{9.6}$$

Here N_A and N_B are the number of A and B atoms in the mixture. Each distinguishable arrangement of the atoms represents a unique state, and to obtain the degeneracy for the single energy state of an ideal solution with a specific composition we need to calculate the number of distinguishable states.

Let us assume that we have N atoms to distribute over N lattice sites ($N = N_A + N_B$). A two-dimensional illustration for $N_A = N_B$ is shown in Figure 9.1. The sites are numbered and we will fill them one by one. We have a choice of N atoms to fill the first site. For the second site there are therefore only $N - 1$ atoms to

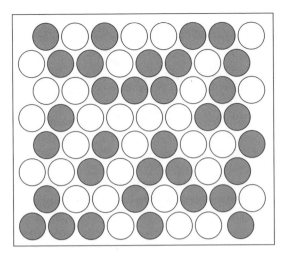

Figure 9.1 Two-dimensional lattice model for a solution of two different atoms of similar radius.

choose from. For the third, there are $N - 2$, etc. The total number of possible arrangements is therefore the product of the number of atoms available to fill all lattice sites one at a time: $N(N - 1)(N - 2)...1 = N!$. Thus, there are $N!$ possible ways to distribute N atoms among N lattice sites. If one type of atom only is distributed on the N sites, all these distributions are indistinguishable. In a lattice containing two types of atom (A and B) a number of distinguishable distributions are obtained. We will term these distinguishable distributions **configurations**.

The number of configurations may be derived indirectly. The switch of two A atoms between two sites containing A atoms leads to an indistinguishable distribution. Therefore if there are N_A sites filled with A atoms there are also $N_A!$ ways of redistributing the A atoms among the N_A sites. Correspondingly, there are $N_B!$ ways of redistributing the B atoms among the N_B sites filled with B atoms. Since the switching of like atoms does not contribute to the distinguishable distributions, the total number of distinguishable arrangements or configurations in an ideal A–B solution is $N!/N_A!N_B!$. The partition function for the ideal solution now can be written as

$$Z = g\exp\left(-\frac{U}{k_B T}\right) = \left(\frac{N!}{N_A!N_B!}\right)\exp\left(-\frac{U_A + U_B}{k_B T}\right) \tag{9.7}$$

where U takes a specific value for a given number of A and B atoms. The Gibbs energy of the ideal solution is given through the partition function as

$$G \approx A = -k_B T \ln Z = -k_B T \ln\left(\frac{N!}{N_A!N_B!}\right) + U_A + U_B \tag{9.8}$$

and the Gibbs energy of mixing is obtained by subtracting the enthalpy contribution from the pure elements, $H \approx U$:

$$\Delta_{mix} G = -k_B T \ln \left(\frac{N!}{N_A! N_B!} \right) \tag{9.9}$$

Using Sterling's approximation ($\ln M! = M \ln M - M$ for M large), the Gibbs energy of mixing becomes

$$\Delta_{mix} G = -k_B T [(N_A + N_B) \ln N - N_A \ln N_A - N_B \ln N_B] + k_B T (N - N_A - N_B)$$

$$= k_B T \left[N_A \ln \left(\frac{N_A}{N} \right) + N_B \ln \left(\frac{N_B}{N} \right) \right] = k_B T [N_A \ln x_A + N_B \ln x_B] \tag{9.10}$$

where x_A and x_B are the mole fractions of A and B respectively. The molar Gibbs energy of mixing for an ideal solution of A and B is obtained by multiplication by $[L/(N_A + N_B)]$:

$$\Delta_{mix} G_m = RT [x_A \ln x_A + x_B \ln x_B] \tag{9.11}$$

The molar entropy of mixing for the ideal solution follows as

$$\Delta_{mix} S_m = -\left(\frac{\partial \Delta_{mix} G_m}{\partial T} \right)_p = -R[x_A \ln x_A + x_B \ln x_B] \tag{9.12}$$

whereas the molar enthalpy of mixing is given through the relation $G = H - TS$ as

$$\Delta_{mix} H_m = \Delta_{mix} G_m + T \Delta_{mix} S_m = 0 \tag{9.13}$$

This is in accordance with the definition of an ideal solution given in Section 3.2.

The ideal solution approximation is well suited for systems where the A and B atoms are of similar size and in general have similar properties. In such systems a given atom has nearly the same interaction with its neighbours, whether in a mixture or in the pure state. If the size and/or chemical nature of the atoms or molecules deviate sufficiently from each other, the deviation from the ideal model may be considerable and other models are needed which allow excess enthalpies and possibly excess entropies of mixing.

Regular solution model

The regular solution model, originally introduced by Hildebrand [2] and further developed by Guggenheim [3], is the most used physical model beside the ideal solution model. In a regular solution A–B, the A and B atoms are assumed to be randomly distributed on the same lattice or quasi-lattice just as was the case for the ideal solution model. Each atom is surrounded by a number of neighbouring atoms, and this nearest neighbour coordination number of the lattice is termed z. There are no vacant lattice sites, and the energy of the system is calculated as a sum of pairwise interactions between atoms on neighbouring sites. For the two-dimensional lattice shown in Figure 9.1, there are six pair interactions to be considered for each atom since the nearest-neighbour coordination number z is 6. The numbers of AA, BB and AB pairs are denoted by N_{AA}, N_{BB} and N_{AB}, and the interaction energies of each pair are u_{AA}, u_{BB} and u_{AB}, respectively. All other contributions to the energy are neglected, including intermediate and long-range interactions. The total number of pairs in the lattice is the product of the number of sites and the coordination number divided by 2 to avoid each pair being counted twice. The average internal energies of the pure substances A and B before mixing, U_A and U_B, can now be given in terms of the interaction energies as $\frac{1}{2} z N_A u_{AA}$ and $\frac{1}{2} z N_B u_{BB}$ respectively. Let us now find an expression for the energy of a solution per atom after mixing a certain number of A and B atoms.

An atom A generates z pairs of either the AA or AB type. Summing over all A atoms gives $z N_A$ numbers of pairs, which is expressed as

$$z N_A = 2 N_{AA} + N_{AB} \tag{9.14}$$

Summing over all B atoms gives similarly

$$z N_B = 2 N_{BB} + N_{AB} \tag{9.15}$$

The total number of pairs $\frac{1}{2} z N$ yields

$$\frac{1}{2} z (N_B + N_A) = N_{AA} + N_{BB} + N_{AB} \tag{9.16}$$

The average internal energy of the system for a given composition is now

$$U = N_{AA} u_{AA} + N_{BB} u_{BB} + N_{AB} u_{AB} \tag{9.17}$$

If N_{AA} and N_{BB} are expressed in terms of eqs. (9.14) and (9.15), U becomes

$$U = \frac{1}{2}(z N_A - N_{AB}) u_{AA} + \frac{1}{2}(z N_B - N_{AB}) u_{BB} + N_{AB} u_{AB} \tag{9.18}$$
$$= \frac{1}{2} z N_A u_{AA} + \frac{1}{2} z N_B u_{BB} + N_{AB}[u_{AB} - \frac{1}{2}(u_{AA} + u_{BB})]$$

It is now convenient to introduce the parameter ω_{AB}, defined as

$$\omega_{AB} = u_{AB} - \frac{1}{2}(u_{AA} + u_{BB}) \tag{9.19}$$

ω_{AB} is closely related to the regular solution constant defined in Section 3.4. Using eq. (9.19), the average internal energy of the system can be expressed as

$$U = U_A + U_B + N_{AB}\omega_{AB} \qquad (9.20)$$

U varies with the number of AB pairs in the solution and is a function of the overall composition. As was the case for the ideal solution, the atoms are assumed to be randomly distributed and N_{AB} can be derived. For a random distribution of the A and B atoms, the probability of finding an atom A at a given lattice site is equal to the mole fraction of A. Consequently, the probability of finding a B atom on a neighbouring site is x_B. The probability of finding an AB pair is therefore $2x_A x_B$, since both AB and BA pairs must be counted. N_{AB} is now the product of the total number of pairs ($\frac{1}{2}zN$) and the probability of finding an AB (or BA) pair among these pairs

$$N_{AB} = \frac{1}{2}zN2x_A x_B = z\left(\frac{N_A N_B}{N}\right) \qquad (9.21)$$

Since the A and B atoms are assumed to be randomly distributed, the partition function becomes

$$Z = g\exp\left(-\frac{U}{k_BT}\right) = \frac{N!}{N_A! N_B!}\exp\left[-\frac{[U_A + U_B + (zN_A N_B/N)\omega_{AB}]}{k_BT}\right] \qquad (9.22)$$

The Gibbs energy for the regular solution of an arbitrary number of A and B atoms follows

$$G \approx A = -k_BT \ln\left[\frac{N!}{N_A! N_B!}\right] + U_A + U_B + \frac{zN_A N_B}{N}\omega_{AB} \qquad (9.23)$$

Note that $U_A (= \frac{1}{2}zN_A u_{AA})$ and $U_B (= \frac{1}{2}zN_B u_{BB})$ depend on the number of A and B atoms in the system. Subtracting the energy for the two pure elements and multiplication by $[L/(N_A + N_B)]$ gives the molar Gibbs energy of mixing of the regular solution A–B:

$$\Delta_{mix}G_m = RT[x_A \ln x_A + x_B \ln x_B] + \Omega_{AB}x_A x_B \qquad (9.24)$$

Here Ω_{AB} is the molar interaction coefficient defined in Chapter 3 (eq. 3.71):

$$\Omega_{AB} = zL\omega_{AB} \qquad (9.25)$$

The excess Gibbs energy of the regular solution, as pointed out in Chapter 3, is a purely enthalpic term:

$$\Delta_{mix} H_m \approx \Delta_{mix} U_m = \Omega_{AB} x_A x_B \tag{9.26}$$

The entropy of mixing is the same as for the ideal solution model.
    The chemical potential of A is given as

$$\mu_A - \mu_A^\circ = RT \ln a_A = RT \ln x_A + \Omega_{AB} x_B^2 \tag{9.27}$$

The activity coefficient for component A is thus

$$RT \ln \gamma_A = \Omega_{AB} x_B^2 \tag{9.28}$$

Similar expressions can be derived for the partial molar Gibbs energy of mixing and the activity coefficient of component B.
    The model was first described in Chapter 3 and the molar Gibbs energy of mixing using different values for Ω_{AB}/RT is shown as a function of composition in Figure 3.10. For large and positive values of Ω_{AB}/RT the solution becomes unstable and segregates into two solutions. The ideal solution model is obtained for $\Omega_{AB}/RT = 0$. Finally, for negative values the solutions is stabilized relative to the pure components, since the unlike atoms attract each other. The activity and activity coefficient for A, using different values for Ω_{AB}/RT, are shown as a function of composition in Figures 9.2(a) and (b). For positive values of Ω_{AB}/RT, the activity coefficient may become larger than unity, while negative values give activity coefficients smaller than one.

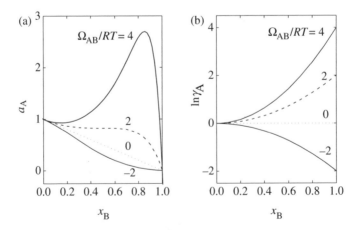

Figure 9.2 (a) a_A and (b) $\ln\gamma_A$ of a regular solution A–B as a function of composition for selected values of Ω_{AB}/RT.

In the derivation of the regular solution model the vibrational contribution to the excess properties has been neglected. However, as a first approximation the vibrational contribution can be taken as independent of the interaction between the different atoms, and this contribution can be factored out of the exponential and taken into account explicitly. The partition function of the solution is then given as

$$Z = g Z_A^{N_A} Z_B^{N_B} \exp\left(-\frac{U'}{k_B T}\right) \tag{9.29}$$

where U' is the average internal energy of eq. (9.20) minus the vibrational contribution and Z_A and Z_B, the vibrational partition functions of the pure atoms A and B. The Gibbs energy now becomes

$$G \approx -k_B T \ln Z \approx U' - k_B T \ln g - k_B T \ln Z_A^{N_A} Z_B^{N_B} \tag{9.30}$$

where the two last terms are entropic contributions to the Gibbs energy of mixing. The entropy of mixing may be divided into two terms; the configurational and the non-configurational contributions. The molar configurational entropy is

$$\Delta_{mix}^{conf} S = L k_B \ln g = R \ln g \tag{9.31}$$

which when assuming a random distribution of atoms corresponds to the ideal entropy of mixing. The second term, the **non-configurational entropy**, is assumed to be proportional to $x_A x_B$ and is given as

$$\Delta_{mix}^{non-conf} S_m = k_B L(\ln Z_A^{N_A} + \ln Z_B^{N_B}) \approx z L x_A x_B \eta_{AB} \tag{9.32}$$

where η_{AB} is a constant. The excess molar Gibbs energy of mixing of the **quasi-regular solution** results:

$$\Delta_{mix}^{exc} G_m = z L x_A x_B (\omega_{AB} - T \eta_{AB}) \tag{9.33}$$

Here the first term is the excess enthalpy of mixing and the second term is the excess entropy of mixing. Just as for the regular solution model, the quasi-regular model is symmetrical about $x_A = x_B = 0.5$ and it is furthermore often convenient to express the excess Gibbs energy through the ratio $\tau = \omega_{AB}/\eta_{AB}$, which for one mole of solution gives

$$\Delta_{mix}^{exc} G_m = x_A x_B \Omega_{AB}\left(1 - \frac{T}{\tau}\right) \tag{9.34}$$

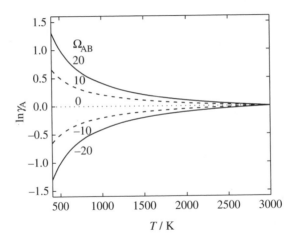

Figure 9.3 $\ln\gamma_A$ of a quasi-regular solution A–B for $x_A = x_B = 0.5$ as a function of temperature for selected values of Ω_{AB}.

It is generally observed that as the temperature increases, real solutions tend to become more ideal and τ can be interpreted as the temperature at which a regular solution becomes ideal. To give a physically meaningful representation of a system τ should be a positive quantity and larger than the temperature of investigation. The activity coefficient of component A for various values of Ω_{AB} is shown as a function of temperature for $\tau = 3000\,\mathrm{K}$ and $x_A = x_B = 0.5$ in Figure 9.3. The model approaches the ideal model as $T \rightarrow \tau$.

Quasi-chemical model

In the models considered until now, the configurational entropy is calculated by assuming that the different species distribute randomly on the relevant lattice sites. While this is a reasonable assumption for the ideal solution case, a large positive or negative interaction energy in the regular solution model suggest that the species are not completely randomly distributed. The quasi-chemical model, developed by Guggenheim [3], aims at giving a more realistic representation of the degeneracy, g. This model has frequently been applied to metallic systems where the deviations from ideality are often limited, but also to salt systems where the interactions are large and lead to the formation of ternary compounds.

While a random distribution of atoms is assumed in the regular solution case, a random distribution of pairs of atoms is assumed in the quasi-chemical approximation. It is not possible to obtain analytical equations for the Gibbs energy from the partition function without making approximations. We will not go into detail, but only give and analyze the resulting analytical expressions.

In the approximate analytical expression for the Gibbs energy of mixing, the pair exchange reaction

$$(A–A) + (B–B) = 2(A–B) \tag{9.35}$$

is an essential parameter [4]. The non-configurational Gibbs energy change for reaction (9.35), termed γ_{AB}, contains both an enthalpic ω'_{AB} and an entropic η'_{AB} term:

$$\gamma_{AB} = \omega'_{AB} - T\eta'_{AB} \tag{9.36}$$

where $\omega'_{AB} = 2\omega_{AB}$ and $\eta'_{AB} = 2\eta_{AB}$. Using eq. (9.36), the Gibbs energy of the solution is expressed as

$$G = N_A\mu_A^o + N_B\mu_B^o - T\Delta_{mix}^{conf}S + \frac{N_{AB}}{2}\gamma_{AB} \tag{9.37}$$

where N_{AB}, as earlier, is the number of A–B pairs in the solution. It is important to note that N_{AB} is no longer equal to the value calculated on the assumption of a random distribution of atoms and thus is not easily derived. Furthermore, the degeneracy is a function of N_{AB}, which again is a function of the interaction energy, γ_{AB}.

Another important feature of the quasi-chemical model is that the model allows for mixing of A and B with different coordination numbers, z_A and z_B. The number of pairs generated by the total number of A and B atoms is then

$$z_A N_A = 2N_{AA} + N_{AB} \tag{9.38}$$

$$z_B N_B = 2N_{BB} + N_{AB} \tag{9.39}$$

When $z_A = z_B$, these two equations reduce to eqs. (9.14) and 9.15).

The configurational entropy of the solution can now be expressed in terms of the **pair fractions**, X_{ij}, defined by

$$X_{ij} = \frac{N_{ij}}{N_{AA} + N_{BB} + N_{AB}} \tag{9.40}$$

and the **coordination equivalent fractions** defined as

$$Y_A = 1 - Y_B = \frac{z_A N_A}{z_A N_A + z_B N_B} = \frac{z_A x_A}{z_A x_A + z_B x_B} \tag{9.41}$$

for solutions where $z_A = z_B$, $Y_A = x_A$ and $Y_B = x_B$.

The resulting configurational entropy is

$$
\begin{aligned}
\Delta_{mix}^{conf}S = &-k_B(N_A \ln x_A + N_B \ln x_B) \\
&-k_B\left[N_{AA}\ln\left(\frac{X_{AA}}{Y_A^2}\right) + N_{BB}\ln\left(\frac{X_{BB}}{Y_B^2}\right) + N_{AB}\ln\left(\frac{X_{AB}}{2Y_A Y_B}\right)\right]
\end{aligned} \tag{9.42}
$$

Equation (9.42) is an approximation frequently used for three-dimensional lattices, but it is exact only in the one-dimensional case [4].

The equilibrium distribution of pairs is calculated by minimizing the Gibbs energy at a given composition. This gives

$$\frac{(\frac{1}{2} X_{AB})^2}{X_{AA} X_{BB}} = \exp\left(-\frac{\gamma_{AB}}{k_B T}\right) \tag{9.43}$$

In eq. (9.43) the relative number of pair interactions is given in terms of the 'equilibrium constant' for the pair exchange reaction we are considering. The name **quasi-chemical model** thus recognizes the equivalence between this mass action type expression and similar expressions for chemical equilibria in general. For a given γ_{AB}, eq. (9.43) gives the relative number of pair interactions which are needed to calculate the Gibbs energy or other thermodynamic properties of the solution using eq. (9.37) as a starting point.

Let us first consider a highly ordered solution. This implies that the Gibbs energy for the pair exchange reaction is large and negative. For a solution with $z_A = z_B$ and using the extreme condition $\gamma_{AB} = -\infty$, the solution should be completely ordered for $x_A = x_B = 0.5$. This implies that the expression for the configuratonal entropy, eq. (9.42), must be zero at this composition. The expression for the configurational entropy at these conditions becomes $\Delta_{mix}^{conf} S_m = -RN_{AB}[1 - (2/z)]\ln 2$, and the configurational entropy is zero only if $z = 2$. For $z > 2$ the configurational entropy becomes negative and thus unphysical. On the other hand it may be argued that the use of coordination number $z = 2$ in three dimensions is non-physical. Still, it is recommended to set z equal to 2 for highly ordered solutions [4].

Using $z_A = z_B = 2$, the molar entropy and enthalpy of mixing for different values of γ_{AB} is shown in Figures 9.4(a) and (b), whereas the corresponding pair distributions are shown as a function of composition in Figure 9.4(c). As γ_{AB} becomes progressively more negative the solution becomes more ordered. The 'm' and 'V' shapes of the entropy and enthalpy of mixing are characteristic features of short-range ordering when the order involves structural entities with composition AB. In cases where other structural entities like AB_2 or A_2B are preferred enthalpically (for $z_A \neq z_B$), the same general features are observed, but the shapes of the curves change.

For solutions with only a small degree of ordering, and particularly for solutions with $\gamma_{AB} > 0$, which exhibits immiscibility, the value of z may be taken from the structure of the components of the solution. A solution of Sn and Cd may for example be modelled using $z_{Sn} = 10$ and $z_{Cd} = 8$ [5].

If the second term in the configurational entropy of mixing, eq. (9.42), is zero, the quasi-chemical model reduces to the regular solution approximation. Here, N_{AB} is given by (eq. (9.21). If in addition $\gamma_{AB} = 0$ the ideal solution model results.

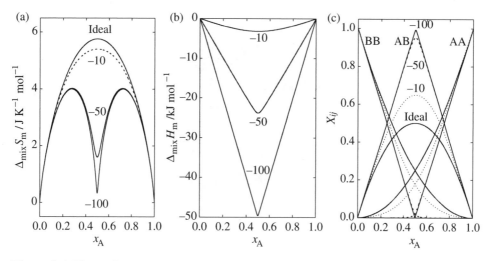

Figure 9.4 The molar entropy (a) and enthalpy (b) of mixing of a quasi-chemical solution A–B at 1000 K for selected values of γ_{AB} with $z_A = z_B = 2$. The distribution of pairs AA, BB and AB for the same conditions is given in (c).

Finally, it should be noted that the model can be extended to multi-component systems [6]. Furthermore, the Gibbs energy of the pair exchange reaction may depend on the relative proportions of the different pairs. In this case γ_{AB} is a polynomial function of the pair fractions X_{AA} and X_{BB} [4].

Flory model for molecules of different sizes

So far we have considered mixtures of atoms or species of similar size and shape. Now we will consider a mixture of a polymeric solute and a solvent of monomers [7, 8]. The ideal entropy of mixing used until now cannot possibly hold for this polymer solution, in which the solute molecule may be thousands or more times the size of the solvent. The long chain polymer may be considered to consist of r chain segments, each of which is equal in size to the solvent molecule. Therefore r is also equal to the ratio of the molar volumes of the solute and the solvent. The solute and the solvent can be distributed in a lattice where each lattice site can contain one solvent molecule. The coordination number of a lattice site is z.

A set of r contiguous sites in the lattice is required for accommodation of the polymer molecule, while the monomers are freely distributed on the remaining lattice sites not occupied by the solute. The situation is illustrated in Figure 9.5, where a large polymeric solute molecule is represented by black circles and the solvent molecules by open circles. The total number of lattice sites, N, is equal to $N_M + rN_P$, where N_M and N_P are the number of monomers and polymer molecules in the solution. The degeneracy of the solution is considered as the g distinguishable ways of arranging the solute molecules in N_P sets of r contiguous sites in the lattice. The $N_M!$ possible arrangements of the solvent on the remaining $(N - rN_P)$ sites do not contribute to the entropy of mixing.

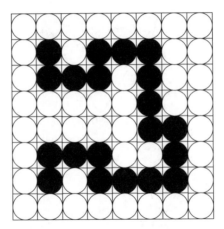

Figure 9.5 Two-dimensional illustration of the lattice model for polymer solutions. Black sites are occupied by the polymer chain, white sites by solvent monomers.

Let v_i be the number of sets of r contiguous lattice sites available to each polymer molecule. If each of the N_P polymer molecules added to the lattice were distinguishable from those already present, the number of ways in which all of them could be arranged in the lattice would be given by the product of the v_i for each molecule added consecutively to the lattice. The degeneracy is then given as

$$g = \frac{1}{N_P!} \prod_{i=1}^{N_P} v_i \tag{9.44}$$

If we distribute N_P polymer molecules over N_P sets of r consecutive lattice sites and then permute the polymer molecules on these fixed sets of consecutive sites, we will overestimate the degeneracy. The factor $1/N_P$ in eq. (9.44) takes into account that the configurations obtained by this kind of permutation are indistinguishable.

We will now calculate v_i. Suppose that i polymer molecules have been previously inserted randomly in the lattice. There then remain a total of $N - ri$ vacant sites in which to place the first segment of molecule $(i + 1)$. The second segment could be assigned to any of z neighbouring sites of the site containing the first segment, unless the site is already occupied by a segment of some of the preceding i molecules. Let f_i represent the probability that a given site adjacent to the site containing the first segment of molecule $(i + 1)$ is occupied. The number of sites available for the second segment is then $z(1 - f_i)$. The probable number of cells available as sites for the third and succeeding segments will be $(z - 1)(1 - f_i)$. Here the possibility that a segment other than the immediately preceding one of the same chain occupies one of the cells in question has been disregarded. The number of sets of r contiguous sites available to molecule v_{i+1} is then

$$v_{i+1} = (N - ri)z(z - 1)^{r-2}(1 - f_i)^{r-1} \tag{9.45}$$

The average probability, $\bar{f}_i$, that a lattice site is occupied by a segment of one of the $(i-1)$ preceding molecules at random is given by the number of vacant sites $1 - \bar{f}_i = (N - ri)/N$. Using this mean field approximation f_i is replaced by $\bar{f}_i$, even though the former is somewhat smaller than the latter. The expression for v_{i+1} then becomes

$$v_{i+1} = \frac{z}{z-1} \left[\frac{(z-1)}{N} \right]^{r-1} (N - ri)^r \tag{9.46}$$

This expression can be further simplified by using

$$(N - ri)^r \approx \frac{(N - ri)!}{[N - r(i+1)]!} \tag{9.47}$$

and

$$\frac{z}{(z-1)} \approx 1 \tag{9.48}$$

By substituting eqs. (9.47) and (9.48) into eq. (9.46) v_{i+1} becomes

$$v_{i+1} \approx \frac{(N-ri)!}{[N-r(i+1)]!} \left[\frac{z-1}{N} \right]^{r-1} \tag{9.49}$$

The degeneracy factor for the arrangement of N_P identical polymer molecules on the lattice consisting of N sites is now obtained by substituting eq. (9.49) in eq. (9.44) [7]:

$$g = \frac{N!}{(N - rN_P)!N_P!} \left[\frac{z-1}{N} \right]^{N_P(r-1)} = \frac{N!}{N_M!N_P!} \left[\frac{z-1}{N} \right]^{N_P(r-1)} \tag{9.50}$$

and the Gibbs energy of the solution can be expressed through the partition function as

$$G = -k_B T \ln Z = U - k_B T \ln \left[\frac{N!}{N_M!N_P!} \left(\frac{z-1}{N} \right)^{N_P(r-1)} \right] \tag{9.51}$$

By using Stirling's approximation, the Gibbs energy of mixing becomes, after some algebra

$$\Delta_{mix}G = k_BTN_M \ln\left[\frac{N_M}{N_M+rN_P}\right]$$
$$+k_BTN_P\left(\ln\left[\frac{N_P}{N_M+rN_P}\right]-(r-1)[\ln(z-1)-1]\right) \tag{9.52}$$

The formation of the polymer solution takes place in two separate steps. In the first the polymer molecules are disordered or disorientated, and subsequently this disordered or disoriented polymer is mixed with the solvent. The entropy of the first step, the disorientation process, is given by eq. (9.52) when we neglect the effect of having solute molecules present by setting $N_M = 0$, thus:

$$\Delta_{dis}S = k_BN_P[\ln r + (r-1)[\ln(z-1)-1]] \tag{9.53}$$

When r is large, the first term becomes negligible compared to the second, and the entropy of disorientation per segment reduces to

$$\frac{\Delta_{dis}S}{rN_P} \approx k_B[\ln(z-1)-1] \tag{9.54}$$

Although the disorientation process might be looked upon as a melting process, eq. (9.54) does not give a good estimate of the entropy of fusion of polymers. A more reasonable approach is to introduce the mean square displacement length of the actual chain in its unperturbed state [7]. The number of such units would then replace r in eq. (9.54), but one still needs to determine the coordination number z in an independent manner.

The entropy of mixing of the disoriented polymer and the solvent, the monomer, is obtained by subtracting eq. (9.54) from eq. (9.52) giving

$$\Delta_{mix}S = -k_B(N_M \ln \phi_M + N_P \ln \phi_P) \tag{9.55}$$

Here ϕ_M and ϕ_P are the volume fraction of the solvent and the solute defined as

$$\phi_M = 1-\phi_P = \frac{N_M}{N_M+rN_P} = 1 - \frac{rN_P}{N_M+rN_P} \tag{9.56}$$

Equation (9.55) is the expression for the entropy of mixing of polymer solutions introduced first by Flory [7]. N_M and N_P can be related through $x_M + x_P = 1$, which for one mole of molecules (polymers + monomers) gives

$$\Delta_{mix}S_m = -(N_M+N_P)k_B(x_M \ln \phi_M + x_P \ln \phi_P)$$
$$= -R(x_M \ln \phi_M + x_P \ln \phi_P) \tag{9.57}$$

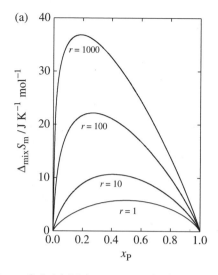 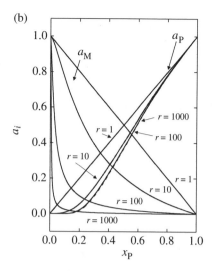

Figure 9.6 (a) Molar entropy of mixing of ideal polymer solutions for $r = 10$, 100 and 1000 plotted as a function of the mole fraction of polymer compared with the entropy of mixing of two atoms of similar size, $r = 1$. (b) Activity of the two components for the same conditions.

Equation (9.57) is similar in form to eq. (9.12); note that whereas the entropy of a mixture of two atoms of similar size is given in terms of the logarithm of the mole fraction in eq. (9.12), the logarithm of the volume fraction is used in eq. (9.57). For $r = 1$, $\phi_i = x_i$ and the two equations become equal.

Alternatively, the entropy of mixing can be expressed as

$$\Delta_{\text{mix}} S = -(N_\text{M} + r N_\text{P}) k_\text{B} (\phi_\text{M} \ln \phi_\text{M} + \phi_\text{P} \ln \phi_\text{P}) \qquad (9.58)$$

The entropies of mixing of hypothetical binary polymer solutions where the solute is characterized by $r = 1$, 10, 100 and 1000 are shown in Figure 9.6(a). The entropy of mixing for one mole of molecules (polymers + monomers) increases with the size of the polymer and it is significantly larger than for an ideal solution of two species of similar size for which $r = 1$. If we instead consider the entropy of mixing per lattice size $(N_\text{M} + r N_\text{P})$, the entropy takes its maximum value for $r = 1$. The entropy of mixing curves when plotted as a function of mole fraction of the polymer are far from symmetrical about $x_\text{P} = 0.5$, while they are symmetrical about $\phi_\text{P} = 0.5$ when plotted as a function of the volume fraction.

For an ideal solution $\Delta_{\text{mix}} G = -T \Delta_{\text{mix}} S$ and the partial molar Gibbs energy of mixing of the solute and solvent is obtained from eq. (9.57) as

$$\mu_\text{M} - \mu_\text{M}^\text{o} = RT \left[\ln(1 - \phi_\text{P}) + \left(1 - \frac{1}{r}\right)\phi_\text{P} \right] \qquad (9.59)$$

and

$$\mu_P - \mu_P^o = RT[\ln \phi_P + (1-r)(1-\phi_P)] \tag{9.60}$$

The activities of the polymer and monomer of the hypothetical solutions given in Figure 9.6(a) are shown in Figure 9.6(b). While $r = 1$ corresponds to Raoult law behaviour, strong negative deviations are observed for $r = 10$, 100 and 1000.

A regular solution type parameter can be added to the ideal polymer model giving [7, 9]

$$\Delta_{mix}G = k_B T(N_M \ln \phi_M + N_P \ln \phi_P) + z\omega_{MP}N_M\phi_P \tag{9.61}$$

The enthalpy of mixing in this case is

$$\Delta_{mix}H = z\omega_{MP}N_M\phi_P \tag{9.62}$$

where ω_{MP} is the difference in pair interaction energy $\omega_{MP} = u_{MP} - \frac{1}{2}(u_{MM} + u_{PP})$ in analogy with the regular solution parameters used earlier. The partial molar Gibbs energies of mixing (eqs. 9.59 and 9.60) now become

$$\mu_M - \mu_M^o = RT\left[\ln(1-\phi_P) + \left(1 - \frac{1}{r}\right)\phi_P\right] + \Omega_{MP}\phi_P^2 \tag{9.63}$$

and

$$\mu_P - \mu_P^o = RT[\ln \phi_P + (1-r)(1-\phi_P)] + \Omega_{MP}r(1-\phi_P)^2 \tag{9.64}$$

The interaction energy $\Omega_{MP} = zL\omega_{MP}$ is often needed to reproduce experimental activity data. One example is shown in Figure 9.7. Here the experimental vapour pressure of benzene over a binary rubber–benzene mixture [10, 11] is compared with activities obtained from the ideal model with $r = \infty$ (eq. 9.59). While mole fractions were used as the compositional variable in Figure 9.6, volume fractions are used in Figure 9.7. The vapour pressure of the solvent using the ideal model cannot reproduce the experimental data and a regular solution term must be added. The dashed line in Figure 9.7 is obtained using eq. (9.63) with $\Omega_{MP}/RT = 0.43$. The dotted line represents the Raoult solution behaviour where the activity of the monomer is assumed to be given by its mole fraction.

Finally, it should be noted that although we have used a model in which the liquid solution conceptually has been divided into a lattice, Hildebrand [12] has shown that a similar expression may be derived without resorting to a hypothetical lattice.

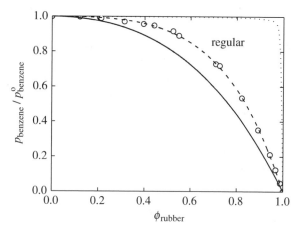

Figure 9.7 Comparison of experimental and calculated vapour pressures of benzene in benzene–rubber mixtures [7].

9.2 Solutions with more than one sub-lattice

Let us now consider a solid or liquid with more than one sub-lattice, typically ionic solutions like NaCl–KBr or compounds like spinels AB_2O_4. In the former case Na^+ and K^+ are distributed at the cation sub-lattice, and Cl^- and Br^- on the anion sub-lattice. In the spinel there are two types of cation that occupy different crystallographic positions. The A atoms in a normal spinel occupy tetrahedral lattice sites, while the B atoms occupy octahedral lattice sites. On heating a certain degree of disordering may take place, but the two sub-lattices remain distinct. Hence we must consider configurational contributions to the entropy from each of the two separate sub-lattices. These can be treated independently of each other. Sub-lattices that contain only one type of atom (and no vacancies) will not contribute to the configurational entropy. This is the case for the third sub-lattice in the spinel case; the oxygen lattice is not affected by disorder.

Ideal solution model for a two sub-lattice system

Temkin was the first to derive the ideal solution model for an ionic solution consisting of more than one sub-lattice [13]. An ionic solution, molten or solid, is considered as completely ionized and to consist of charged atoms: anions and cations. These anions and cations are distributed on separate sub-lattices. There are strong Coulombic interactions between the ions, and in the solid state the positively charged cations are surrounded by negatively charged anions and vice versa. In the Temkin model, the local chemical order present in the solid state is assumed to be present also in the molten state, and an ionic liquid is considered using a quasi-lattice approach. If the different anions and the different cations have similar physical properties, it is assumed that the cations mix randomly at the cation sub-lattice and the anions randomly at the anion sub-lattice.

A binary ionic solution must contain at least three kinds of species. One example is a solution of AC and BC. Here we have two cation species A^+ and B^+ and one common anion species C^-. The sum of the charge of the cations and the anions must be equal to satisfy electro-neutrality. Hence $N_{A^+} + N_{B^+} = N_{C^-} = N$ where N_{A^+}, N_{B^+} and N_{C^-} are the total number of each of the ions and N is the total number of sites in each sub-lattice. The total number of distinguishable arrangements of A^+ and B^+ cations on the cation sub-lattice is $N!/N_{A^+}!N_{B^+}!$. The expression for the molar Gibbs energy of mixing of the ideal ionic solution AC-BC is thus analogous to that derived in Section 9.1 and can be expressed as

$$\Delta_{mix}G_m = -RT \ln\left(\frac{N!}{N_{A^+}!N_{B^+}!}\right) = RT[X_{A^+} \ln X_{A^+} + X_{B^+} \ln X_{B^+}] \quad (9.65)$$

Here X_{A^+} and X_{B^+} are the ionic fraction of A^+ and B^+ respectively defined as

$$X_{B^+} = 1 - X_{A^+} = \frac{N_{A^+}}{N_{A^+} + N_{B^+}} \quad (9.66)$$

The entropy of mixing is obtained by differentiation of eq. (9.65) with regard to temperature.

$$\Delta_{mix}S_m = -R[X_{A^+} \ln X_{A^+} + X_{B^+} \ln X_{B^+}] \quad (9.67)$$

An analogous expression can be derived for an ionic solution with a common cation, and the ideal entropy for a system AC–BD is twice as large as that for the AC–BC system. This approach can also be used for an alloy $A_{1-x}B_xC$, where the atoms A and B are randomly distributed on one sub-lattice and C fills completely the second separate sub-lattice.

Regular solution model for a two sub-lattice system

The regular model for an ionic solution is similarly analogous to the regular solution derived in Section 9.1. Recall that the energy of the regular solution model was calculated as a sum of pairwise interactions. With two sub-lattices, pair interactions between species in one sub-lattice with species in the other sub-lattice (nearest neighbour interactions) and pair interactions within each sub-lattice (next nearest neighbour interactions), must be accounted for.

Let us first derive the regular solution model for the system AC–BC considered above. The coordination numbers for the nearest and next nearest neighbours are both assumed to be equal to z for simplicity. The number of sites in the anion and cation sub-lattice is N, and there are $\frac{1}{2}zN$ nearest and next nearest neighbour interactions. The former are cation–anion interactions, the latter cation–cation and anion–anion interactions. A random distribution of cations and anions on each of the two sub-lattices is assumed, and the average internal energy of the AC–BC mixture expressed in terms of pairwise interactions is

$$U = N_{A^+A^+} u_{A^+A^+} + N_{B^+B^+} u_{B^+B^+} + N_{A^+B^+} u_{A^+B^+} + N_{C^-C^-} u_{C^-C^-}$$
$$+ N_{A^+C^-} u_{A^+C^-} + N_{B^+C^-} u_{B^+C^-} \tag{9.68}$$

where N_{ij} and u_{ij} are the number and internal energy of pair interaction ij. The number of C^-C^- interactions is simply $\frac{1}{2} zN$, while the number of A^+C^- and B^+C^- interactions are $\frac{1}{2} zN_{A^+}$ and $\frac{1}{2} zN_{B^+}$. The number of A^+A^+ and B^+B^+ interactions are $\frac{1}{2}(zN_{A^+} - N_{A^+B^+})$ and $\frac{1}{2}(zN_{B^+} - N_{A^+B^+})$, by analogy with the regular solution model for a single lattice system derived in Section 9.1. Substituting these expressions in eq. (9.68), the internal energy for the ionic solution becomes

$$U = \frac{1}{2} z[N_{A^+} u_{A^+A^+} + N_{B^+} u_{B^+B^+} + N u_{C^-C^-} + N_{A^+} u_{A^+C^-} + N_{B^+} u_{B^+C^-}]$$
$$+ N_{A^+B^+} [u_{A^+B^+} - \frac{1}{2}(u_{A^+A^+} + u_{B^+B^+})] \tag{9.69}$$

Introducing once more a pairwise interaction parameter $\omega_{A^+B^+}$

$$\omega_{A^+B^+} = u_{A^+B^+} - \frac{1}{2}(u_{A^+A^+} + u_{B^+B^+}) \tag{9.70}$$

and the internal energy for the pure neutral components AC and BC

$$U_{AC} = \frac{1}{2} zN_A [u_{A^+A^+} + u_{C^-C^-} + u_{A^+C^-}] \tag{9.71}$$

$$U_{BC} = \frac{1}{2} zN_B [u_{B^+B^+} + u_{C^-C^-} + u_{B^+C^-}] \tag{9.72}$$

eq. (9.69) becomes

$$U = U_{AC} + U_{BC} + N_{A^+B^+} \omega_{A^+B^+} \tag{9.73}$$

$N_{A^+B^+}$ is easily derived when the cations are assumed to be randomly distributed on the cation sub-lattice. The probability of finding an AB (or BA) pair is $2X_{A^+} X_{B^+}$ in analogy with the derivation of the regular solution in Section 9.1. $N_{A^+B^+}$ is then the product of the total number of cation–cation pairs multiplied by this probability

$$N_{AB} = \frac{1}{2} zN2 X_{A^+} X_{B^+} = z\left(\frac{N_{A^+} N_{B^+}}{N}\right) \tag{9.74}$$

The partition function for the regular ionic solution model is now

$$Z = \frac{N!}{N_{A^+}! N_{B^+}!} \exp\left[-\frac{[U_{AC} + U_{BC} + (zN_{A^+} N_{B^+}/N)\omega_{A^+B^+}]}{k_B T}\right] \tag{9.75}$$

and the Gibbs energy of the solution results:

$$G \approx A = -k_B T \ln \left[\frac{N!}{N_{A^+}! N_{B^+}!} \right] + U_{AC} + U_{BC} + \frac{z N_{A^+} N_{B^+}}{N} \omega_{A^+B^+} \qquad (9.76)$$

By subtracting the enthalpy $(H \approx U)$ for the pure compounds AC and BC and multiplying by $[L/(N_{A^+} + N_{B^+})]$ the molar Gibbs energy of mixing is obtained:

$$\Delta_{mix} G_m = RT[X_{A^+} \ln X_{A^+} + X_{B^+} \ln X_{B^+}] + \Omega_{A^+B^+} X_{A^+} X_{B^+} \qquad (9.77)$$

The first term is the ideal entropy of mixing while the second term is the enthalpy of mixing in the regular solution approximation:

$$\Delta_{mix} H_m \approx \Delta_{mix} U_m = \Omega_{A^+B^+} X_{A^+} X_{B^+} \qquad (9.78)$$

where $\Omega_{A^+B^+} = zL\omega_{A^+B^+}$. The partial Gibbs energy of mixing of one of the components, e.g. AC, can be derived by differentiation with respect to the number of AC neutral entities, which is equal to the number of A^+ cations:

$$\mu_{AC} - \mu_{AC}^o = RT \ln X_{A^+} + \Omega_{A^+B^+} X_{B^+}^2 \qquad (9.79)$$
$$= RT \ln x_{AC} + \Omega_{A^+B^+} x_{BC}^2$$

The equations for the regular solution model for a binary mixture with two sub-lattices are quite similar to the equations derived for a regular solution with a single lattice only. The main difference is that the mole fractions have been replaced by ionic fractions, and that while the pair interaction is between nearest neighbours in the single lattice case, it is between next nearest neighbours in the case of a two sub-lattice solution.

Reciprocal ionic solution

Let us now consider a slightly more complex system, the system AC–BD. The ideal configurational entropy of a system like this that contains two cations A^+ and B^+ and two anions C^- and D^- is readily derived as

$$\Delta_{mix} S_m = -R[X_{A^+} \ln X_{A^+} + X_{B^+} \ln X_{B^+} + X_{C^-} \ln X_{C^-} + X_{D^-} \ln X_{D^-}] \qquad (9.80)$$

With $\Delta_{mix} H_m = 0$ the ideal Temkin model for ionic solutions [13] is obtained. If deviations from ideality are observed, a regular solution expression for this mixture that contains two species on each of the two sub-lattices can be derived using the general procedures already discussed. The internal energy is again calculated from the sum of the pairwise interactions between nearest and next nearest neighbours. The numbers of each interaction can be derived since the different species on the different sub-lattices are assumed to be randomly distributed. The expression for the average molar internal energy of the AB–CD solution becomes

$$U_m = X_{A^+} X_{C^-} U_{AC,m} + X_{A^+} X_{D^-} U_{AD,m} + X_{B^+} X_{C^-} U_{BC,m}$$
$$+ X_{B^+} X_{D^-} U_{BD,m} + (X_{A^+} X_{B^+} \Omega_{A^+B^+} + X_{C^-} X_{D^-} \Omega_{C^-D^-}) \tag{9.81}$$

where $\Omega_{C^-D^-}$ is defined by analogy with $\Omega_{A^+B^+}$. $U_{ij,m}$ is the molar internal energy of the pure salts ij. The corresponding molar Gibbs energy of the solution is

$$G_m \approx A_m = X_{A^+} X_{C^-} U_{AC,m} + X_{A^+} X_{D^-} U_{AD,m} + X_{B^+} X_{C^-} U_{BC,m}$$
$$+ X_{B^+} X_{D^-} U_{BD,m} + (X_{A^+} X_{B^+} \Omega_{A^+B^+} + X_{C^-} X_{D^-} \Omega_{C^-D^-})$$
$$+ RT (X_{A^+} \ln X_{A^+} + X_{B^+} \ln X_{B^+} + X_{C^-} \ln X_{C^-} + X_{D^-} \ln X_{D^-}) \tag{9.82}$$

In the preceding treatments we have neglected the difference between enthalpy and internal energy and between Gibbs energy and Helmholtz energy. More importantly, we have neglected the non-configurational entropy, since we have assumed this contribution not to be affected by the mixing. The different components have been described in terms of internal energy/enthalpy only, since we have focused on the mixing properties and thus subtracted the properties of the pure components. In the following example, the energetics of the components are included as a parameter in the model and we can no longer neglect the vibrational entropy. A modified version of eq. (9.82) in which the internal energies of the four salts are exchanged with the Gibbs energies is

$$G_m = X_{A^+} X_{C^-} G_{AC,m} + X_{A^+} X_{D^-} G_{AD,m} + X_{B^+} X_{C^-} G_{BC,m} + X_{B^+} X_{D^-} G_{BD,m}$$
$$+ (X_{A^+} X_{B^+} \Omega_{A^+B^+} + X_{C^-} X_{D^-} \Omega_{C^-D^-})$$
$$+ RT (X_{A^+} \ln X_{A^+} + X_{B^+} \ln X_{B^+} + X_{C^-} \ln X_{C^-} + X_{D^-} \ln X_{D^-}) \tag{9.83}$$

The regular solution parameters are still assumed to be enthalpic in nature. In other words, the vibrational entropy is, as earlier, considered not to be affected by the mixing. The last term of eq. (9.83), the configurational entropy, is also unaffected by this modification.

Let us now look at this slightly more complex case where the Gibbs energy of the components are needed. Until now we have mixed one salt like AC with another like BD. This implies that the fraction of A atoms on the cation sub-lattice has been equal to the fraction of C atoms on the anion sub-lattice. Let us consider a composition like that marked with a cross in Figure 9.8. There are several possible

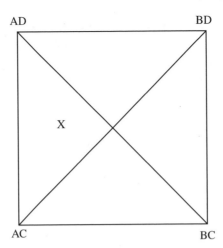

Figure 9.8 Ternary reciprocal system AC–AD–BD–BC.

combinations of the four neutral components AC, AD, BC and BD that can represent this particular composition in this ternary reciprocal system. The particular mixture can for example be made from the three components AC–AD–BC or alternatively from AC–AD–BD. This is important in experimental studies of such systems, where the most suitable choice of standard state is given by the characteristics of the particular experiment in question. In a thermodynamic analysis of the system on the other hand, all four components are often considered.

In the case of reciprocal systems, the modelling of the solution can be simplified to some degree. The partial molar Gibbs energy of mixing of a neutral component, for example AC, is obtained by differentiation with respect to the number of AC neutral entities. In general, the partial derivative of any thermodynamic function Y for a component A_aC_c is given by

$$\left(\frac{\partial Y}{\partial n_{A_aC_c}}\right)_{n \neq n_{A_aC_c}} = a\left(\frac{\partial Y}{\partial n_A}\right) + c\left(\frac{\partial Y}{\partial n_C}\right) \tag{9.84}$$

Using AC as an example, the chemical potential relative to the standard state (pure AC) can be shown to be

$$\mu_{AC} - \mu^o_{AC} = RT \ln X_{A^+} X_{C^-} + (X_{B^+} X_{B^+} \Omega_{A^+B^+} + X_{D^-} X_{D^-} \Omega_{C^-D^-}) \tag{9.85}$$
$$+ X_{B^+D^-} [G_{BC,m} + G_{AD,m} - G_{AC,m} - G_{BD,m}]$$

This expression can be simplified by introducing, $\Delta_r G$, the Gibbs energy of the reciprocal reaction

$$AC + BD = AD + BC \tag{9.86}$$

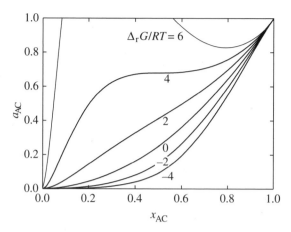

Figure 9.9 Activity of AC in the hypothetical ternary reciprocal system AC–BD for different values of $\Delta_r G/RT$.

for which eq. (9.85) becomes

$$\mu_{AC} - \mu_{AC}^\circ = RT \ln X_{A^+} X_{C^-} + (X_{B^+} X_{B^+} \Omega_{A^+B^+} + X_{D^-} X_{D^-} \Omega_{C^-D^-})$$
$$+ X_{B^+} X_{D^-} \Delta_r G \qquad (9.87)$$

For the ideal solution $\Omega_{A^+B^+} = \Omega_{C^-D^-} = 0$ and the partial Gibbs energy of AC is

$$\mu_{AC} - \mu_{AC}^\circ = RT \ln a_{AC} = RT \ln X_{A^+} X_{C^-} + X_{B^+} X_{D^-} \Delta_r G \qquad (9.88)$$

The activity of AC in the ionic mixture AC–BD using eq. (9.88) is shown in Figure 9.9 for different values of $\Delta_r G$. The curves for large positive values for the reciprocal reaction show that immiscibility in these cases must be expected at low temperatures.

The reciprocal lattice model as derived above is the basis for many different variants. For simplicity we have assumed the interactions between the next nearest neighbours $A^+ - B^+$ and $C^- - D^-$ to be independent of composition, even though experiments have shown that this is often not the case. It is relatively simple to introduce parameters which allow the interaction energy, for example between A^+ and B^+, to depend on the concentration of C^- and D^- [14]. One may also include other terms that take into account excess enthalpies that are asymmetric with regard to composition and the effects of temperature and pressure.

Equation (9.83) is also the basis for the compound energy model. The excess energy of the mixture is here represented by any type of equation, for example a power series [15, 16]. Equation (9.83) has also been derived using the conformal solution theory after Blander [14] and as an extension of the molten salts models presented by Flood, Førland and Grjotheim [17].

9.3 Order–disorder

Bragg–Williams treatment of convergent ordering in solid solutions

In cases where Ω_{AB} is negative, the enthalpy of mixing is also negative and the solution is stable over its entire compositional region. Since such a negative $\Delta_{mix}H_m$ implies that the attraction between unlike species is larger than between like atoms, the formation of fully ordered structures is often observed in these systems on cooling. This tendency to order often causes crystallographic sites, which are equivalent in the disordered state, to transform into two or more crystallographically distinct sites in the ordered low-temperature state. This situation is called convergent ordering, since the two sites are identical when randomly occupied. Bragg and Williams used the regular solution model to describe low-temperature convergent ordering in solid solutions [18, 19]. The model considers an AB solution consisting of two sub-lattices termed a and b, where the pairwise interactions between nearest neighbours give rise to long-range order in the stoichiometric compound AB at low temperatures. In the fully ordered structure, all the a sites are occupied by A atoms and all the b sites are occupied by B atoms. An order parameter, σ, is defined so that it is 1 in the long-range ordered structure and 0 in the disordered structure. The fraction of atoms A on the correct sites in the ordered state is now given by $\frac{1}{2}(1 + \sigma)$, and the fraction on incorrect sites by $\frac{1}{2}(1 - \sigma)$.

Long-range ordering on cooling suggests a cooperative mechanism in which the behaviour of the different atoms is correlated. Considerable short-range order may exist in real materials close to the ordering temperature, but here the domains are not correlated. Furthermore, the size of the domains will decrease with increasing temperature. This type of short-range order may be present even far above the transition temperature and is governed by the balance between the enthalpic and entropic terms, minimizing the Gibbs energy. Such short-range order is often not considered in theoretical models like the Bragg–Williams approach, which we will now use to describe the variation in the degree of order, σ, in the low-temperature phase with temperature.

Let us assume that there are N sites in the two lattices a and b. Thus $N_A + N_B = 2N$ and there are a total of zN interactions between nearest neighbours. The energy of the system in the disordered state is for a given composition given by eq. (9.17):

$$U = N_{AA}u_{AA} + N_{BB}u_{BB} + N_{AB}u_{AB} \tag{9.89}$$

where N_{ij} is the number of ij pairs and u_{ij} is the energy of the ij interaction, respectively. The numbers of each interaction can be calculated by the occupancy in the two lattices, as for the regular solution model. However, in the present case the A and B atoms are not randomly distributed. Instead, the number of AA interactions is given by product of the occupancy of A atoms on the two lattice sites multiplied by the total number of interactions. The occupancy is described by the degree of order σ and thus

$$N_{AA} = zN\left[\frac{1}{2}(1-\sigma)\frac{1}{2}(1+\sigma)\right] = zN\left[\frac{1}{4}(1-\sigma^2)\right] \qquad (9.90)$$

Taking into account that $N_{AA} = N_{BB}$ and $N_{AB} = zN - N_{AA} - N_{BB}$ the energy of the disordered solid solution becomes

$$H(\sigma) \approx U(\sigma) = zN\left[\frac{1}{4}(1-\sigma^2)u_{AA} + \frac{1}{4}(1-\sigma^2)u_{BB} + \frac{1}{2}(1+\sigma^2)u_{AB}\right] \qquad (9.91)$$

Since the enthalpy of the fully ordered state corresponding to $\sigma = 1$ is zNu_{AB}, the enthalpy for the disordering process is

$$\begin{aligned}
\Delta_{dis}H &= H(\sigma) - H(\sigma = 1) = H(\sigma) - zNu_{AB} \\
&= \frac{1}{4}zN[(1-\sigma^2)(u_{AA} + u_{BB} - 2u_{AB})] = \frac{1}{2}zN\omega_{AB}(1-\sigma^2)
\end{aligned} \qquad (9.92)$$

The ideal entropy of mixing of A and B atoms on the a lattice is

$$\Delta_{mix}S_a = -k_B N\left(\frac{1}{2}(1+\sigma)\ln\left[\frac{1}{2}(1+\sigma)\right] + \frac{1}{2}(1-\sigma)\ln\left[\frac{1}{2}(1-\sigma)\right]\right) \qquad (9.93)$$

The configurational entropy on the b lattice is given by an identical equation. For the perfect long-range ordered modification the configurational entropy is zero, and the entropy change due to disorder is thus

$$\Delta_{dis}S = -k_B N[(1-\sigma)\ln(1-\sigma) + (1+\sigma)\ln(1+\sigma) - 2\ln 2] \qquad (9.94)$$

Combining the entropy and enthalpy of the disordering process, the Gibbs energy of disordering is

$$\begin{aligned}
\Delta_{dis}G = \Delta_{dis}H - T\Delta_{dis}S = \tfrac{1}{2}zN\omega_{AB}(1-\sigma^2) + k_B TN[(1-\sigma)\ln(1-\sigma) \\
+ (1+\sigma)\ln(1+\sigma) - 2\ln 2]
\end{aligned} \qquad (9.95)$$

For $\Delta_{mix}H < 0$ the Gibbs energy is reduced by ordering, and the degree of order is obtained by $d(\Delta_{mix}G)/d\sigma = 0$ which yields

$$\ln\frac{1+\sigma}{1-\sigma} = -\frac{\sigma z\omega_{AB}}{k_B T} = \frac{2\sigma T_{trs}}{T} \qquad (9.96)$$

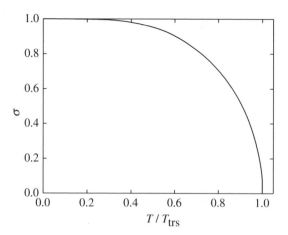

Figure 9.10 Order parameter σ for the Bragg–Williams model as a function of reduced temperature.

The variation of order with temperature is shown in Figure 9.10. As the temperature is raised, a high degree of order remains until close to the transition temperature, where the disorder is induced rapidly. This cooperative behaviour is due to the fact that the energy associated with the disordering becomes progressively less as the disorder takes place. The transition is of second order. No short-range order remains above the transition temperature.

Non-convergent disordering in spinels

Non-convergent disordering is, in contrast to convergent disordering, a process where the crystallographic sites are distinct even when randomly occupied. Let us use spinels as an example. The spinel-type structure contains one tetrahedral and two octahedral cations per formula unit, AB_2O_4. While some A atoms prefer a tetrahedral local environment, others prefer an octahedral environment, and all sorts of distributions of the two cations on the two cation sub-lattices may result. The composition of a spinel is thus in general represented by $A_{1-x}B_x[A_xB_{2-x}]O_4$, where x may vary from 0 to 1. $x = 0$ corresponds to what is termed a normal spinel $(A)^{\text{tetr}}(B_2)^{\text{octa}}O_4$, while $x = 1$ represents an inverse spinel $(B)^{\text{tetr}}(A,B)^{\text{octa}}O_4$. For $x = 2/3$ the A and B atoms are randomly distributed on the two sub-lattices and the spinel is said to be a random spinel.

The molar configurational entropy of a spinel can be expressed in terms of the composition parameter x as

$$\Delta_{\text{config}} S_m = -R\left[x\ln x + (1-x)\ln(1-x) + x\ln\left(\frac{x}{2}\right) + (2-x)\ln\left(1-\frac{x}{2}\right)\right] \quad (9.97)$$

While the configurational entropy of the normal spinel $(x = 0)$ is 0, a large configurational entropy is obtained for the random spinel $(x = 2/3)$; see Figure 9.11(a). The

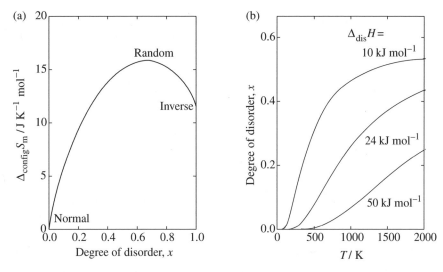

Figure 9.11 (a) Configurational entropy of a spinel AB_2O_4 as a function of the composition parameter, x. (b) Degree of disorder, x, as a function of temperature for selected values of $\Delta_{dis}H$.

inverse spinel is characterized by a single type of atom on the tetrahedral lattice, while equal numbers of A and B atoms are distributed on the octahedral sites and the entropy for $x = 1$ is, although lower than for the random distribution, substantial.

Both ordered and normal spinels should disorder at high temperature, since the entropy of the random spinel is larger. The disordering reaction for a normal spinel can be described by a quasi-chemical reaction

$$(A)^{tetr} + (B)^{octa} = (A)^{octa} + (B)^{tetr} \tag{9.98}$$

The Gibbs energy of this defect reaction defines the degree of disorder through the equilibrium constant.

$$\Delta_{dis}G = \Delta_{dis}H - T\Delta_{dis}S = -RT \ln K \tag{9.99}$$

$$K = \frac{x^2}{(1-x)(2-x)} \tag{9.100}$$

Since the entropy of a normal spinel is zero, $\Delta_{dis}S = \Delta_{config}S$. The corresponding enthalpy term can be interpreted in terms of site preference enthalpies; different cations prefer different crystallographic positions. It follows that the degree of disorder is a function of the site preference enthalpy and temperature [20]. Equilibrium distributions of A and B atoms are given as a function of the enthalpy of reaction (9.98) and temperature in Figure 9.11(b). Although the enthalpy term supports ordering, while the entropy term gives disordering, solid state equilibria of this type may of course be kinetically hindered at lower temperatures.

Similar methods have been used in other cases as well, and a recent example is an analysis of the cation distribution in the complex oxide, $LaSr_{2-x}Ca_xCu_2GaO_7$. Here site preference enthalpies for La, Sr and Ca have been derived [21].

9.4 Non-stoichiometric compounds

Mass action law treatment of defect equilibria

Analyses of the defect chemistry and thermodynamics of non-stoichiometric phases that are predominately ionic in nature (i.e. halides and oxides) are most often made using quasi-chemical reactions. The concentrations of the point defects are considered to be low, and defect–defect interactions as such are most often disregarded, although defect clusters often are incorporated. The resulting mass action equations give the relationship between the concentrations of point defects and partial pressure or chemical activity of the species involved in the defect reactions.

Consider a simple non-stoichiometric perovskite-type oxide, $ABO_{3-\delta}$. In the perovskite crystal structure the A and B cations occupy two different cation sublattices with coordination numbers 12 and 6, respectively; see Figure 7.16. The oxygen anions occupy a third sub-lattice, and the oxygen atoms and vacancies are considered randomly distributed on this oxygen lattice. The oxygen non-stoichiometry leads to valence defects on the B sub-lattice. If the A site ion is trivalent, all the B atoms are trivalent for $\delta = 0$ (i.e. in ABO_3) and divalent for $\delta = 0.5$ (i.e. in $ABO_{2.5}$). We disregard any effect of intrinsic disorder, of ionic or electronic type. The removal of oxygen atoms results in the reduction of B atoms, and for many systems this can be expressed in terms of a defect chemical reaction using the Kröger–Vink notation [22] as

$$2O_O^x + 4B_B^x = 2V_O^{\cdot\cdot} + 4B_B' + O_2(g) \tag{9.101}$$

where O_O^x and $V_O^{\cdot\cdot}$ are an oxygen ion and an oxygen vacancy (with effective charge +2) on the oxygen sub-lattice and B_B^x and B_B' are trivalent and divalent B-ions on the B sub-lattice. The equilibrium constant (K) for the quasi-chemical reaction is

$$K = \frac{[V_O^{\cdot\cdot}]^2 [B_B']^4}{[O_O^x]^2 [B_B^x]^4} \cdot pO_2(g) \tag{9.102}$$

where [i] denotes the concentration, usually given as site fractions, of defect i. The temperature dependence of the equilibrium constant is, as always, given through

$$\Delta G^\circ = \Delta H^\circ - T\Delta S^\circ = -RT \ln K \tag{9.103}$$

from which the oxygen partial pressure corresponding to a certain value of δ can be deduced:

$$\log pO_2 (g) = \log K + 4[\log(1 - 2\delta) - \log(2\delta)] - 2 \log\left(\frac{\delta}{3 - \delta}\right) \qquad (9.104)$$

Here the site fractions of the defects in eq. (9.102) are expressed in terms of the oxygen non-stoichiometry parameter δ.

This type of defect equilibrium treatment has been used extensively to model the defect chemistry and non-stoichiometry of inorganic substances and has the great advantage that it easily takes several simultaneous defect equilibria into account [22]. On the other hand, the way the mass action laws are normally used they are focused on partial thermodynamic properties and not on the integral Gibbs energy. The latter is often preferred in other types of thermodynamic analyses. In such cases the following solid solution approach is an alternative.

Solid solution approach

For thermochemical uses, an expression for the integral Gibbs energy of formation of the compound $ABO_{3-\delta}$ can be derived by integration of eq. (9.104), but in order to show clearly some of the main implications of the model a more detailed analysis starting from the partition function is preferred [23].

It can be shown that the partition function in this case can be expressed as

$$Z = \sum_c \sum_s \exp\left(-\frac{H_{c,s}}{k_B T}\right) = \sum_c \exp\left(-\frac{G_c}{k_B T}\right) \qquad (9.105)$$

The summation over the enthalpy, $H_{c,s}$, is running over all the vibrational states, s, for all the different atomic configurations, c, of the system. The summation over the enthalpy for a particular atomic configuration is related to the Gibbs energy of that particular configuration, G_c and the total partition function, Z, is given as the summation of the Gibbs energy of all the different atomic configurations of the system. The Gibbs energy of formation for a given composition can now be derived by summation over all configurations that have a certain Gibbs energy of formation, $\Delta_f G_c$, and an associated degeneracy, g_c.

$$\Delta_f G = -k_B T \ln Z = -k_B T \ln \sum_c g_c \exp\left(-\frac{\Delta_f G_c}{k_B T}\right) \qquad (9.106)$$

The configurational entropy term, given by the degeneracy, g_c, is included in $\Delta_f G$ but not in $\Delta_f G_c$. Let us assume the existence of two compounds with different formal oxidation states for the B atom, ABO_3 and $ABO_{2.5}$. The two compounds have the same (perovskite-type) structure and the non-stoichiometric phase

$ABO_{3-\delta}$ is seen as a solution of these two end members. Often only one of the limiting compounds is physically realizable.

If we assume that the non-stoichiometric $ABO_{3-\delta}$ can be described as an ideal solution of the two limiting compounds, ABO_3 and $ABO_{2.5}$, all configurations with a certain composition have the same Gibbs energy of formation since there are no defect–defect interactions. The Gibbs energy of formation of a configuration, $\Delta_f G_c$, is for a certain composition given as

$$\Delta_f G_c (ABO_{3-\delta}) = (1 - 2\delta)\Delta_f G_m^o (ABO_3) + 2\delta \Delta_f G_m^o (ABO_{2.5}) \qquad (9.107)$$

The pure elements at one bar and at a particular temperature are chosen as the standard state. Since all configurations with a given composition have the same Gibbs energy of formation, the total Gibbs energy of formation of a material with a specific composition is given by taking the number of configurations for that composition into consideration. In the ideal solution approach a random distribution of the different species on the different sub-lattices are assumed. Let us assume that oxygen atoms and oxygen vacancies on the oxygen sub-lattice and B^{2+} and B^{3+} (thus $A = A^{3+}$) on the B sub-lattice are randomly distributed. In this case, the degeneracy, g_c, is

$$g_c = \frac{(3N)!}{N_{V_O}!(3N - N_{V_O})!} \cdot \frac{N!}{N_{B^{2+}}!(N - N_{B^{2+}})!} \qquad (9.108)$$

where N is the number of B atoms, N_{V_O} is the number of oxygen vacancies and $N_{B^{2+}}$ is the number of B^{2+} in $ABO_{3-\delta}$.

By substitution of eqs. (9.107) and (9.108) in eq. (9.106) an expression for the total Gibbs energy of formation of the oxide in the ideal solution approximation is obtained:

$$\Delta_f G_m (ABO_{3-\delta}) = (1 - 2\delta)\Delta_f G_m^o (ABO_3) + 2\delta \Delta_f G_m^o (ABO_{2.5})$$
$$+ RT\left[(1 - 2\delta)\ln(1 - 2\delta) + 2\delta \ln(2\delta) + \delta \ln\left(\frac{\delta}{3}\right) + (3 - \delta)\ln\left(1 - \frac{\delta}{3}\right)\right] \qquad (9.109)$$

$\Delta_f G^o (ABO_{2.5})$ is here the standard Gibbs energy of formation of each of all possible configurations of perovskite-type $ABO_{2.5}$. The fact that we have a number of configurations with this Gibbs energy of formation gives rise to the contribution of configurational origin given in the square brackets. For $\delta = 0.5$ this term represents (in the present ideal solution approximation) the entropy connected with disordering of ordered $ABO_{2.5}$ giving a completely random distribution of oxygen atoms and oxygen vacancies on the oxygen sub-lattice. Hence the total Gibbs energy of formation of an oxide with a certain composition is given as a sum of a non-configurational term and a configurational term.

The chemical potential of oxygen can now be derived and the related quantity $\log pO_2$ expressed as a function of δ:

$$\log pO_2 = \left(\frac{1}{RT \ln 10}\right)[4\Delta_f G_m^o(ABO_3) - 4\Delta_f G_m^o(ABO_{2.5})]$$

$$+ 4[\log(1 - 2\delta) - \log(2\delta)] - 2\log\left(\frac{\delta}{3 - \delta}\right) \qquad (9.110)$$

The first term on the right-hand side is in this ideal solution approach given by the standard Gibbs energy of oxidation

$$4\Delta_f G_m^o(ABO_3) - 4\Delta_f G_m^o(ABO_{2.5}) \equiv \Delta_{ox} G^o = \Delta_{ox} H^o - T\Delta_{ox} S^o \quad (9.111)$$

which corresponds to the reaction

$$4ABO_{2.5}(\text{perovskite}) + O_2(g) = 4ABO_{3.00}(\text{perovskite}) \qquad (9.112)$$

In the ideal solid solution model used, the enthalpy and entropy of oxidation are independent of composition.

By comparing eq. (9.110) with eq. (9.104), a thermodynamic expression for the equilibrium constant of the defect reaction, eq. (9.101), is obtained:

$$\log K = \left(\frac{1}{RT \ln 10}\right)[4\Delta_f G_m^o(ABO_3) - 4\Delta_f G_m^o(ABO_{2.5})] \qquad (9.113)$$

In the ideal limit, the two models (the mass action law and the solid solution model) are identical: one focuses on the partial thermodynamic properties, the other on the integral properties. Non-ideal terms giving an excess enthalpy of mixing can be incorporated into the solid solution model [23]. This will enable us to take compositional effects on the Gibbs energy of oxidation into consideration and the Gibbs energy of oxidation will no longer be directly related to eq. (9.112). The mass action law treatment does not usually consider interaction terms explicitly.

Both the reduced and the oxidized compounds in eq. (9.112) take the same structure. In the present case (the perovskite-type), the enthalpy of formation of $ABO_{2.50}$ is the enthalpy of formation of a disordered phase with many possible configurations that, however, all have the same enthalpy of formation (the ideal solution approach). With regard to the redox entropy it should be noted that the last two terms in eq. (9.110) represent the partial configurational entropy of oxygen. Hence the entropic contribution to the Gibbs energy of the redox reaction (9.112) should not include the structural configurational contribution, since this term is included explicitly in the configurational part of eq. (9.110). Thus, when comparing calorimetric entropies with entropies deduced from equilibration studies, the configurational entropy should be subtracted from the calorimetric entropy. This is in many ways analogous to the treatment of polymer solutions considered in Section 9.1 where the entropy of disorientation of the polymer and the entropy of

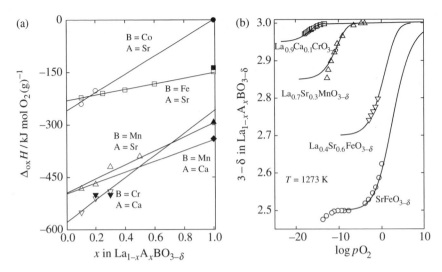

Figure 9.12 (a) Enthalpy of oxidation of $La_{1-x}A_xBO_{3-\delta}$ as a function of x. Open symbols represents values deduced from non-stoichiometry versus partial pressure isotherms. Closed symbols represent calorimetric values. (b) Comparison of experimental and calculated non-stoichiometry versus partial pressure isotherms [23]. Reproduced by permission of the Royal Society of Chemistry.

the subsequent mixing of the monomer and the disoriented polymer were treated separately. It should also be noted that similar solution models may be applied for less ionic compounds like sulfides, selenides and tellurides. Here, the electrons are considered to be delocalized and the configurational entropy due to valence defects is thus negligible [24].

The enthalpies of oxidation of a number of perovskite-type oxides, $La_{1-x}A_xBO_{3-\delta}$ (A = Sr, Ca, B = Cr, Mn, Fe, Co), deduced by applying eq. (9.110) to experimental data for the variation of the non-stoichiometry with the partial pressure of oxygen are given in Figure 9.12(a) [23]. Calorimetric data are also given where available. Note that the redox reactions considered involve B^{3+} and B^{4+}. Examples of the agreement between experimental and calculated non-stoichiometry data are given in Figure 9.12(b) [23].

Non-stoichiometry in solid solutions may also be handled by the compound energy model: see for example a recent review by Hillert [16]. In this approach the end-member corresponding to vacancies is an empty sub-lattice and it may be argued that the model loses its physical significance. Nevertheless, this model represents a mathematically efficient description that is often incorporated in thermodynamic representations of phase diagrams.

References

[1]   K. Huang, *Statistical Mechanics*, 2nd edn. New York: John Wiley, 1987.
[2]   J. H. Hildebrand, *J. Am. Chem. Soc.* 1929, **51**, 66.

[3]   E. A. Guggenheim, *Mixtures*. Oxford: Oxford University Press, 1952.
[4]   A. D. Pelton, S. A. Degterov, G. Erikson, C. Robelin and Y. Dessureault, *Metall. Mater. Trans. B.* 2000, **31B**, 651.
[5]   C. H. P. Lupis, *Chemical Thermodynamics of Materials*. Amsterdam: Elsevier, 1983.
[6]   A. D. Pelton and P. Chartrand, *Metall. Mater. Trans. A.* 2001, **32A**, 1355.
[7]   P. J. Flory, *J. Chem. Phys.* 1942, **10**, 51.
[8]   M. L. Huggins, *J. Am. Chem. Soc.* 1942, **64**, 1712.
[9]   E. A. Guggenheim, *Proc. Roy. Soc. (London) A.* 1944, **183**, 213.
[10]   G. Gee and Treloar, *Trans. Faraday Soc.* 1942, **38**, 147.
[11]   G. Gee and W. J. C. Orr, *Trans. Faraday Soc.* 1946, **42**, 507.
[12]   J. H. Hildebrand, *J. Chem. Phys.* 1947, **15**, 225.
[13]   M. Temkin, *Acta Phys. Chim. USSR.* 1945, **20**, 411.
[14]   M. Blander and S. J. Yosim, *J. Chem. Phys.* 1963, **39**, 2610.
[15]   M. Hillert and L.-I. Staffansson, *Acta Chem. Scand.* 1970, **24**, 3618.
[16]   M. Hillert, *J. Alloys Comp.* 2001, **320**, 161.
[17]   H. Flood, T. Førland and K. Grjotheim, *Z. Anorg. Allg. Chem.* 1954, **276**, 289.
[18]   W. L. Bragg and E. J. Williams, *Proc. Roy. Soc. London, Ser. A.* 1934, **145**, 699.
[19]   W. L. Bragg and E. J. Williams, *Proc. Roy. Soc. London, Ser. A.* 1935, **151**, 540.
[20]   H. St C. O'Neill and A. Navrotsky, *Am. Mineral.* 1983, **68**, 181.
[21]   K. B. Greenwood, D. Ko, A. V. Griend, G. M. Sarjeant, J. W. Milgram, E. S. Garrity, D. I. DeLoach, K. R. Poeppelmeier, P. A. Salvador and T. O. Mason, *Inorg. Chem.* 2000, **39**, 3386.
[22]   F. A. Kröger, *The Chemistry of Imperfect Crystals*. Amsterdam: North-Holand, 1974.
[23]   E. Bakken, T. Norby and S. Stølen, *J. Mater. Chem.* 2002, **12**, 317.
[24]   S. Stølen and F. Grønvold, *J. Phys. Chem. Solids.* 1987, **48**, 1213.

Further reading

P. J. Flory, *Principles of Polymer Chemistry*. Ithaca, NY: Cornell University Press, 1953.
E. A. Guggenheim, *Mixtures*. Oxford: Clarendon Press, 1952.
N. A. Gocken, *Statistical Thermodynamics of Alloys*. New York: Plenum Press, 1986.
T. L. Hill, *Introduction to Statistical Thermodynamics*. New York: Dover Publications, 1986.
F. A. Kröger, *The Chemistry of Imperfect Crystals*. Amsterdam: North-Holland, 1974.
C. H. P. Lupis, *Chemical Thermodynamics of Materials*. Amsterdam: Elsevier, 1983.
A. D. Pelton, Solution Models: Chapter 3 in *Advanced Physical Chemistry for Process Metallurgy*. London: Academic Press, 1997.
F. Reif, *Fundamentals of Statistical and Thermal Physics*. Singapore: McGraw-Hill, 1965.

10

Experimental
thermodynamics

Thermodynamics is largely an experimental science, although computational and theoretical methods are in rapid development (see Chapter 11). The term *experimental thermodynamics* encompasses all experimental investigations that allow the determination of parameters from which thermodynamic data can be extracted. The field of experimental thermodynamics covers direct determination of enthalpy, entropy and Gibbs energy of substances, phase transitions or reactions, but also thermodynamically directed studies of phase equilibria and volumetric properties. In this chapter, most attention is given to experimental techniques for direct determination of thermodynamic properties, but methods for studying phase equilibria and volumetric properties are also treated shortly. In addition brief treatments of the measurement of temperature and pressure are given. These two intensive properties are of special importance in thermodynamics and profoundly affect and often completely determine the state of a system.

10.1 Determination of temperature and pressure

Techniques for accurate and reproducible measurement of temperature and temperature differences are essential to all experimental studies of thermodynamic properties. Ideal gas thermometers give temperatures that correspond to the fundamental thermodynamic temperature scale. These, however, are not convenient in most applications and practical measurement of temperature is based on the definition of a temperature scale that describes the thermodynamic temperature as accurately as possible. The analytical equations describing the latest of the international temperature scales, **the temperature scale of 1990 (ITS–90)** [1, 2]

Chemical Thermodynamics of Materials by Svein Stølen and Tor Grande
© 2004 John Wiley & Sons, Ltd ISBN 0 471 492320 2

contains a number of parameters that are determined by calibration of the measurement probe using the fixed points on ITS–90. These fixed points correspond to situations where two or three phases at a specific pressure are in equilibrium at a
constant temperature. The fixed points for the ITS–90 are given in Table 10.1.

Between the triple point of equilibrium hydrogen (13.8033 K) and the freezing
point of silver (1234.93 K), T_{90} is defined by means of platinum resistance thermometers calibrated at specific sets of defining fixed points. The temperatures are
given in terms of the ratio of the resistance of the thermometer at temperature T_{90} to
the resistance at the triple point of water:

$$W(T_{90}) = \frac{R(T_{90})}{R(273.16\,\text{K})} \tag{10.1}$$

The temperature T_{90} for a specific resistance thermometer is calculated from the
equation

$$W(T_{90}) = W_r(T_{90}) + \Delta W(T_{90}) \tag{10.2}$$

where $W(T_{90})$ is the resistance ratio of the thermometer measured at the temperature T_{90}, $W_r(T_{90})$ is the thermometer-independent resistance ratio calculated from a
reference function [1, 2], and $\Delta W(T_{90})$ is a difference function obtained by calibration of the thermometer using the temperature fixed points. The reference function
is given as complex polynomials, with different polynomials being used in different temperature regimes. For example, the difference function for the temperature region from 273.15 to 933.473 K is

$$\Delta W(T_{90}) = a[W(T_{90}) - 1] + b[W(T_{90}) - 1]^2 + c[W(T_{90}) - 1]^3 \tag{10.3}$$

Table 10.1 Temperature fixed points for ITS-90.

Type of transition	Compound	T in K	Type of transition	Compound	T in K
T_{trp}	e-H_2	13.8033	T_{fre}	In	429.7485
T_{trp}	Ne	24.5561	T_{fre}	Sn	505.078
T_{trp}	O_2	54.3584	T_{fre}	Zn	692.677
T_{trp}	Ar	83.8058	T_{fre}	Al	933.473
T_{trp}	Hg	234.3156	T_{fre}	Ag	1234.93
T_{trp}	H_2O	273.16	T_{fre}	Au	1337.33
T_{fus}	Ga	302.9146	T_{fre}	Cu	1357.77

T_{trp} = triple point temperature
T_{fus} = fusion temperature
T_{fre} = freezing temperature

where a, b and c are determined from the triple point of water and the freezing points of Sn, Zn and Al. For high accuracy the effect of trace impurities on the temperature fixed points should be remembered [3] and the calibration must be performed with very pure metals using procedures recommended by the Comité Consultatif de Thermométrie [2, 4].

The resistance of platinum increases with temperature. Above 30 K the resistance–temperature slope is high, giving a high sensitivity. Below 30 K the slope of resistance versus temperature is much lower, a fact that make accurate and sensitive temperature measurements difficult, and alternative thermometers like germanium and rhodium with 0.5 at. % iron are frequently used. The resistivity of an intrinsic semiconductor like germanium follows an exponential law. However, since trace impurities greatly affect the transport properties, donor- or acceptor-doped germanium is used for thermometry. The typical resistance versus temperature response for Pt, n-type Ge and Rh–Fe resistance thermometers are compared in Figure 10.1. The slope of the curve gives the sensitivity of the thermometer.

At relatively high temperatures thermocouple thermometers are most commonly used to measure temperature. The thermoelectric power of three frequently used thermocouples is compared in Figure 10.2. The choice of thermocouple depends on the temperature range, the chemistry of the problem in question, sensitivity requirements and resistance towards thermal cycling. The temperature range and typical uncertainty of some of the most commonly used thermocouple thermometers are given in Table 10.2.

Above the freezing point of silver, T_{90} is defined in terms of a defining fixed point and the Planck radiation law, and optical pyrometers are frequently used as temperature probes. The Comité Consultatif de Thermométrie gives a thorough discussion of the different techniques for approximation of the international temperature scale of 1990 [2, 4].

Pressure has dimensions of force per unit area, and pressure multiplied by volume has the dimensions of energy. The SI unit of pressure is the pascal (Pa) and the standard pressure is set to 1 bar (equal to 10^5 Pa or 0.1 MPa). Methods for measuring pressure are outside the scope of this book. A thorough discussion of pressure measurements in different pressure regimes are found in Goodwin *et al.*, *Measurements of the Thermodynamic Properties of Single Phases* (see Further reading). Pressure measurements in the GPa range are briefly mentioned in relation to the high-pressure techniques described in Section 10.2.

10.2 Phase equilibria

The strategy that is followed for mapping a phase diagram depends on the specific problem considered. In general there are two complementary approaches that must be used. In the first, samples prepared at a particular temperature–pressure condition and subsequently quenched from these conditions are analyzed, typically by optical or electron microscopy and X-ray diffraction. This approach does not

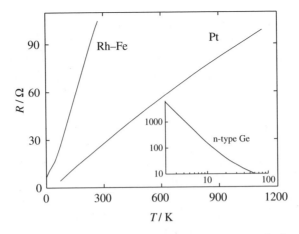

Figure 10.1 Resistance versus temperature response of some typical resistance thermometers.

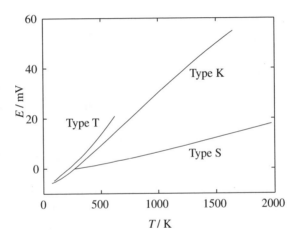

Figure 10.2 Thermoelectric power versus temperature response of three frequently used thermocouples.

necessarily, however, give the equilibrium phase or phase mixture at the annealing conditions and must therefore be combined with dynamic techniques that allows for mapping phase changes with temperature and pressure, such as thermal analysis and *in situ* diffraction and spectroscopy. Thermal analysis is convenient and frequently used to determine liquidus/solidus curves as well as solid–solid transformation temperatures [5]. A short description of the main techniques, differential thermal analysis (DTA) and differential scanning calorimetry (DSC), is given in Section 10.3. The development of high-intensity synchrotron radiation facilities has also made *in situ* X-ray diffraction studies of reactions more applicable [6].

   Annealing at temperatures up to and above 2000 K is easily achieved due to the availability of different furnaces. Studies at high pressure demand more complex

Table 10.2 Characteristics of some frequently used thermometers.

Thermometer	Usual T-range	Typical uncertainty
Resistance thermometers		
Pt	14–1235 K	0.5 mK
Ge	1–100 K	$\Delta T/T < 2\cdot10^{-4}$
Rh–Fe	0.5–30 K	0.3 mK
C	0.5–30 K	$\Delta T/T < 5\cdot10^{-3}$
Thermocouples		
S: Pt10%Rh/Pt	223–1873 K	0.3 K < 1273 K
		1 K > 1273 K
R: Pt13%Rh/Pt	223–1873 K	0.3 K < 1273 K
		1 K > 1273 K
B: Pt30%Rh/Pt	573–2073 K	0.5 to 2 K
T: Cu/Cu–Ni	73–623 K	0.1 K
E: Ni–Cr/Cu–Ni	73–1143 K	1 K < 300 K
		1 K > 300 K
J: Fe/Cu–Ni	273–1033 K	0.5 K < 300 K
		2 K > 300 K
K: Ni–Cr/Ni–Al	73–1533 K	1 K < 1273 K
		3 K > 1273 K
N: Ni–Cr–Si/Ni–Si	273–1573 K	0.5 K < 1273 K
		3 K > 1273 K
W5%Re/W20%Rh	1273–2473 K	3–10 K
Radiation		
	373–3273 K	1 K < 1273 K
		5 K > 1273 K

and specialized equipment. Some of the main techniques with characteristic pressure ranges are listed in Table 10.3. For static compression the sample is contained in a pressure vessel. The maximum pressure is limited by the construction of the vessel and apparatus and the mechanical properties of the materials used.

Pressures up to 4–6 GPa can be obtained by **piston–cylinder** devices, where the pressure is generated by pushing a piston against the sample cell [7]. Pressure is measured directly from the force applied and the cross-sectional area of the piston, but calibration of the pressure scale using the phase transition pressure between polymorphs of, for example, Bi is some times necessary. Large sample masses can be used; in industrial diamond synthesis tens of cm^3 samples are used in belt devices. The method allows internal probes for acoustic velocity measurements, but the sample is inaccessible to X-ray and spectroscopic probes. Higher pressures are obtained in **multi-anvil presses** [8] that contain anvils made of high-strength materials like tungsten, boron carbide, sapphire or sintered diamond. With an anvil base to tip ratio of 100, 30 GPa can be reached with tungsten carbide, 60 GPa with sintered diamond and 140 GPa with single crystal diamond [9]. In addition to the

Table 10.3 Characteristics of the main high-pressure devices.

Devices	Approx. sample volume cm^3	Approx. p-range GPa	Approx. T-range K
Autoclaves	10^3	0–0.1	200–600
Sealed pressure vessels	10^2	0–0.5	300–1200
Piston cylinder	10^{-1}	0–5	300–2000
Multi-anvil	10^{-2}	0–25	300–2000
Paris–Edinburgh	10^{-3}	0–8	200–2000
Diamond anvil	10^{-4}	0–100	200–3000
Shock apparatus		0–300	300–3000

exceptional strength of single-crystal diamond, diamonds are essentially transparent to radiation in the UV–visible–IR region (below 5 eV), and to X-ray radiation above 10 keV, and *in situ* spectroscopy and X-ray diffraction are facilitated.

In general, the sample volume is proportional to the size of the anvil and the press. Above 15 GPa large sample devices refer to millimetre and centimetre sized samples. This will require anvils of the order of decimetres to metres and presses from 200 to 50 000 tons.

In situ studies at high pressure and temperature are to a large degree performed using **diamond anvil cells** [10] and the **Paris–Edinburgh device** [11]. A schematic illustration of a piston–cylinder type diamond anvil cell is given in Figure 10.3. In diamond anvil cells two small opposing single crystal diamonds compress a gasket that contain the sample chamber in centre. The gasket experiences a gradient in pressure from ambient to the peak pressure. The sample is within the sample chamber contained in a pressure-transmitting medium that under ideal conditions gives approximately hydrostatic pressure conditions. Pressure is often measured with an internal calibration substance with an accurately determined equation of state or from the pressure shift of the ruby fluorescence wavelength of a tiny ruby grain added to the sample [12]. The ruby scale has been calibrated to 180 GPa [13].

10.3 Energetic properties

A large number of techniques have been used to investigate the thermodynamic properties of solids, and in this section an overview is given that covers all the major experimental methods. Most of these techniques have been treated in specialized reviews and references to these are given. This section will focus on the main principles of the different techniques, the main precautions to be taken and the main sources of possible systematic errors. The experimental methods are rather well developed and the main problem is to apply the different techniques to systems with various chemical and physical properties. For example, the thermal stability of the material to be studied may restrict the experimental approach to be used.

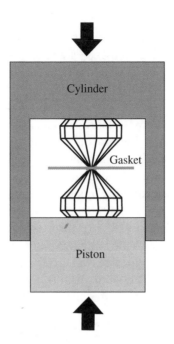

Figure 10.3 Schematic illustration of a piston–cylinder type arrangement of the diamond anvil cell. The sample is contained in a hole drilled in the gasket.

Calorimetric, electrochemical and vapour pressure methods are treated separately. The different techniques are to a large extent complementary. In general, enthalpy and entropy are measured most accurately by calorimetry, while electrochemical and vapour pressure techniques represent efficient direct methods for determination of activities and Gibbs energies.

Thermophysical calorimetry

The entropy of a compound can only be derived directly by calorimetry through integration of the heat capacity of the compound, and accurate entropy data rely on accurate determination of the heat capacity from 0 K. Different techniques are preferred in different temperature regimes. Calorimeters with electrically heated shields, which follow the surface temperature of the calorimeter with its contained sample (**adiabatic calorimeters**), offer unique possibilities for accurate determination of heat capacity and enthalpies of transition [14] and are the most accurate instruments below 1000 K. Absolute values are determined through accurate determination of the electrical energy provided and the corresponding temperature rise of the calorimeter. Minimization of the heat leak between the calorimeter and the immediate surroundings, the adiabatic shields, are of primary importance to the accuracy of the measurements. The high-temperature calorimeter used at the University of Oslo since the early 1960s [15, 16] has two shields: an inner adiabatic shield and an outer guard shield as shown in Figure 10.4. Both shields are divided into three different parts that are regulated independently. For adiabatic calorimeters, instrumental temperature drift rates, characterizing the heat leak, of the order of < 0.15 μK·s^{-1} and < 1.5 μK·s^{-1} are readily obtainable at 100 K [17] and 800 K [16], respectively. Sophisticated low-temperature instruments have an accuracy of 0.1% or less below 300 K [17], whereas corresponding high-temperature calorimeters may be accurate to 0.2% from 300 to 1000 K [16]. These calorimeters can be operated either with stepwise energy inputs preceded and followed by equilibration periods as illustrated in Figure 10.5, or alternatively with continuous input of energy. Stepwise heating secures thermal equilibrium in the sample and these instruments may be used to follow exothermal reactions taking place in a sample.

Adiabatic calorimeters are complex home-made instruments, and the measurements are time-consuming. Less accurate but easy to use commercial **differential scanning calorimeters** (DSCs) [18, 19] are a frequently used alternative. The method involves measurement of the temperature of both a sample and a reference sample and the 'differential' emphasizes the difference between the sample and the reference. The two main types of DSC are heat flux and power-compensated instruments. In a heat flux DSC, as in the older differential thermal analyzers (DTA), the

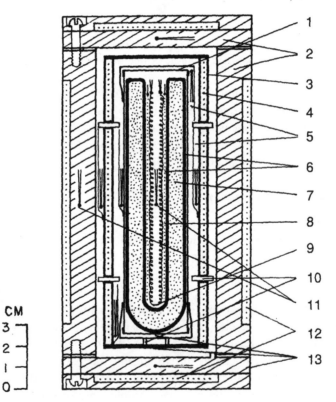

Figure 10.4 Adiabatic high-temperature calorimeter [15]. 1: Calorimeter proper; 2: Silver guard; 3: Silver shield; 4: Shield heater; 5: Thermocouple sleeve; 6: Silica glass container; 7: Sample; 8: Calorimeter heater; 9: Pt resistance thermometer; 10: Silica ring spacer; 11: Type S thermocouple; 12: Guard heater; 13: Removable bottom. Reproduced by permission of F. Grønvold.

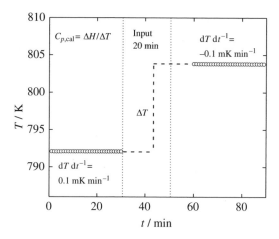

Figure 10.5 Schematic representation of the stepwise heating mode of operation of an adiabatic calorimeter.

temperature difference between sample and reference material is measured and there is no clear distinction between DSCs of the heat flux type and DTA instruments. Only a part of the heat released or absorbed during a phase transition is detected, and the thermocouple system is of crucial importance for the quality of the instrument. The relationship between heat flux DSCs and classical DTAs is seen in a number of DSC constructions, where both the sample and the reference material are contained within a single furnace chamber: see Figure 10.6(a). The temperature sensors in these constructions are typically placed in a disc of good thermal conductivity. A quite different approach is used in Calvet type apparatuses [20] where two cylindrical sample chambers are placed in a common calorimeter block and the heat flux in or out of the two chambers is determined by differential thermopiles.

Whereas the heat flux DSC measures the temperature difference between the sample and the reference sample, power-compensated DSCs are based on compensation of the heat to be measured by electrical energy. Here the sample and the reference are contained in separate micro-furnaces, as shown in Figure 10.6(b). The time integral over the compensating heating power is proportional to the enthalpy absorbed by or released from the sample.

Proper calibration of the DSC instruments is crucial. The basis of the enthalpy calibration is generally the enthalpy of fusion of a standard material [21, 22], but electrical calibration is an alternative. A resistor is placed in or attached to the calorimeter cell and heat peaks are produced by electrical means just before and after a comparable effect caused by the sample. The different heat transfer conditions during calibration and measurement put limits on the improvement. DSCs are usually limited to temperatures from liquid nitrogen to 873 K, but recent instrumentation with maximum temperatures close to 1800 K is now commercially available. The accuracy of these instruments depends heavily on the instrumentation, on the calibration procedures, on the type of measurements to be performed, on the temperature regime and on the

(a)

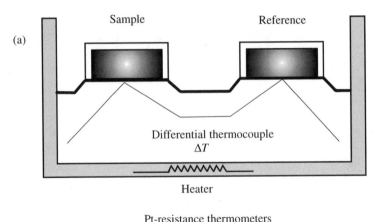

(b)

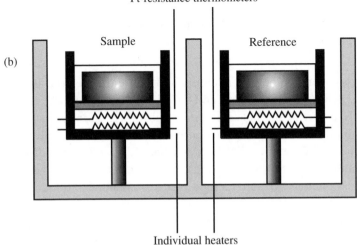

Figure 10.6 Schematic representation of (a) heat flux DSC and (b) power-compensated DSC.

properties of the material to be investigated. On careful calibration, heat capacities can be determined with an accuracy of 1–2% in the optimal temperature regime (300–600 K), while enthalpies of first-order transitions can be determined even more accurately [21].

Heat capacities at high temperatures, $T > 1000$ K, are most accurately determined by **drop calorimetry** [23, 24]. Here a sample is heated to a known temperature and is then dropped into a receiving calorimeter, which is usually operated around room temperature. The calorimeter measures the heat evolved in cooling the sample to the calorimeter temperature. The main sources of error relate to temperature measurement and the attainment of equilibrium in the furnace, to evaluation of heat losses during drop, to the measurements of the heat release in the calorimeter, and to the reproducibility of the initial and final states of the sample. This type of calorimeter is in principle unsurpassed for enthalpy increment determinations of substances with negligible intrinsic or extrinsic defect concentrations in the region of interest, like high-purity synthetic sapphire, α-Al_2O_3, except in the vicinity of the melting point.

A special form of drop calorimetry has been developed for high-temperature studies of metals and alloys. At high temperatures reactions between the sample and the container may lead to serious errors. They may be avoided by electromagnetic levitation of the sample in a vacuum furnace, and **levitation drop calorimetry** is a very powerful high-temperature technique [25]. Conventional steady state and quasi-steady state techniques for accurate measurement of heat capacity are generally limited to temperatures below 2000 K. A special technique for studies of metals at very high temperature is the so-called **pulse calorimetry**, in which the sample reaches high temperatures and different thermophysical properties are recorded in short times – sub-microsecond techniques have been reported [26]. The accuracy is lower than obtained by drop calorimetry and it seems reasonable to assume an uncertainty of about 3–5%. **Modulation calorimetry** has proven extremely useful for studies of phase transitions [27]. Two popular modulated techniques are AC calorimetry [28], and temperature-modulated DSC [29]. The latter technique proves especially useful in characterization of glasses.

Thermochemical calorimetry

Most thermochemical calorimetric methods are used to determine enthalpy changes of chemical reactions. The reaction may give the enthalpy of interest directly or may represent a step in a thermodynamic cycle needed to obtain an enthalpy of interest. These techniques are also very suitable for direct determination of enthalpy of mixing in the liquid state or indirect determination of enthalpy of mixing in the solid state. Calorimetric methods for studies of chemical reactions involving solids can be divided into three main categories:

- solution calorimetry
- combustion calorimetry
- direct reaction calorimetry

The measurement of an enthalpy change is based either on the law of conservation of energy or on the Newton and Stefan–Boltzmann laws for the rate of heat transfer. In the latter case, the heat flow between a sample and a heat sink maintained at isothermal conditions is measured. Most of these isoperibol **heat flux calorimeters** are of the twin type with two sample chambers, each surrounded by a thermopile linking it to a constant temperature metal block or another type of heat reservoir. A reaction is initiated in one sample chamber after obtaining a stable stationary state defining the baseline from the thermopiles. The other sample chamber acts as a reference. As the reaction proceeds, the thermopile measures the temperature difference between the sample chamber and the reference cell. The rate of heat flow between the calorimeter and its surroundings is proportional to the temperature difference between the sample and the heat sink and the total heat effect is proportional to the integrated area under the calorimetric peak. A calibration is thus needed to transform the calorimetric signal into an enthalpy. Calibration can be obtained by dropping a substance with known heat capacity into the calorimeter or by electrical calibration. A thermopile system and a typical calorimetric signal are shown in Figure 10.7.

Measurements based on the law of conservation of energy are of two main types. In **phase change calorimetry** the enthalpy of the reaction is exactly balanced by the enthalpy of a phase change of a contained compound surrounded by a larger reservoir of the same compound used to maintain isothermal conditions in the calorimeter. The latter enthalpy, the measurand, is often displayed indirectly through the change in the volumetric properties of the heat reservoir compound, e.g. ice/water.

Adiabatic calorimetry uses the temperature change as the measurand at nearly adiabatic conditions. When a reaction occurs in the sample chamber, or energy is supplied electrically to the sample (i.e. in heat capacity calorimetry), the temperature rise of the sample chamber is balanced by an identical temperature rise of the adiabatic shield. The heat capacity or enthalpy of a reaction can be determined directly without calibration, but corrections for heat exchange between the calorimeter and the surroundings must be applied. For a large number of isoperibol

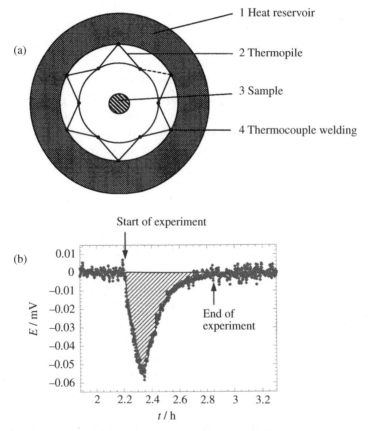

Figure 10.7 (a) Horizontal cross-section of a thermopile showing a star-like arrangement of the thermocouples and (b) voltage signal from a typical experimental run.

solution and combustion calorimeters the temperature change of the calorimeter during a reaction is recorded. From the corrected temperature change and the energy equivalent of the calorimeter (determined by electrical calibration in a separate experiment) the enthalpy of a reaction can be calculated [30, 31].

The basic principle of **solution calorimetry** is simple. In one experiment the enthalpy of solution of, for example, $LaAlO_3$(s) [32] is measured in a particular solvent. In order to convert this enthalpy of solution to an enthalpy of formation, a thermodynamic cycle, which gives the formation reaction

$$\tfrac{1}{2}La_2O_3(s) + \tfrac{1}{2}Al_2O_3(s) = LaAlO_3(s) \tag{10.4}$$

must be set up. The enthalpy of formation of $LaAlO_3$(s) can be obtained from three enthalpy of solution measurements that correspond to the following reactions:

$$LaAlO_3\,(s,T) + \text{solvent}\,(T) = \text{solution}\,(T) \tag{10.5}$$

$$\tfrac{1}{2}La_2O_3\,(s,T) + \text{solvent}\,(T) = \text{solution}\,(T) \tag{10.6}$$

$$\tfrac{1}{2}Al_2O_3\,(s,T) + \text{solvent}\,(T) = \text{solution}\,(T) \tag{10.7}$$

The enthalpy of formation of the ternary compound is given by

$$\Delta_f H_m^o\,(LaAlO_3,\ s,T) = \Delta_r H_m^o\,(10.5) - [\Delta_r H_m^o\,(10.6) + \Delta_r H_m^o\,(10.7)]$$
$$+ \tfrac{1}{2}[\Delta_f H_m^o\,(La_2O_3, s, T) + \Delta_f H_m^o\,(Al_2O_3, s, T)] \tag{10.8}$$

Integral and partial molar enthalpies of mixing in solid solutions may be derived by similar investigations of a series of solid solutions with systematic variation in composition.

The solution experiments may be made in aqueous media at around ambient temperatures, or in metallic or inorganic melts at high temperatures. Two main types of **ambient temperature solution calorimeter** are used: adiabatic and isoperibol. While the adiabatic ones tend to be more accurate, they are quite complex instruments. Thus most solution calorimeters are of the isoperibol type [33]. The choice of solvent is obviously crucial and aqueous hydrofluoric acid or mixtures of HF and HCl are often-used solvents in materials applications. Very precise enthalpies of solution, with uncertainties approaching ±0.1% are obtained. The effect of dilution and of changes in solvent composition must be considered. Whereas low temperature solution calorimetry is well suited for hydrous phases, its ability to handle refractory oxides like Al_2O_3 and MgO is limited.

High-temperature solution calorimeters [34–36] are in general of the twin heat flux type. They are applicable from around 900 K to around 1500 K and a precision of about ±1% is often obtained. Although of lower precision than lower temperature solution calorimeters, the effect of dilution is small and a large range of materials can be studied. The main development of recent years is related to measurements on smaller samples. Samples with masses of the order of 10–15 mg are routinely used, and sensitive calorimeters for samples present only in very small quantities, 1–5 mg [37] have been developed. Solution calorimetry has been used to study metals and alloys since the early 1950s [38, 39], and has been applied to oxides since 1964 [40].

A large number of solvents are used. Solubility, kinetics, thermal history, particle size, stirring and heat flow sizes are all factors that must be considered. For solution of oxides, buffer-type systems are used [41, 42]. Lead borate ($2PbO \cdot B_2O_3$), alkali tungstates or molybdates ($3Na_2O \cdot 4MoO_3$) and alkali borates ($LiBO_3 \cdot NaBO_3$) are all solvents for oxides [34]. Lead borate is the usual solvent, but a number of oxides dissolve slowly (oxides of Ti, Zr and rare earth elements) and other solvents have in such cases proven more useful. Alkali molybdates and tungstates are used for relatively basic oxides and also for nitrides. The redox properties of the solvent are of particular importance for transition metal compounds, since the oxidation state of the transition metal after solution depends on the solvent and the atmosphere. While Cr_2O_3 dissolves in molten lead borate near 1000 K giving Cr(VI) in O_2, Cr(III) is formed in Ar. MnO_2 will, in the same solvent, give Mn(III) in O_2 [34].

Low melting metals (Sn and also Bi, In, Pb, and Cd) are extensively used as solvents in calorimetric studies of metallic phases [35]. Transition metals do not, however, dissolve readily in tin [43] and other solvents such as Cu and Al have been used. An experimental probe for high-temperature solution calorimetry is shown in Figure 10.8.

In **drop solution calorimetry** the sample may not necessarily be equilibrated at the calorimeter temperature prior to dissolution. It is often useful to drop the sample into the calorimeter from room temperature [34]. This method is preferred when the sample might decompose at the calorimeter temperature. In drop solution calorimetry the enthalpy increment of the sample and its enthalpy of solution at the calorimetric temperature are determined. The difference in the enthalpy of solution of products and reactants gives the enthalpy of reaction at room temperature. This method has been used for determination of the enthalpy of formation of carbonates and hydrates [44, 45].

In **solute–solvent calorimetry** the compound to be studied is present as a mixture with another element or compound in solid form at room temperature and dropped into a hot calorimeter with resulting formation of a liquid product [35]. In order to determine the enthalpy of formation of LaB_6, Pt was added in a proportion that gave the composition of a low melting eutectic. The liquid phase formed enhanced the reaction rate and enabled the energetic parameters to be extracted [46].

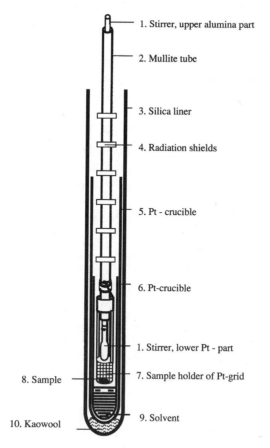

1. Stirrer, upper alumina part

2. Mullite tube

3. Silica liner

4. Radiation shields

5. Pt - crucible

6. Pt-crucible

1. Stirrer, lower Pt - part

8. Sample   7. Sample holder of Pt-grid

10. Kaowool   9. Solvent

Figure 10.8 Experimental setup for measurement of the enthalpy of solution at high temperatures.

In **combustion calorimetry** [47, 48] the enthalpies of chemical reactions of elements and compounds with reactive gases like oxygen or fluorine are determined. Examples are [49, 50]:

$$GeS_2(cr) + 8F_2(g) = GeF_4(g) + 2SF_6(g) \tag{10.9}$$

$$TiN(s) + 2O_2(g) = TiO_2(s) + NO_2(g) \tag{10.10}$$

The technique is versatile and can be used for a large number of compounds, and recent examples are given in [47, 51, 52].

In many cases the measured enthalpies must be corrected for impurities present in the original sample, often in an ill-defined state. A carbon impurity present in an Si_3N_4 sample may reasonably be assumed to be present as SiC [53]. The carbon-containing species in VSi_2, for example, is more uncertain: it may be SiC, but it may also be present in the form of vanadium carbide or may even be present in its elemental form [47].

Incomplete combustion and several reaction products represent another difficulty, which may increase the uncertainty. Whereas sulfur-containing compounds invariably produce $SF_6(g)$ at room temperature and the combustion yield when fluorinating GeS_2 is high, oxygen combustion is often less ideal and the combustion yield rather low. Furthermore, several different oxides are often present in the reaction product. Combustion calorimetry seems in general to work better for fluorine compared with oxygen combustion for inorganic solids. Metal fluorides are often stoichiometric, and moreover the combustion yield is generally higher than for oxygen combustion. However, difficulties are also observed for fluorides. Whereas S, Se, Te, P, As, I, Ge and Si invariably produce only one reaction species at room temperature [47], uranium-containing compounds may produce not only UF_6, but also UF_5, UF_4, UF_3, U_2F_9 and U_4F_{17} [54].

In combustion calorimetry the chemical reactions are usually ignited and the calorimeters are most often of the isoperibol type working at around room temperature. The energetics of solid–gas reactions may also be studied at high temperatures in heat flux or in adiabatic calorimeters, these techniques may be described under the heading **direct reaction calorimetry**. The enthalpy of formation of nitrides (e.g. Na_3WN_3) has been determined indirectly through measurement of the enthalpy of oxidation of the nitride at high temperature [55].

$$Na_3WN_3(s,\ 298\ K) + 9/4O_2(g,\ 977\ K)$$
$$= 3/2Na_2O(soln,\ 977\ K) + WO_3(soln,\ 977\ K) + 3/2N_2(g,\ 977\ K) \quad (10.11)$$

Enthalpies of oxidation of stoichiometric and even non-stoichiometric oxides may similarly be obtained by heating reduced oxides in air in an adiabatic calorimeter to a temperature at which the oxidation proceeds sufficiently fast [56]:

$$SrMnO_{2.50}(s) + 0.25\ O_2(g) = SrMnO_3(s) \quad (10.12)$$

By controlling the partial pressure of the volatile species, partial enthalpies of solid–gas reactions such as oxidation and hydrogenation can also be obtained. The partial enthalpies of solution of hydrogen and deuterium in metals and alloys have been measured by high-temperature heat flux calorimetry [57, 58]. In both cases the calorimetric technique was combined with the determination of pressure–composition isotherms. A recent construction used for determination of the partial molar enthalpy of oxygen for the high-temperature superconductor $YBa_2Cu_3O_z$ simultaneously measures the heat effect owing to a small change in oxygen partial pressure and the corresponding change in stoichiometry [59]. Two separate instruments, a calorimeter and a thermobalance, are placed in series in the gas path. The obvious advantage of this is that the mass change is measured directly and not derived from the partial derivative of the compositional variation of the equilibrium partial pressure.

Enthalpies of formation of metallic compounds and alloys can also be determined directly through reactions such as [60]:

$$Al(s, T_1) + Ce(s, T_1) = AlCe(liq, T) \tag{10.13}$$

$$AlCe(s, T_1) = AlCe(liq, T) \tag{10.14}$$

$$\Delta_f H_m (AlCe, s, T_1) = \Delta_f H(10.13) - \Delta_f H(10.14) \tag{10.15}$$

In recent years a range of refractory intermetallic compounds have been studied [35, 61]. The main criterion for a good result is that the reaction should go to completion in a reasonable time.

Electrochemical methods

Wagner pioneered the use of solid electrolytes for thermochemical studies of solids [62]. Electrochemical methods for the determination of the Gibbs energy of solids utilize the measurement of the electromotive force set up across an electrolyte in a chemical potential gradient. The electrochemical potential of an electrochemical cell is given by:

$$E = \frac{RT}{q_i F} \int_{a_1}^{a_2} t_{ion} \, d \ln a_i \tag{10.16}$$

where

$$t_{ion} = \sigma_{ion} / \sigma_{tot} \tag{10.17}$$

is the transference number of the mobile ion in the electrolyte, σ_{ion} and σ_{tot} are the ionic and total (ionic + electronic) conductivity, and q_i is the ionic charge of species i and F is Faraday's constant. a_1 and a_2 are the activities of the ionic conducting species at the two electrodes. If the activity of a species at one electrode is fixed, the activity of the species on the other electrode is determined from the observed electrochemical potential.

Solid electrolytes are frequently used in studies of solid compounds and solid solutions. The establishment of cell equilibrium ideally requires that the electrolyte is a pure ionic conductor of only one particular type of cation or anion. If such an ideal electrolyte is available, the activity of that species can be determined and the Gibbs energy of formation of a compound may, if an appropriate cell is constructed, be derived. A simple example is a cell for the determination of the Gibbs energy of formation of NiO:

$$Pt \mid Ni, NiO \mid\mid ZrO_2 (CaO) \mid Pt, air \tag{10.18}$$

The left-hand side of the cell, the working electrode, has its p_{O_2} fixed by the Ni + NiO equilibrium pressure, while on the right-hand side the reference electrode has p_{O_2} given by the air atmosphere. Alternatively, a buffer may be used on the reference electrode side. The left- and right-hand side half-cell reactions are respectively

$$O^{2-} + Ni(s) = NiO(s) + 2e^- \tag{10.19}$$

$$\tfrac{1}{2}O_2(g) + 2e^- = O^{2-} \tag{10.20}$$

and the total cell reaction is:

$$Ni(s) + O_2(g) = NiO(s) \tag{10.21}$$

When a perfect ionic conductor electrolyte is used,

$$\Delta_f G_m^o(NiO) = -nFE^o = -2FE^o \tag{10.22}$$

where n is the number of electrons involved in the cell reaction and E^o is the standard potential or the electromotive force of the cell reaction. The electromotive force (EMF) should be independent of time and whether the cell temperature is approached from above and below. The EMF should furthermore be measured at zero current and the same value should be obtained when passing a small current in either direction.

The choice of electrolyte is crucial and a few are widely used for galvanic cell studies. In order to be used for thermochemical experiments, a solid electrolyte should have an ionic conductivity exceeding 10^{-4} S·m^{-1} [63] and a transference number larger than 0.99. In other words, only pure ionic fast-ion conductors are suitable for EMF measurements. Outside of the purely ionic regime of the electrolyte electron or hole conduction becomes significant, and although corrected, electrochemical potentials may be obtained if the transference numbers are known. These corrections are however complex [64, 65]. In addition, the presence of electronic conduction allows the cell reaction to proceed spontaneously, even with an open external circuit. This gives a net transfer of material between the electrodes and often precludes stable EMFs. Hence both the electrolyte and the temperature and partial pressure regimes of the measurements should be considered. The ionic conductivity for selected oxygen, silver, fluorine and proton-conducting electrolytes are shown in Figure 10.9(a). The partial pressure ranges of pure ionic conduction for some of the more usual solid electrolytes are shown in Figure 10.9(b) [66].

Among the **oxygen ion conductors**, CaO or Y_2O_3 stabilized ZrO_2 (CSZ and YSZ) [67, 68], and Y_2O_3 or La_2O_3 stabilized ThO_2 [69] are frequently used. CSZ and YSZ are limited to oxygen partial pressures in the range from 10^{-13} to 10^{10} Pa at 1273 K [68]. Lower partial pressures are allowed with the thoria-based

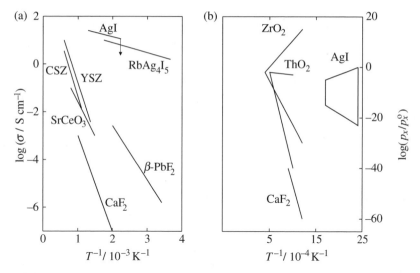

Figure 10.9 (a) Conductivity and (b) pressure range of pure (> 99%) ionic conductivity for some of the more usual solid electrolytes.

electrolytes, which are limited to the range from 10^{-20} to 10^{-11} Pa at 1273 K [69]. The electronic properties of the thoria-based electrolytes depend to a large extent on trace impurities in the particular material. The most widely used fluoride electrolyte is CaF_2 [67]. The transference number for fluorine in CaF_2 is > 0.99 for calcium activities below $6 \cdot 10^{-6}$ at 1113 K [70] and fluorine partial pressures down to 10^{-43} Pa at 1073 K and to 10^{-65} Pa at 767 K [71]. A third group of much used solid electrolytes are the β-aluminas. In Na-β-alumina [72] Na^+ is the ion-conducting species situated in oxygen-deficient layers separating the four oxide layers thick 'spinel blocks' of the structure.

Although the experimental setup is simple in principle, several factors make construction of electrochemical cells difficult: sealing of the cell, gas permeability, and materials stability and compatibility in general. Experimental cell designs have been discussed by Kleykamp [73] and Pratt [67]. In single-compartment cells, a cell stack is kept in a common protective atmosphere. Transport of the ionic conducting species via the gas phase should be insignificant; it can be obtained by choosing a reference electrode with potential similar to that of the working electrode. An electrolyte may also be used in the form of a crucible in order to increase the gas phase path distance between the electrodes. However, in cases where the difference in equilibrium pressure is large, sealing of the cell is necessary. This can be done by sealing one electrode with an alumina-based cement or with a fused gasket of a high melting glass, or by using O-rings outside the furnace.

Commonly used cells use Pt/O_2 or Pt/air as reference electrodes. At very low partial pressures of oxygen, care must be taken to avoid direct permeation of oxygen through stabilized zirconia from the air (or reference electrode) [74, 75]. The effect may be avoided by applying reference electrodes with activity near that observed at the working electrode. A well-defined buffer system like a metal–metal oxide or a metal–metal fluoride mixture is one solution to the problem. The thermodynamic properties of these buffer systems are well known over extended temperature ranges. Direct permeation of oxygen through the electrolyte may also be avoided by controlled continuous adjustment of activity at the reference electrodes. By containing the sample in the inner of two concentric zirconia tubes, the electrodes on the outer tube may be used to measure and control the oxygen partial pressure between the two tubes and the reference electrode potential [76].

It is also important to consider the materials' stability and compatibility. If interfacial reactions between the electrode and electrolyte are known to occur, direct contact between the electrolyte and the electrode must be avoided by using, for example, a compatible intermediate material. Liquid silver has been used to separate an $Fe-O-SiO_2$ electrode from the ZrO_2 electrolyte [77]. Correspondingly, the choice of reference and working electrode is important and knowledge of the phase equilibria in the system of interest is essential for a valid interpretation of electrochemical cell measurement. The species present at the working electrode should be coexisting phases in the system of interest. Care must be taken since the coexisting phases may change with temperature. In the Fe–Mo–O system Fe and Mo_2Fe_3 are in equilibrium with $Fe_2Mo_3O_8$ below 1189 K, and with Fe_2MoO_4 above 1189 K; see [78].

Many materials have been used as electrical connectors. By far the most common is Pt, but Pt forms very stable intermediate phases with actinides and lanthanides, and in these cases W or Mo electrical connectors have been used. Correspondingly, W may react with refractory oxides like Rh_2O_3 forming WO_2, and an intermediate layer of Rh has been proposed to prevent this [73].

The activity of the mobile species in a closed electrolyte cell can be controlled [79, 80]. Ions can be pumped in or out of a closed compartment and the change in composition determined accurately. The composition of a non-stoichiometric compound can therefore be changed in small steps and both the composition and the activity of the species are simultaneously determined. Detailed information about the compositional dependence of the partial Gibbs energy under isothermal conditions can thus be obtained by this technique, termed **coulometric titration**. The technique is most commonly used to study phases with properties that are highly dependent on small changes in stoichiometry. Early examples are studies of sulfides, selenides and tellurides [80, 81], while more recent ones include careful studies of the partial Gibbs energy of oxygen in $U_{1-y}Gd_yO_{2-x}$ [82]. In order to reach equilibrium in a reasonable time the material to be studied must have a high diffusion coefficient for the conducting ion of the electrolyte. An additional advantage of the technique is that phase boundaries can be determined accurately, and this has been used for simultaneous mapping of phase equilibria and thermodynamic properties.

Vapour pressure methods

A heterogeneous phase equilibrium involving a gas phase represents a convenient way of determining the Gibbs energy of a substance. A substance may evaporate congruently:

$$AB_2(s) = AB_2(g) \tag{10.23}$$

or non-congruently

$$AB_2(s) = AB(s) + B(g) \tag{10.24}$$

Vapour pressure methods are used to determine the pressure p_i of the volatile species i in equilibrium with a solid compound with a well-defined composition. The activity can then be deduced through

$$a_i = \frac{f_i}{f_i^{\,o}} \approx \frac{p_i}{p_i^{\,o}} \tag{10.25}$$

where f_i and p_i are the fugacity and the partial pressure of the species i, and $f_i^{\,o}$ and $p_i^{\,o}$ are the fugacity and activity of the species i in its standard state. The activity is determined directly by measurement of the vapour pressure of an element or a compound at a certain temperature (static or effusion methods) or indirectly, through equilibration of the sample with a well-defined gas phase. The techniques are here considered under two main headings: effusion and equilibration methods.

The most usual **effusion methods** are based on equilibration of a substance in a Knudsen cell. A small fraction of the vapour molecules effuse through a small effusion orifice in the lid of the cell (diameter from 0.1 to 1 mm), ideally without

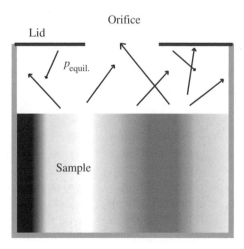

Figure 10.10 Schematic representation of a Knudsen effusion cell.

disturbing the equilibrium in the cell; see Figure 10.10. The equilibrium partial vapour pressure of species i is given by the steady state evaporation rate [83]:

$$p_i = \frac{\mathrm{d}m_i}{\mathrm{d}t} \frac{1}{Af} \sqrt{\frac{2\pi k_{\mathrm{B}}T}{M_i}} \qquad (10.26)$$

where $(\mathrm{d}m_i/\mathrm{d}t)$ and M_i are the mass rate of effusion and the molar mass of the effusing species, A is the area of Knudsen cell orifice and f is a correction factor, the Clausing factor [83]. The methods are used successfully in the vapour pressure range between 10^{-7} and 10 Pa for temperatures up to 2800 K [84].

The **mass loss technique** [83] gives the total vapour pressure only and is suitable mainly for samples that evaporate congruently, or compounds that vaporize incongruently but with one dominating vapour species. The evaporation rate is deduced from the mass loss or through collecting and analyzing the vapour effusing from the cell.

The **momentum sensor techniques** [85] are based on the force transferred from a gas to a surface on impact or recoil. Impact momentum sensors [86] are generally not very sensitive, partly because molecules simultaneously condense and revaporize from the target. Recoil-based techniques are hence preferred. In one version, the vapour pressure is deduced from the change in mass of a Knudsen cell that is observed on opening/shutting the orifice at the measuring temperature [87]. In the torsion recoil method [88] the Knudsen cell is suspended on a fibre. Two orifices are made in the cell perpendicular to the fibre and in opposite directions. The vapour pressure is deduced from the torsion force that results from the vapour effusing through the two orifices. The recoil of the anti-parallel effusing vapour twists the supporting torsion fibre to a degree determined by the elastic torsion momentum of the fibre. The deflection angle is the measurand. A third variant is based on measurement of the recoil momentum of a linear pendulum [89].

Mass spectrometry techniques are the most usual and versatile methods for analysis of the gas [90]. Here the effusing vapour is ionized by an ionization source and the product analyzed with a mass spectrometer. The different vapour species are identified and the partial pressures of all species determined. The partial pressure of species i of a compound or a solution with a specific composition is at a specific temperature:

$$p_i = k \frac{1}{\sigma_i} I_i T \qquad (10.27)$$

where k is a pressure calibration factor, σ_i is the ionization cross-section of species i, and I_i is the intensity of species i. The pressure calibration factor may be determined *in situ* by use of a twin-type Knudsen cell with the sample in one cell and the reference materials in the other or through separate experiments on the reference material [91]. For binary alloys the pure metal whose activity is measured is the natural reference. Since high temperatures may be used, the choice of the Knudsen cell material is important and reactions between the sample and the cell must be avoided. Care must also be taken to avoid fragmentation of the gas molecules on ionization [90]. The mass spectrometric analysis allows detailed thermodynamic studies of compounds where the vapour consists of more than one species, e.g. for $NaDyI_4(s)$, where the main gaseous species are $NaI(g)$ and $DyI_3(g)$ [92].

A number of techniques are based on direct measurement of the total vapour pressure in equilibrium with a compound at a given temperature, i.e. **equilibration methods**. The most usual methods are based on the use of pressure gauges covering pressures from 10^{-7} to 100 kPa [93]. Methods based on thermogravimetric determination of the mass of the vapour [94] and on atomic absorption spectroscopy have also been reported [95].

The equilibrium vapour pressures may also be determined indirectly, for example through measurement of the exact composition of a non-stoichiometric compound in equilibrium with a gas with a well-defined activity of the volatile species. While Knudsen effusion studies by mass spectrometry depends on complex and expensive instrumentation, some equilibration studies are readily performed in rather simple experimental setups. For example, the composition of a non-stoichiometric compound can be determined as a function of the vapour pressure of a volatile species such as oxygen in the case of $La_{2-x}Sr_xCuO_{4-\delta}$ by thermogravimetry [96] and the technique is complementary to coulometric titration. While only certain discrete partial pressures of oxygen in practice are feasible by thermogravimetry, oxygen permeability and materials compatibility problems in general is less of a problem.

The temperature of decomposition of carbonates of the YBCO high-temperature superconductor to oxides [97]:

$$2YBa_2Cu_3O_{6.3}(CO_2)_{0.19}(s) + 2.62CO_2(g) + 0.2O_2(g)$$
$$= 5CuO(s) + 3BaCO_3(g) + Y_2BaCuO_5(s) \qquad (10.28)$$

at different well-defined partial pressures of $CO_2(g)$ also facilitates determination of thermodynamic properties through second or third law treatments of the equilibrium pressure data. The main systematic error is often related to inadequate equilibration, and it is important that the equilibrium pressure is obtained both on decomposition (i.e. on heating) and on carbonatization (i.e. on cooling). It is often advantageous to start out with a partly decomposed sample in order to reduce nucleation problems.

A range of different methods measures the solubility of hydrogen in metals and alloys. **Manometric methods** [98] and gas volumetric methods [99] have been used to determine pressure–composition isotherms at selected temperatures for a range of alloys [100–103].

In the **isopiestic method** two condensed phases are equilibrated via the gas phase [104,105]. The composition and pressure of the gas phase are determined by use of a reference compound for which the partial pressure of a volatile component

Table 10.4 Selected experimental determinations of the standard enthalpy of formation of LaNi$_5$ at 298.15 K.

Compound	$\Delta_f H_m$ / kJ mol^{-1}	Method	Reference
LaNi$_5$	-126.3 ± 7.5	HCl solution	Semenko [107]
	-159.1 ± 8.3	HCl solution	O'Hare [108]
	-165.6 ± 10.2	Al solution	Colinet [109]
	-161.4 ± 10.8	Al solution	Colinet [110]
	-157.8 ± 18.1	Liquid reaction	Kleppa [111]

is known as a function of temperature and composition. Experiments can be performed isothermally by equilibration of one sample with the reference sample [106]. The sample is taken out and its composition determined analytically. Alternatively, several samples are equilibrated at the same time in a temperature gradient. This method is well suited for studies of non-stoichiometric compounds and alloys. Various binary and ternary systems with Zn, Cd, As, Sb or Te as volatile components have been studied [105].

In the **dew point method**, the sample is kept in an evacuated silica glass tube, which is placed in a temperature gradient [93]. The sample is contained in the hot end and the temperature of the cold end controlled to the temperature where the vapour of the volatile component just starts to condense. The activity of the volatile species in the compound is given from the dew point temperature.

Some words on measurement uncertainty

It is evident that the accuracy of an enthalpy determined by direct reaction calorimetry will depend largely on the completeness of the reaction and on the corrections made in order to take this source of systematic error into consideration. Local saturation and precipitation are similarly possible sources for systematic errors in solution calorimetry. Correspondingly, obvious and less obvious sources of systematic errors may be found for all experimental techniques. It is difficult to give a definite common uncertainty to a particular measurement technique and the uncertainty is to a large extent determined not only by the technique itself but also by the temperature of the reaction, the type of compound studied and so on. Therefore it is difficult to estimate the uncertainty of an experiment, and results obtained by different methods often do not agree within the stated uncertainties or reproducibilities. The enthalpy of formation, determined by calorimetry, of LaNi$_5$ obtained by leading scientific groups using combustion calorimetry, solution calorimetry and direct reaction calorimetry are given in Table 10.4. Four of the five data for LaNi$_5$ are equal within the stated estimate of the uncertainty. In other cases, like GeSe$_2$ and Si$_3$N$_4$, larger systematic errors in some data are inferred. The F-combustion mean value for GeSe$_2$ [112, 113] is 18.7 kJ·mol^{-1}, 22% more negative than the value obtained by direct reaction calorimetry [114]. The combined uncertainty of the F-combustion and direct reaction values is 4.8 kJ ·mol^{-1}. For β-Si$_3$N$_4$, the

Table 10.5 Selected experimental determinations of the enthalpy of transition of AgI (α-AgI = β-AgI).

$\Delta_{trs}H_m^o$ / J mol^{-1}	Year	Method	Reference
6319	1954	Adiabatic	118
6402	1957	Adiabatic	119
5920	1963	Adiabatic	120
6319	1968	Adiabatic	121
6277	1969	Drop	122
6153	1969	Adiabatic	123
6308	1989	Adiabatic	124
6302	1989	Adiabatic	124
5072	1958	Clapeyron	125
6319	1966	DTA	126
6485	1967	DTA	127
8398	1970	DTA	128
6061	1981	DSC	129
5404	1983	EMF	130

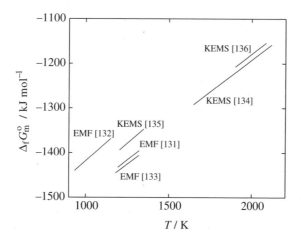

Figure 10.11 Gibbs energy of formation of BaZrO$_3$ reported in literature.

F-combustion values are 24.2 kJ·mol^{-1} [115, 116], 3% more positive than those obtained by solution calorimetry [117] and again larger than the combined estimated uncertainties.

Considerable spread is also observed in reported enthalpies of transition in single-component systems. As an example, the reported enthalpy of the first-order transition giving the fast ionic conductor phase of AgI at 420 K are compared in Table 10.5. In general, the agreement between the results obtained by adiabatic or

Table 10.6 Selected experimental determinations of the standard enthalpy of formation of $BaZrO_3$ from oxides.

Compound	$\Delta_{f,o}H_m$ / kJ mol^{-1}	Method	Reference
$BaZrO_3$	-89.3 ± 13	EMF	[131]
	-90.4 ± 0.3	EMF	[132]
	-103.9 ± 2.4	EMF	[133]
	-106.7 ± 20.9	KEMS	[134]
	-76.1 ± 0.7	KEMS	[135]
	-97.5 ± 0.8	KEMS	[136]
	-117.3 ± 3.7	HF/HNO$_3$ solution	[137]
	-119.7 ± 0.4	F-combustion	[138]

drop calorimetry is good, while considerable spread is observed for other techniques. This result is also true for enthalpy and entropy increments due to a change in temperature.

The next question is: how do the results of calorimetry, electrochemical and vapour pressure methods compare? The formation properties of $BaZrO_3$ have been extensively studied. The directly measured Gibbs energy of the reaction

$$BaO(s) + ZrO_2(s) = BaZrO_3(s) \qquad (10.29)$$

is given in Figure 10.11. Some data are obtained by electrochemical measurements [131–133] and others by Knudsen effusion mass spectrometry [134–136]. Although deconvolution of directly measured Gibbs energies to enthalpic and entropic contributions is in general difficult, the agreement with two sets of calorimetric determinations [137, 138] is reasonable; see Table 10.6.

10.4 Volumetric techniques

A number of different techniques are used for density measurements and the method of choice will depend largely on the physical and chemical properties of the material to be studied [139]. The **Archimedes method** is often used. For solid materials the volume can be derived by measuring the weight difference between a solid body in air and immersed in a liquid with known density. The apparent loss of weight yields the volume when using large single crystals, but also for polycrystalline materials that do not contain closed porosity. The method may also be applied to measure the density of liquids. In this case a solid body with known density and volume is immersed in the liquid in question and the weight difference between the solid body in air and immersed in the liquid is measured.

In **pycnometry**, the sample is weighted in a calibrated pycnometer before and after this is filled with a liquid of known density. The volume of the sample and thus its mass is determined. Powders may be used, and it is often advantageous to use a liquid with low surface tension, since this facilitates the removal of gas bubbles that frequently occur on filling the liquid.

The **gradient tube technique** is based on immersion of the solid in a liquid with a gradient in density. When one liquid is placed on another of higher density, a linear density gradient develops near the interface. If convection is not allowed the diffusion is taking place very slowly and the gradient remains virtually constant for long periods of time (of the order of months). A crystal introduced into the tube will sink until it reaches a level of density corresponding to its own density. At this level the crystals remain stationary.

Alternatively, the density of a crystal can be determined indirectly by **diffraction** [140]. Here the unit cell, defined as the smallest repeating unit which shows the full symmetry of the crystal structure, is determined from the observed diffraction pattern. The density is subsequently obtained from the unit cell volume by taking into consideration the number of formula units in the unit cell. X-ray diffraction is frequently used also to determine the density or rather the molar volume as a function of temperature. This implicitly gives the isobaric expansivity, also termed the thermal volume expansion coefficient, α. Accurate data are only obtained through accurate temperature determination. Temperature can be monitored either directly by placing a thermocouple in the vicinity of the sample or by mixing the sample with an inert standard for which the unit cell volume is known as a function of temperature. High-temperature X-ray diffraction to above 1500 K is readily performed using both commercial diffractometers and in large-scale X-ray synchrotron facilities. Diffraction is also frequently used to measure the volume change connected with first-order phase transitions through careful measurement of the unit cell volume of the phases above and below the phase transition temperature.

Besides high-temperature X-ray diffraction, **dilatometry** is a common technique for determination of the isobaric expansivity and for volumetric changes of phase transition [141]. In this technique one may use a single crystal or a polycrystalline sample with near theoretical density (negligible porosity). While the linear expansion along each of the unit cell axes can be determined in the case of single crystals, only the average linear expansion is obtained for polycrystalline materials. The material is typically heated at a constant heating rate and the elongation of the sample is measured as a function of temperature. A dilatometer is usually calibrated with regard to expansion using the volumetric properties of known substances such as sapphire, while temperature is calibrated by using accurately known phase transition temperatures such as the melting temperatures of Au or Ag. Dilatometers that allow good control of the sample atmosphere may also be used to study the volumetric properties (isothermally) as a function of the chemical potential of a volatile species such as oxygen. At high temperatures reactions between sample and container may lead to serious systematic errors.

Other methods are obviously needed for liquids. In the simplest approach the thermal expansivity is derived by measurements of the density as a function of temperature. It is then necessary to correct for the thermal expansion of the solid body used in the measurements. The Archimedes method is difficult to apply in cases when the vapour pressure above the liquid becomes substantial. In these cases the liquid can be contained in a closed container (e.g. of quartz), where the liquid is allowed to expand into a narrow cylinder. This technique allows for direct reading of the liquid volume as a function of temperature [142]. The volume of the container is calibrated before the measurement by using a liquid with known density. At high temperatures the main obstacle to proper measurements of volumetric properties of liquids is the lack of inert container materials.

Several methods are also available for determination of the isothermal compressibility of materials. High pressures and temperatures can for example be obtained through the use of diamond anvil cells in combination with X-ray diffraction techniques [10]. κ_T is obtained by fitting the unit cell volumes measured as a function of pressure to an equation of state. Very high pressures in excess of 100 GPa can be obtained, but the disadvantage is that the compressed sample volume is small and that both temperature and pressure gradients may be present across the sample.

The compressibility may alternatively be measured by **shock compression** [143] or **sound velocity experiments** [144]. In the latter, acoustic sound is propagated through a medium by longitudinal waves with a wave velocity that is related to the compressibility and density of the material. In the former case a shock wave is produced by, for example, a piston propelled toward the sample. The shock wave travels through the material. Behind the shock wave the material is compressed and a series of subsequent shock waves having increasingly higher velocity is generated. This train of shock waves eventually coalesces into a single wave front that proceeds at the speed of sound in the material. Application of the principles of conservation of mass, momentum and energy across the wavefront relates the characteristics of the waves to the thermodynamic variables of interest, pressure, volume and internal energy through the Hugoniot equations [144].

References

[1]   H. Preston-Thomas, *Metrologia*. 1990, **27**, 3.
[2]   *Supplementary Information for the International Temperature Scale of 1990*. Paris: Bureau International des Poids et Mesures, 1990.
[3]   S. Stølen and F. Grønvold, *J. Chem. Thermodyn.* 1999, **31**, 379.
[4]   *Techniques for Approximating the International Temperature Scale of 1990*. Paris: Bureau International des Poids et Mesures, 1990.
[5]   W. W. Wendlandt, *Thermal Analysis*, 3rd edn. Chichester: John Wiley, 1986.
[6]   D. D. L. Chung, *X-ray Diffraction At Elevated Temperatures: a Method for in Situ Process Analysis*. Chichester: John Wiley, 1993.
[7]   F. R. Boyd and J. L. England, *J. Geophys. Res.* 1960, **65**, 741.
[8]   J. M. Besson, in *High Pressure Techniques in Chemistry and Physics: A Practical Approach* (W. B. Holtzapfel and N. S. Isaacs eds.). Oxford: Oxford University Press, 1997, p. 1.

[9] H. K. Mao and R. J. Hemley, in *Ultrahigh Pressure Mineralogy: Physics and Chemistry of the Earth's Deep Interior* (R. J. Hemley ed.). Reviews in Mineralogy, Mineralogical Society of America, vol. 37, 1998, p. 1.

[10] H. K. Mao, R. J. Hemley and A. L. Mao, in *High Pressure Science and Technology* (S.C. Schmidt, J. W. Shaner, G. A. Samara and M. Ross eds.). New York: American Institute of Physics, 1994, p. 1613.

[11] J. M. Besson, G. Hamel, T. Grima, R. J. Nelmes, J. S. Loveday, S. Hull, and D. Häusermann, *High Press. Res.* 1992, **8**, 625.

[12] H. K. Mao, P. M. Bell, J. Shaner and D. Steinberg, *J. Appl. Phys.* 1978, **49**, 3276.

[13] P. M. Bell, J. Xu and H. K. Mao, in *Shock Waves in Condensed Matter* (Y. Gupta ed.). New York: Plenum Press, 1986, p. 125.

[14] E. D. West and E. F. Westrum Jr, in *Calorimetry of Non-Reacting Systems* (J.P. McCullough and D. W. Scott eds.). New York: Plenum Press, 1968–1975, p. 333.

[15] F. Grønvold, *Acta Chem. Scand.* 1967, **21**, 1695.

[16] S. Stølen, R. Glöckner and F. Grønvold, *J. Chem. Thermodyn.* 1996, **28**, 1263.

[17] M. Sorai, K. Kaji and Y. Kaneko, *J. Chem. Thermodyn.* 1992, **24**, 167.

[18] M. J. Richardson, in *Compendium of Thermophysical property Measurement Methods. I. Survey of Measurements Techniques* (K. G. Maglic, A. Cezairliyan and V. E. Peletsky eds.). New York: Plenum Press, 1984, p. 669.

[19] G. Höhne, W. Hemminger and H.J. Flammersheim, *Differential Scanning Calorimetry: An Introduction for Practitioners*. Berlin: Springer-Verlag, 1996.

[20] E. Calvet and H. Prat, *Recent Progress in Microcalorimetry*. London: Pergamon, 1963.

[21] E. Gmelin and S. M. Sarge, *Pure & Appl. Chem.* 1995, **67**, 1789.

[22] S. Stølen and F. Grønvold, *Thermochimica Acta* 1999, **327**, 1.

[23] T. B. Douglas and E. G. King, in *Experimental Thermodynamics, Volume. I, Calorimetry of Non-Reacting Systems* (J. P. McCullough and D. W. Scott eds.). New York: Plenum Press, 1968–1975, p. 293.

[24] D. A. Ditmars, in *Compendium of Thermophysical Property Measurement Methods 1. Survey of Measurement Techniques* (K. G. Maglic, A. Cezairliyan and V. E. Peletsky eds.). New York: Plenum Press, 1984, p. 527.

[25] V. Ya. Chekhovskoi, in *Compendium of Thermophysical Property Measurement Methods 1. Survey of Measurement Techniques* (K. G. Maglic, A. Cezairliyan and V. E. Peletsky eds.). New York: Plenum Press, 1984, p. 555.

[26] A. Cezairliyan, in *Compendium of Thermophysical Property Measurement Methods 1. Survey of Measurement Techniques* (K. G. Maglic, A. Cezairliyan and V. E. Peletsky eds.). New York: Plenum Press, 1984, p. 643.

[27] Ya. A. Kraftmakher, in *Compendium of Thermophysical Property Measurement Methods 1. Survey of Measurement Techniques* (K. G. Maglic, A. Cezairliyan and V. E. Peletsky eds.). New York: Plenum Press, 1984, p. 591.

[28] I. Hatta, *Thermochimica Acta.* 1997, **300**, 7.

[29] M. Reading, D. Elliott and V. L. Hill, *J. Thermal Anal.* 1993, **40**, 949.

[30] A. King and H. Grover, *J. Appl. Phys.* 1941, **12**, 557.

[31] A. C. MacLeod, *Trans. Faraday Soc.* 1967, **63**, 289.

[32] Y. Kanke and A. Navrotsky, *J. Solid State Chem.* 1998, **141**, 424.

[33]   E. H. P. Cordfunke and W. Ouweltjes, in *Experimental Thermodynamics, Volume IV, Solution calorimetry* (K. N. Marsh and P. A. G. O'Hare eds.). Blackwell Scientific Publications, 1994, p. 25.

[34]   A. Navrotsky, *Phys. Chem. Minerals.* 1997, **24**, 222.

[35]   O. J. Kleppa, *J. Phase Equil.* 1994, **15**, 240.

[36]   C. Colinet and A. Pasturel, in *Experimental Thermodynamics, Volume IV. Solution Calorimetry* (K. N. Marsh and P. A. G. O'Hare eds.). Blackwell Scientific Publications, 1994, p. 89.

[37]   L. Topor, A. Navrotsky, Y. Zhao and D. J. Weidner, *EOS Trans Am Geophys. Union.* 1992, **73**, 300.

[38]   L. B. Ticknor and M. B. Bever, *J. Metals.* 1952, **4**, 941.

[39]   O. J. Kleppa, *J. Amer. Chem. Soc.* 1955, **59**, 175.

[40]   T. Yokokawa and O. J. Kleppa, *J. Phys. Chem.* 1964, **68**, 3246; *Inorg. Chem.* 1965, **3**, 954.

[41]   O. J. Kleppa, *Colloq. Intern. CNRS.* 1972, No. 201, 119.

[42]   A. Navrotsky, *Phys. Chem. Minerals.* 1977, **2**, 89.

[43]   R. Boom, *Scr. Metall.* 1974, **8**, 1277.

[44]   A. Navrotsky and C. Capobianco, *Am. Mineral.* 1987, **72**, 782.

[45]   S. Fritsch, J. E. Post and A. Navrotsky, *Geochim. Cosmochim. Acta.* 1997, **61**, 2613.

[46]   L. Topor and O. J. Kleppa, *J. Chem. Thermodynam.* 1984, **16**, 993.

[47]   P. A. G. O'Hare, in *Energetics of Stable Molecules and Reactive Intermediates* (M. E. Minas da Piedade ed.). New York: Kluwer Academic, 1999, p. 55.

[48]   S. Sunner and M. Månsson (eds.), *Experimental Chemical Thermodynamics, Volume 1: Combustion Calorimetry.* Oxford: Pergamon Press, 1979.

[49]   P. A. G. O'Hare and L. A. Curtiss, *J. Chem. Thermodyn.* 1995, **27**, 643.

[50]   G. L. Humphrey, *J. Am. Chem. Soc.*, 1951, **73**, 2261.

[51]   V. Ya. Leonidov and P. A. G. O'Hare, *Pure Appl. Chem.* 1992, **64**, 103.

[52]   P. A. G. O'Hare, *Pure & Appl. Chem.* 1999, **71**, 1243.

[53]   P. A. G. O'Hare, I. Tomaszkiewicz and H. J. Seifert, *J. Mater. Res.* 1997, **12**, 3203.

[54]   G. K. Johnson, *J. Chem. Thermodyn.* 1979, **11**, 483.

[55]   S. Elder, F. J. DiSalvo, L. Topor and A. Navrotsky, *Chem. Mater.* 1993, **5**, 1545.

[56]   L. Rørmark, A. B. Mørch, K. Wiik, S. Stølen, and T. Grande, *Chem. Mater.* 2001, **13**, 4005.

[57]   O. J. Kleppa, *Ber. Bunsenges. Phys. Chem.* 1983, **87**, 741.

[58]   Y. Sakamoto, M. Imoto, K. Takai, T. Yanaru and K. Ohshima, *J. Phys.: Cond. Matter* 1996, **8**, 3229.

[59]   V. Pagot, C. Picard, P. Gerdanian, R. Tetot and C. Legros, *J. Chem. Thermodyn.* 1998, **30**, 403.

[60]   G. Borzone, G. Cacciamani and R. Ferro, *Metall. Trans.* 1991, **22A**, 2119.

[61]   J. C. Gachon, J. Charles and J. Hertz, *CALPHAD* 1985, **9**, 29.

[62]   K. Kiukkola and C. Wagner, *J. Electrochem. Soc.* 1957, **104**, 379.

[63]   H. Schmalzried, in *Metallurgical Chemstry, Proc. Symp. NPL, Teddington, England* (O. Kubaschewski ed.). London: HMSO, 1972, p. 39.

[64]   H. Schmalzried, *Z. Phys. Chem. NF.* 1963, **38**, 87.

[65]   H. Schmalzried, *Ber. Bunsenges. Phys. Chem.* 1962, **66**, 572.

[66]   I. Katayama and Z. Kozuka, *Nippon Kinzoku Gakkai Kaiho.* 1986, **25**, 528.

[67]   J. N. Pratt, *Met. Trans.* 1990, **21A**, 1223.

[68]   H. Schmalzried, *Z. Phys. Chem.* 1960, **25**, 178.

[69]   J. W. Patterson, *J. Electrochem. Soc.* 1971, **118**, 1033.

[70]   C. Wagner, *J. Electrochem. Soc.* 1968, **115**, 933.

[71]   J. W. Hinze and J. W. Patterson, *J. Electrochem. Soc.* 1973, **120**, 96.

[72]   Y. F. Y. Yao and J. T. Kummer, *J. Inorg. Nucl. Chem.* 1967, **29**, 2453.

[73]   H. Kleykamp, *Ber. Bunsenges. Phys. Chem.* 1987, **87**, 777.

[74]   J. Foultier, P. Fabry and M. Kleitz, *J. Electrochem. Soc.* 1976, **123**, 204.

[75]   A. M. Anthony, J. F. Baumard and J. Corish, *Pure Appl. Chem.* 1984, **56**, 1069.

[76]   W. Piekarczyk, W. Weppner and A. Rabenau, *Z. Naturforsch.* 1979, **34A**, 430.

[77]   T. Oishi, T. Goto, Y. Kayahara, K. Ono and J. Moriyama, *Metall. Trans.* 1982, **13B**, 423.

[78]   H. Kleykamp and V. Schauer, *J. Less-Common Met.* 1981, **81**, 229.

[79]   C. Wagner, *Progr. Solid State Chem.* 1977, **6**, 1.

[80]   C. Wagner, *J. Chem. Phys.* 1953, **21**, 1819.

[81]   K. Kiukkola and C. Wagner, *J. Electrochem. Soc.* 1957, **104**, 308.

[82]   A. Nakamura, *Z. Phys. Chem.* 1998, **207**, 223.

[83]   M. Knudsen, *Ann. Phys.* 1909, **28**, 999; 1909, **29**, 179.

[84]   K. Hilpert, *Structure and Bonding.* 1990, **73**, 97.

[85]   J. Tomiska, in *Thermochemistry of Alloys* (H. Brodowsky and H.-J. Schaller eds.). Dordrecht: Kluwer Academic, 1989, p. 247.

[86]   J. L. Margrave, *The Characterization of High-Temperature Vapors*, New York: Wiley, 1967.

[87]   J. L. Margrave, *J. Chem. Phys.* 1957, **27**, 1412.

[88]   M. Volmer, *Z. Phys. Chem.* 1931, **156A**, 863.

[89]   J. Tomiska, *J. Phys. E: Sci. Instrum.* 1981, **14**, 420; *Rev. Sci. Instrum.* 1981, **5**, 750.

[90]   K. Hilpert, *Rap. Commun. Mass Spectr.* 1991, **5**, 175.

[91]   A. Neckel, in *Thermochemistry of Alloys* (H. Brodowsky and H.-J. Schaller eds.). Dordrecht: Kluwer Academic, 1989, p. 221.

[92]   K. Hilpert, M. Miller, H. Gerads and B. Saha, *Ber. Bunsenges. Phys. Chem.* 1990, **94**, 35.

[93]   A. N. Nesmayanov, *Vapour Pressure of the Chemical Elements*. Amsterdam, New York: Elsevier, 1963; T. Boublik, F. Vojtech and H. Eduard, *The Vapor Pressure of Pure Substances*. Amsterdam, New York: Elsevier Scientific, 1973.

[94]   Y. G. Sha and H. Wiedemeier, *J. Electron Mater.* 1990, **19**, 159; *J. Electron Mater.* 1990, **19**, 761.

[95]   E. Samuelsson and A. Mitchell, *Met. Trans.* 1992, **23B**, 805.

[96]   H. Kanai, J. Mizusaki, H. Tagawa, S. Hoshiyama, K. Hirano, K. Fujita, M. Tezuka and T. Hashimoto, *J. Solid State Chem.* 1997, **131**, 150.

[97]   B. E. Vigeland, *Dissertation*, Institutt for Uorganisk Kjemi, Norges Tekniske Høgskole, Universitetet i Trondheim, Norway, 1996, nr. 83.

[98]   M. Mrowietz and A. Weiss, *Ber. Bunsenges. Phys. Chem.* 1985, **89**, 49.

[99]   R. Kadel and A. Weiss, *Ber. Bunsenges. Phys. Chem.* 1978, **82**, 1290.

[100] Y. Sakamoto, K. Ohira, M. Kokubu and T. B. Flanagan, *J. Alloys Comp.* 1997, **253**, 212.

[101] Y. Sakamoto, F. L. Chen, M. Ura and T. B. Flanagan, *Ber. Bunsenges. Phys. Chem.* 1995, **99**, 807.

[102] S. Ramaprabhu, R. Leiberich and A. Weiss, *Z. Phys. Chem. NF.* 1989, **161**, 83.

[103] R. Sivakumar, S. Ramaprabhu, K. V. S. Rama Rao, H. Anton and P. C. Schmidt, *J. Alloys Comp.* 1999, **285**, 143.

[104] D. A. Sinclair, *J. Phys. Chem.* 1933, **37**, 495.

[105] H. Ipser, *Ber. Bunsenges. Phys. Chem.* 1998, **102**, 1217.

[106] G. Wnuk, T. Pomianek, J. Romanowska and M. Rychlewski, *J. Chem. Thermodyn.* 1997, **29**, 931.

[107] K. N. Semenko, R. A. Sirotina and R. A. Savchenkova, *Russ. J. Phys. Chem.* 1979, **53**, 1356.

[108] W. N. Hubbard, P. L. Rawlins, P. A. Connick, R. E. Stedwell and P. A. G. O'Hare, *J Chem. Thermodyn.* 1983, **15**, 785.

[109] C. Colinet, A. Pasturel, A. Percheron-Guegan and J. C. Achard, *J. Less-Common Met.* 1987, **134**, 109.

[110] A. Pasturel, F. Liautaud, C. Colinet, C. Allibert, A. Percheron-Guegan and J. C. Achard, *J. Less-Common Met.* 1984, **96**, 93.

[111] S. Watanabe and O. J. Kleppa, *J. Chem. Thermodyn.* 1983, **15**, 633.

[112] P. A. G. O'Hare, *J. Chem. Thermodyn.* 1986, **18**, 555.

[113] P. A. G. O'Hare, S. Susman and K. J. Violin, *J. Non-Cryst. Solids* 1987, **89**, 24.

[114] S. Boone and O. J. Kleppa, *Thermochim. Acta.* 1992, **197**, 109.

[115] J. L. Wood, G. P. Adams, J. Mukerji and J. L. Margrave, *Third International Conference on Chemical Thermodynamics*, Baden-bei-Wien, Austria, 1973, p. 115.

[116] P. A. G. O'Hare, I. Tomaszkiewicz, C. M. Beck and H. J. Seifert, *J. Chem. Thermodyn.* 1999, **31**, 303.

[117] J. J. Liang, L. Topor, A. Navrotsky and M. Mitomo, *J. Mater. Res.* 1999, **14**, 1959.

[118] K. H. Lieser, *Z. Phys. Chem. NF.* 1954, **2**, 238.

[119] S. Hoshino, *J. Phys. Soc. Jpn.* 1957, **12**, 315.

[120] J. Nölting, *Ber. Bunsenges. Phys. Chem.* 1963, **67**, 172.

[121] C. M. Perrott and N. H. Fletcher, *J. Chem. Phys.* 1968, **48**, 2143.

[122] J. Carré, H. Pham and M. Rolin, *Bull. Soc. Chim. France.* 1969, 2322.

[123] J. Nölting and D. Rein, *Z. Phys. Chem. NF.* 1969, **66**, 150.

[124] R. Shaviv, E. F. Westrum Jr, F. Grønvold, S. Stølen, A. Inaba, H. Fujii and H. Chihara, *J. Chem. Thermodyn.* 1989, **21**, 631.

[125] A. J. Majumdar and R. Roy, *J. Phys. Chem.* 1959, **63**, 1858.

[126] K. J. Rao and C. N. R. Rao, *J. Mater. Sci.* 1966, **1**, 23.

[127] C. Berger, R. Raynaud, M. Richard and L. Eyraud, *Compt. Rend. B (Paris).* 1967, **265**, 716.

[128] M. Natarajan and C. N. R. Rao, *J. Chem. Soc. A.* 1970, 3087.

[129] B. E. Mellander, B. Baranowski and A. Lunden, *Phys. Rev. B.* 1981, **23**, 3770.

[130] N. E. Quaranta and J. C. Bazan, *Solid State Ionics.* 1983, **11**, 71.

[131] V.A. Levitskii, Yu. Ya. Skolis, Yu. Khekimov and N. N. Shevchenko, *Russ. J. Phys. Chem.* 1974, **48**, 24.

[132] B. Deo, J. S. Kachhawaha and V. B. Tare, *Mater. Res. Bull.* 1976, **11**, 653.

[133] V. A. Levitskii, *J. Solid State Chem.* 1978, **25**, 9.

[134] R. Odoj and K. Hilpert, *Z. Phys. Chem. NF.* 1976, **102**, 191.

[135] S. Dash, Z. Singh, R. Prasad and D. D. Sood, *J. Chem. Thermodyn.* 1990, **22**, 557.

[136] T. Matsui, S. Stølen and H. Yokoi, *J. Nucl. Mater.* 1994, **209**, 174.

[137] M. E. Huntelaar, A. S. Booji and E. H. P. Cordfunke, *J. Chem. Thermodyn.* 1994, **26**, 1095.

[138]   A. S. L'vova and N. N. Feodosev, *Russ. J. Inorg. Chem.* 1964, **38**, 14.
[139]   F. M. Richards, *Determination of the Density of Solids,* International Tables of Crystallography, Volume C. Dordrecht: Kluwer Academic, 1999, p. 141.
[140]   A. R. West, *Solid State Chemistry and its Application.* Chichester: John Wiley & Sons, 1998.
[141]   B. Wunderlich, *Thermal Analysis.* New York: Academic Press, 1990.
[142]   L. A. King and D. W. Seegmiller, *J. Chem. Eng. Data.* 1971, **16**, 23.
[143]   T. S. Duffy and Y. Wang, in *Ultrahigh Pressure Mineralogy: Physics and Chemistry of the Earth's Deep Interior* (R. J. Hemley ed.). Reviews in Mineralogy, Mineralogical Society of America, vol. 37, 1998, p. 425.
[144]   R. C. Liebermann and B. Li, in *Ultrahigh Pressure Mineralogy: Physics and Chemistry of the Earth's Deep Interior* (R. J. Hemley ed.). Reviews in Mineralogy, Mineralogical Society of America, vol. 37, 1998, p. 459.

Further reading

E. Calvet and H. Prat, *Recent Progress in Microcalorimetry.* London: Pergamon, 1963.
P. J. Gellings and H. J. M. Bouwmeester (eds.) *The CRC Handbook of Solid State Electrochemistry.* Boca Raton, FL: CRC Press, 1996.
A. R. H. Goodwin, K. N. Marsh and W. A. Wakeham (eds.) *Measurement of the Thermodynamic Properties of Single Phases.* Amsterdam: Elsevier Science, 2003.
S. V. Gupta, *Practical Density Measurement and Hydrometry.* Bristol: Institute of Physics, 2002.
Y. M. Gupta (ed.) *Shock Waves in Condensed Matter.* New York: Plenum Press, 1986.
R. J. Hemley (ed.) *Ultrahigh Pressure Mineralogy: Physics and Chemistry of the Earth's Deep Interior.* Reviews in Mineralogy, Mineralogical Society of America, vol. 37, 1998.
W. B. Holtzapfel and N. S. Isaacs (eds.) *High Pressure Techniques in Chemistry and Physics: A Practical Approach.* Oxford: Oxford University Press, 1997.
G. W. H. Höhne, W. Hemminger and H. J. Flammersheim, *Differential Scanning Calorimetry,* 2nd edn. Berlin: Springer-Verlag, 2003.
J. P. McCullough and D. W. Scott (eds.) *Calorimetry of Non-Reacting Systems.* New York: Plenum Press, 1968.
K. D. Maglic, A. Cezairliyan and V. E. Peletsky (eds.) *Compendium of Thermophysical Property Measurement Methods.* New York: Plenum Press, 1984.
K. N. Marsh and P. A. G. O'Hare (eds.) *Solution Calorimetry.* Oxford: Blackwell Scientific, 1994.
S. C. Schmidt, J. W. Shaner, G. A. Samara and M. Ross (eds.) *High Pressure Science and Technology.* New York: American Institute of Physics, 1994.
S. Sunner and M. Månsson (eds.) *Combustion Calorimetry.* Oxford: Pergamon Press, 1979.
G. K. White and P. J. Meeson, *Experimental Techniques in Low-Temperature Physics.,* 4th edn. Oxford: Clarendon Press, 2002.

11

Thermodynamics and materials modelling

Neil L. Allan, *University of Bristol*

The previous chapter discussed *experimental* methods for determining thermodynamic properties of materials. An increasingly attractive option is to calculate these quantities directly. Thanks to both the tremendous increase in computer power – the ratio of performance to price increasing typically by an order of magnitude every five years – and the development of powerful software, theoretical prediction of thermodynamic properties may rival experimental measurement in some specific areas. Values can be obtained for properties either under conditions inaccessible to experiment (e.g. the elevated temperatures and high pressures deep in the Earth's mantle) or too dangerous for experiment (e.g. radioactive materials). In addition, molecular modelling can provide unique insights into the behaviour of the material at the atomic level, enabling us to examine the underlying reasons for the trends in properties from one material to another. Why, for example, does a particular compound adopt a particular structure rather than the myriad of other possibilities? Again, why do some materials *contract* on heating while most *expand*? Much of this book has been concerned with macroscopic thermodynamic quantities – enthalpies, entropies and so on – and what information about atoms and molecules can be obtained from these data. Here we go in the opposite direction, and obtain macroscopic quantities from the calculation of atomic properties.

There are many computational techniques available, covering many orders of magnitude of length and time-scales, as shown schematically in Figure 11.1. **Potential-based** methods depend on the use of analytical expressions for the interaction energies between the atoms in the molecule or solid under study. These are parametrized by fitting either to experiment or to the results of quantum mechanical

Published in *Chemical Thermodynamics of Materials* by Svein Stølen and Tor Grande
© 2004 John Wiley & Sons, Ltd ISBN 0 471 492320 2

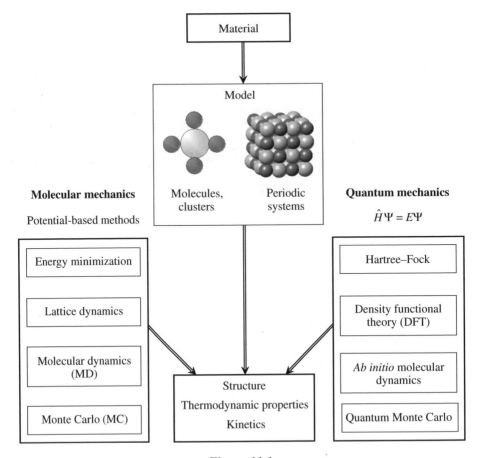

Figure 11.1

calculations. Classical **energy minimization** (also known as **molecular mechanics**), **lattice dynamics**, **molecular dynamics** and **Monte Carlo** techniques all use these potentials. **Molecular dynamics** is unique in following explicitly the evolution in time of a system over the pico- and nanosecond time-scales, by using the potentials together with Newton's laws of motion. More accurate are first principles quantum mechanical methods which solve the Schrödinger equation directly to obtain the energy of the molecule or periodic system with no recourse to interatomic potentials; but they are *much* more computationally expensive.

There is thus an inevitable trade-off between speed and accuracy. This often presents an acute problem for materials modelling, where we are normally interested in assemblies of large numbers of atoms or molecules! In particular, for glasses and other amorphous systems we must use large simulation cells to represent the long-range disorder. A further challenge is the difficulty of calculating entropies and Gibbs energies, and allowing for the effects of temperature. We are often forced to ignore atomic vibrations entirely and work in the **static limit**, which refers to a temperature of absolute zero and in the absence of vibrations.

We start in this chapter with potential-based methods, the computationally cheapest approach, which can be applied to large assemblies of molecules. We then move on to the use of quantum mechanical techniques, as used for problems involving smaller numbers of atoms. The aim is to give a brief overview of the subject and its applications, and to show what type of information can be obtained from the different methods. The reader is referred to specialist texts for fuller details.

11.1 Interatomic potentials and energy minimization

We would ideally tackle all problems in molecular-level modelling by using quantum mechanics to calculate the wavefunction and the energy of the system. However, the size of many systems is such that the computer requirements (computer time and memory) make this totally unfeasible and we have to resort to a different approach. All **potential-based** methods (often referred to in pharmaceutical modelling as **force-field** methods) take no explicit account of the electronic motion. Instead, making the Born–Oppenheimer approximation, they calculate the energy of the system as a function of the nuclear positions only. Essentially, we view the solid or liquid as comprised of a set of interacting spheres,[1] the motions of which can then be treated using the laws of *classical* rather than *quantum* physics. Of course, these methods are incapable of providing any information about bonding, charge transfer, reaction pathways or electronic properties; but such details are often not required *per se* for calculation of thermodynamic properties, and potential-based methods can sometimes provide answers to an accuracy comparable with that of the highest-level quantum mechanical methods. But such success depends crucially on the accuracy of the potentials describing the interactions between the atomic 'spheres'. If these potentials are poor, then so will be the results of the simulation.

Intermolecular potentials

In general there is *no* analytic expression for the energy of a molecule or solid in terms of the positions of the atoms. Instead we *assume* that the potential energy Φ of the system can be written as a *sum* of various interactions (Figure 11.2), as follows:

$$\Phi = \underbrace{\sum_{i<j} V_{ij}}_{\substack{\text{over} \\ \text{all pairs of} \\ \text{atoms} \\ \text{2-body potential}}} + \underbrace{\sum_{i<j<k} V_{ijk}}_{\substack{\text{over all} \\ \text{3 atom} \\ \text{combinations} \\ \text{3-body potential}}} + \underbrace{\sum_{i<j<k<l} V_{ijkl}}_{\substack{\text{over all} \\ \text{4 atom} \\ \text{combinations} \\ \text{4-body potential}}} + \dots$$

$$(11.1)$$

---

1 Einstein's comments on quantum mechanics include his belief that God does not play dice. God may have a weakness for snooker and billiards.

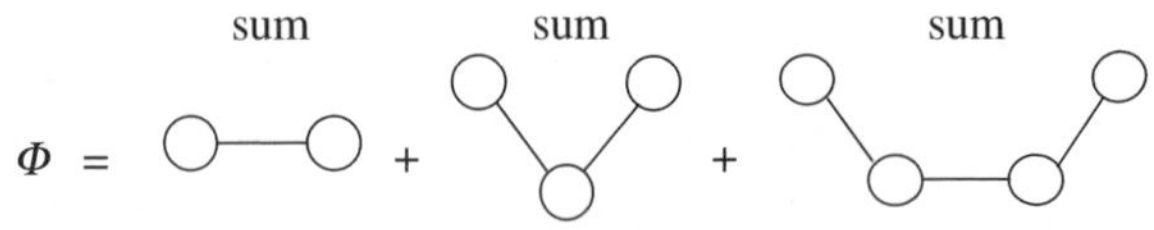

$$\Phi = \quad \overset{\text{sum}}{\bigcirc\!\!-\!\!\bigcirc} \; + \; \overset{\text{sum}}{\bigvee} \; + \; \overset{\text{sum}}{\diagup\!\!-\!\!\diagdown}$$

Figure 11.2

where each subscript label specifies a particular atom.

Consider, for example, Figure 11.3, which shows four atoms in a molecule or a solid. Using eq. (11.1) the potential energy for these four atoms is given by

$$\Phi = \underbrace{V_{12} + V_{13} + V_{14} + V_{23} + V_{24} + V_{34}}_{\text{2-body}} + \underbrace{V_{123} + V_{124} + V_{134} + V_{234}}_{\text{3-body}} + \underbrace{V_{1234}}_{\text{4-body}} \quad (11.2)$$

Extension to a molecule with more than four atoms or to a solid is straightforward. Usually the two-body terms are much larger than the three-body terms, which in turn are greater than the four-body. For ionic solids, for example, the three-body and four-body terms are often neglected. In contrast, for metals and semiconductors including only two-body terms leads to very poor results (see Sutton (Further reading)).

The next step is to assume particular functional forms for the various terms. These can be extremely elaborate, but most are usually based on a simple, chemically intuitive model of the interactions, e.g. stretching of bonds, changes in bond and torsion angles, Coulomb forces and van der Waals intermolecular interactions. Thus, for a **molecular solid** comprised of discrete molecules, we might well use

$$\Phi = \sum_{\substack{\text{bond lengths} \\ \text{in molecules}}} \frac{k_r}{2}(r - r_0)^2 + \sum_{\substack{\text{bond lengths} \\ \text{in molecules}}} \frac{k_\theta}{2}(\theta - \theta_0)^2$$

$$+ \sum_{\substack{\text{torsion angles} \\ \text{in molecules}}} \frac{V_n}{2}(1 + \cos(n\omega - \xi)) + \sum_{\substack{\text{atoms } i \text{ and } j}} \frac{q_i q_j}{4\pi\varepsilon_0 r_{ij}} \quad (11.3)$$

$$+ \sum_{\substack{\text{non-bonded} \\ \text{atoms } i \text{ and } j}} 4\varepsilon_{ij}\left\{ \left(\frac{\sigma_{ij}}{r_{ij}}\right)^{12} - \left(\frac{\sigma_{ij}}{r_{ij}}\right)^{6} \right\}$$

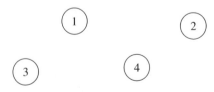

Figure 11.3

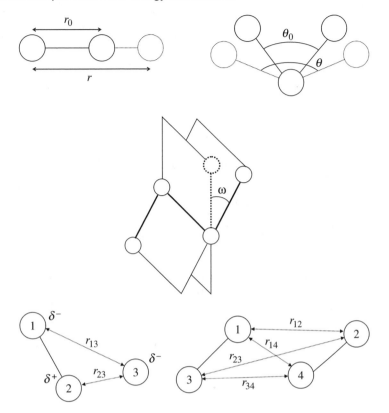

Figure 11.4 Schematic representation of the different potentials used in eq. (11.3).

The first three of the terms on the right-hand side of eq. (11.3) all involve solely **intramolecular** interactions (Figure 11.4). The first is the (two-body) interaction between pairs of directly bonded atoms in the molecules. The particular form chosen here is harmonic. The force constant for each bond is k_r, the bond length is r and r_0 is a constant (the bond length adopted by the bond when all other terms in the force field are zero). The second term is the three-body contribution, again harmonic, involving the summation over all the bond (valence) angles θ in the molecules. The third is the four-body torsion term, which changes with rotation about the various bonds in the molecules. There is a contribution from each quartet of bonded atoms A–B–C–D: ω is the torsion angle, n is a constant reflecting the periodicity of rotation (3 for ethane), and ξ is a phase factor determining the particular values of the torsion angle for which this term is a minimum.

The fourth term on the right-hand side of eq. (11.3) is the electrostatic interaction (Coulomb's law) between pairs of charged atoms i and j, separated by distance r_{ij}. Since electrostatic interactions fall off slowly with r (only as r^{-1}) they are referred to as **long-range** and, for an infinite system such as a periodic solid, special techniques, such as the Ewald method, are required to sum up all the electrostatic interactions (cf. Section 7.1) (see e.g. Leach, Jensen (Further reading)). The

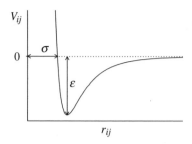

Figure 11.5 The Lennard-Jones 6–12 potential (the fifth term on the right-hand side of eq. (11.3)).

final term represents the van der Waals interactions between pairs of non-bonded atoms, both those in different molecules (intermolecular forces) and also those in the same molecule. In eq. (11.3) we have chosen the well-known Lennard-Jones functional form for this van der Waals term (Figure 11.5), in which the attractive part representing the dispersion interactions varies as r^{-6} and a short-range repulsive part (Pauli repulsion) varies as r^{-12}.

For **ionic solids**, where directional bonding is largely absent, three- and four-body terms are often neglected, and eq. (11.1) becomes

$$\Phi = \sum_{\substack{i<j \\ \text{over} \\ \text{all pairs of ions}}} V_{ij} \tag{11.4}$$

where the interaction V_{ij} between any pair of ions i and j is given by the sum of the electrostatic interaction and the van der Waals term, V_{vdw}:

$$V_{ij} = \frac{q_i q_j}{4\pi\varepsilon_0 r_{ij}} + V_{vdw}(r_{ij}) \tag{11.5}$$

V_{vdw} may be represented by a potential of Lennard-Jones form (as in the last term of eq. (11.3)). A common alternative is the Buckingham form:

$$V_{vdw}(r_{ij}) = A \exp(-r_{ij}/\rho) - \frac{C}{r_{ij}^6} \tag{11.6}$$

where an exponential replaces the r^{-12} term in the Lennard-Jones potential and A, ρ and C are constants. Increasingly, ionic models of this type also include extra terms to allow for the polarization (distortion) of the electron cloud around ions in low-symmetry environments; one used for many years is the **shell** model [1].

The quality of any force field and set of interatomic potentials depends crucially on the values chosen for all the parameters in the various potential functions and the assignment of charges to the individual atoms or ions. This task is far from trivial, and much care is needed. Even for a binary oxide such as CaO, using Buckingham potentials, we must assign charges to the Ca and O ions (usually +2 and −2 in agreement with the usual valence rules) and decide values for A, ρ and C for the Ca–Ca potential, the Ca–O potential and the O–O potential. One general approach is to obtain these values by fitting to observed quantities such as enthalpies of formation, experimental unit cell dimensions, bond lengths, angles, vibrational frequencies and so on, for a small set of compounds. A second approach growing considerably in importance is to fit to potential energy surfaces generated by quantum mechanical techniques. This is useful when it is necessary to ensure that the potentials are accurate over wide ranges of internuclear separations, since only a few distances may be sampled in the molecules or solids used in the fitting method. This is often crucial when simulating interfaces or defects in materials where the interatomic distances at equilibrium may be very different from those in the perfect bulk crystal. Transferability is generally a key requirement of any set of potentials, for it is often vital that a set of parameters obtained from a small number of cases can be applied to a much wider range of problems, often involving much more complex systems.

Given the potentials we can evaluate the potential energy Φ of the system for any input structure using eq. (11.1). We now turn to examine a range of techniques all of which use this quantity.

Energy minimization, molecular mechanics and lattice statics

Given an input structure for our periodic solid we calculate Φ (typically per unit cell) using the interatomic potentials. We are usually interested in minimum points on the potential energy hypersurface. Comparison of the relative energies of two or more possible structures following **energy minimization** enables us to predict the likely structure for a given material. Such energy minimization is often referred to in pharmaceutical modelling as **molecular mechanics**.

In **lattice statics** simulations all vibrational effects are neglected[2] and the internal energy of the solid U is simply equal to Φ, and the entropy is zero. Such minimizations give the crystal structure and internal energy (often referred to as the lattice energy) of the low-temperature phase. In the **static limit** at 0 K and zero pressure[3] the crystal structure is thus determined by the equation

$$\partial U / \partial Z_i = \partial \Phi / \partial Z_i = 0 \tag{11.7}$$

---

2 The inclusion of vibrational terms is dealt with later. We note in passing the fascinating example of solid He, where lattice statics is completely inappropriate. The binding forces are so weak that even at the lowest temperatures solidification occurs only at pressures of at least 2.5 MPa, and it is the zero-point vibrational energy that stabilizes the structure.

3 For all practical purposes in modelling condensed phases the difference between zero pressure and one atmosphere can be ignored. At zero pressure, also, U and H are equal.

where the Z_i are **all** the variables that define the structure, namely the lengths of the three lattice vectors, the angles between these vectors, and the positions of the atoms in the unit cell.

Considerable effort has been made to develop efficient algorithms for quick and efficient minimization; there is a vast literature on the subject. Minimization methods are divided into two classes – those that use derivatives of the energy with respect to the variables defining the structure (useful for providing information about the shape of the energy surface and thus enhancing the efficiency of the minimization), and those that do not. Considerable care is often needed in the choice of minimizer.

A typical example of energy minimization using interatomic potentials is a study [2] of the ternary fluorides AMF_3 (A = Li^+–Cs^+, M = Mg^{2+}–Ba^{2+}). Not all these compounds have been reported experimentally. The computational study was based on structures adopted by AMO_3 oxides, since the oxide and fluoride ions have similar ionic radii. The possible structures, shown in Figure 11.6, fall into two classes:

1. The first arises when A is large enough for the formation of close-packed layers AF_3, which can be stacked in various ways. The simplest such structure is the cubic perovskite (Figure 11.6) in which the AF_3 layers are cubic close-packed. Known fluorides with this structure include $KMgF_3$, $RbMgF_3$, $RbCaF_3$, $CsCaF_3$ and $LiBaF_3$. In all except the last of these the larger univalent ion is 12-coordinate occupying the position labelled A in the centre of the unit cell in Figure 11.6(a); the distance to its nearest anion neighbours is $a_0/\sqrt{2}$, where a_0 is the cubic unit cell parameter. The divalent ions occupy the position marked M and are 6-coordinate with a smaller nearest-neighbour distance of $a_0/2$. In contrast, $LiBaF_3$ has an 'inverse perovskite' structure in that the large Ba^{2+} ion is 12-coordinate and the smaller Li^+ ion 6-coordinate. Orthorhombic perovskites, in which the M–F–M bridges linking the MF_6 octahedra are not linear (Figure 11.6(b)), are also common. In the three hexagonal fluoride structures known – $RbNiF_3$, $CsCoF_3$ and $CsNiF_3$ – there are other stacking sequences of the AF_3 layers (Figure 11.6(c)).

2. In ternary *oxides* AMO_3 the second class of structures arises when A and M are the same size and the size is suitable for octahedral co-ordination. These adopt structures in which both ions are 6-coordinate. An example is the lithium niobate structure, which contains hexagonally packed anion layers (Figure 11.6(d)). Surprisingly, no known fluoride adopts such a structure.

In [2] all the possible structures listed above were considered for each ternary fluoride. The potentials for A^+–F^- and M^{2+}–F^- were exactly those derived for the binary systems AF and MF_2, all of which were based on a *single* F^-–F^- potential. These potentials were *assumed* to be transferable unchanged to the ternary fluorides.

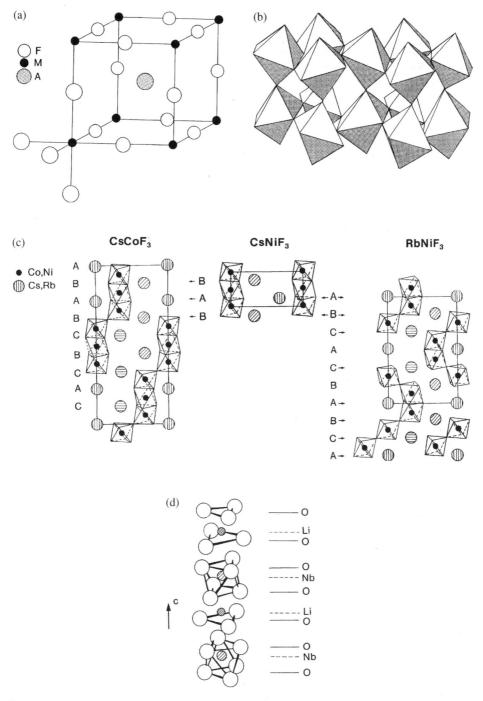

Figure 11.6 AMF$_3$ crystal structures. (a) 'Ideal' cubic perovskite structure. (b) Tilting of MX$_6$ octahedra in orthorhombically distorted AMF$_3$ perovskites. (c) RbNiF$_3$, CsCoF$_3$ and CsNiF$_3$ crystal structures. (d) Crystal structure of lithium niobate.

Table 11.1 Calculated low-temperature phases of AMF_3 compounds. The * denotes compounds for which there are experimental crystallographic data.

Ion A	Ion M			
	Mg^{2+}	Ca^{2+}	Sr^{2+}	Ba^{2+}
Li^+	$LiNbO_3$	Orthorhombic (Inverse)	Orthorhombic (Inverse)	Cubic* (Inverse)
Na^+	Orthorhombic*	$LiNbO_3$	Orthorhombic (Inverse)	Orthorhombic (Inverse)
K^+	Cubic*	Orthorhombic*	$LiNbO_3$	$LiNbO_3$
Rb^+	$RbNiF_3$*	Cubic*	Orthorhombic	$LiNbO_3$
Cs^+	$CsNiF_3$*	Cubic*	Cubic	Orthorhombic

Table 11.1 lists the resulting low-temperature phases calculated for this set of compounds. Where experimental data are available (marked with a star) the predicted structures are those observed at low temperatures. 'Inverse' denotes a perovskite structure in which a large divalent ion is 12-coordinate and a smaller univalent ion 6-coordinate. Unit cell dimensions are predicted to within 1% of the measured values.

Five of the compounds in which the univalent and divalent cations are of comparable size ($LiMgF_3$, $NaCaF_3$, $KSrF_3$, $KBaF_3$ and $RbBaF_3$) are predicted to adopt the lithium niobate structure, in agreement with simple ion size arguments. Lithium niobate itself is an important ferroelectric material, so the question of a possible fluoride analogue is of particular interest.

To investigate this further the enthalpies of formation $\Delta_{f,flu}H$ of all the ternary fluorides from the binary fluorides,

$$AF + MF_2 \rightarrow AMF_3 \tag{11.8}$$

were calculated using the lattice energies obtained for each compound from lattice statics minimizations. The calculated values of ΔH for all the known ternary fluorides marked with a star in Table 11.1 are negative, with the exception of $KCaF_3$ which is small and positive. Of those that are apparently unknown, the enthalpies of formation of $LiMgF_3$ and $CsSrF_3$ from the binary fluorides are calculated to be negative, while $LiCaF_3$ has a positive value close to that of $KCaF_3$. The remaining systems $NaCaF_3$, $KSrF_3$ and $KBaF_3$, which are predicted to have the lithium niobate structure, all have large positive enthalpies of formation from binary fluorides. These values of $\Delta_{f,flu}H$ suggest in general why fluorides with the lithium niobate structure have not been reported, but leave open the tantalizing question of why $LiMgF_3$, also predicted to adopt this structure but with a negative enthalpy of formation from LiF and MgF_2, is unknown.

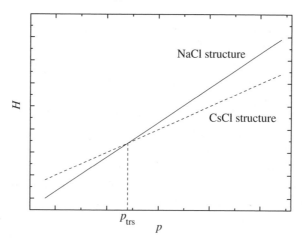

Figure 11.7 H vs. p for the NaCl and CsCl phases of CaO.

High pressure

Thermodynamic properties at high pressures are of great interest for instance to Earth scientists who wish to understand the behaviour of the Earth's mantle, where pressures reach 100 GPa. To carry out energy minimizations in the static limit at non-zero pressures we minimize the enthalpy $H = U + pV$ with respect to all the variables that define the structure, where p is the applied pressure and V the volume. When p is zero we regain eq. (11.7).

Figure 11.7 shows schematically the resulting calculated variation of H with p for the NaCl-type and the CsCl-type phases of CaO. The NaCl-type structure, which is stable at low pressures, is the rock salt structure in which the Ca and O atoms are 6-coordinate. In the CsCl structure, stable at high pressures, both cation and anion are 8-coordinate. In the static limit where the entropy is set to zero, the thermodynamically most stable phase at any pressure is that with the lowest value of H; at the thermodynamic transition pressure, p_{trs}, the enthalpies of the two phases are equal. For CaO the particular set of potentials used in Figure 11.7 indicates a transition pressure of 75 GPa between the NaCl-type and CsCl-type structures, which compares with experimental values in the range 60–70 GPa.

Changes in phase have important consequences for other thermodynamic properties and thus geophysical implications. For example, the bulk modulus at any pressure p in the static limit is given by the value of $V(d^2U/dV^2)$ (at that pressure); for CaO this increases markedly across the phase boundary.

Elevated temperatures and thermal expansion: Helmholtz, Gibbs energies and lattice dynamics

How can these calculations be extended to finite temperatures? How can we calculate, for example, how a material expands with temperature? Temperature can be included in simulations in several ways. Two of these, Monte Carlo and molecular dynamics, will be deferred until the next section. They avoid the *direct* calculation of the Helmholtz energy, which is much more challenging than that of the internal energy. Helmholtz and Gibbs energies cannot be determined accurately from Monte Carlo and molecular dynamics simulations because they do not sample adequately high-energy regions of phase space which make important contributions to these energies. For periodic solids, particularly at temperatures some way below the melting temperature, a valuable alternative is the use of **lattice dynamics** to calculate vibrational frequencies and hence, combined with the lattice statics contributions, to give absolute Helmholtz and Gibbs energies and their various derivatives directly. Such calculations are computationally much more expensive than the static energy minimizations, but are still nevertheless orders of magnitude cheaper than molecular dynamics or Monte Carlo simulations, which require long runs for similar precision.

The equilibrium structure at applied pressure p and temperature T can be found by minimizing the Gibbs energy, G, given by

$$G = U - TS + pV \tag{11.9}$$

simultaneously with respect to all the variables that define the structure. At zero pressure this reduces to minimizing the Helmholtz energy A:

$$A = U - TS \tag{11.10}$$

The Helmholtz energy thus plays a key role. In the **quasiharmonic** approximation it is assumed that at temperature T this can be written as the sum of static and vibrational contributions

$$A = \Phi_{\text{stat}} + A_{\text{vib}} \tag{11.11}$$

where Φ_{stat} is the potential energy of the static lattice calculated in the previous section and A_{vib} the vibrational contribution obtained from statistical thermodynamics and the partition function for the simple harmonic oscillator (see e.g. Atkins and de Paula (Further reading)). This is given by

$$A_{\text{vib}} = \sum_{q,j} \{\tfrac{1}{2} h v_j(q) + k_B T \ln[1 - \exp(-h v_j(q) / k_B T)]\} \tag{11.12}$$

where the $v_j(q)$ are the vibrational (phonon) frequencies for wave vector q (see Chapter 8). The first term on the right of eq. (11.12) is the **zero point energy**.

The vibrational frequencies are obtained from the force-field using the second derivatives of the potential energy with respect to displacements of the atoms (in a more elaborate version of the argument used in Section 8.2 for the one-dimensional chain); the calculation is analogous to the calculation of normal mode frequencies for molecules. The resulting vibrational frequencies can be compared with those obtained experimentally by inelastic neutron scattering, and infrared and Raman spectroscopy. In practice for an ionic solid, it is essential to include ionic polarization using, for example, the shell model; otherwise the frequencies are too high. We must sum over sufficient wave vectors q in the Brillouin zone to ensure convergence in the desired thermodynamic property. For efficient minimization of the Helmholtz energy we also require its derivatives, as obtained from the derivatives of the vibrational frequencies with respect to the variables that define the structure; the calculation of these can be a formidable task.

Given A_{vib} and Φ_{stat}, we thus obtain the equilibrium structure (and the corresponding volume) at any pressure and temperature, and the corresponding values of the Gibbs and Helmholtz energies. The quasiharmonic approximation implicit in eqs. (11.11) and (11.12) usually holds at low temperatures, and often up to one-half or two-thirds of the melting temperature. Above this the vibrations become strongly anharmonic and eq. (11.12) breaks down.

Given A, V and T (or G, p and T), all other thermodynamic properties can be obtained by appropriate thermodynamic manipulations. For example, given the variation of volume with temperature (at given pressure) it is straightforward to calculate the isobaric expansivity α given by

$$\alpha = \frac{1}{V}\left(\frac{\partial V}{\partial T}\right)_p \tag{11.13}$$

By way of example, Figure 11.8 shows a comparison [3] of calculated and experimental expansion coefficients for MgF_2, which has the rutile structure.

Furthermore, this type of simulation provides a convenient route to the calculation of key quantities such as the entropies and heat capacities discussed throughout this book. Since $S = -(\partial A/\partial T)_V$, and $C_V = (\partial U/\partial T)_V$ it follows that

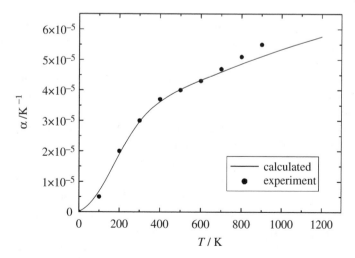

Figure 11.8 Calculated and experimental thermal expansion of MgF_2 [3].

$$S = \sum_{q,j} \frac{(h\nu_j(q)/T)}{\exp(h\nu_j(q)/k_BT)-1} - k_B \ln[1 - \exp(-h\nu_j(q)/k_BT)] \qquad (11.14)$$

and

$$C_V = \sum_{q,j} k_B \left(\frac{h\nu_j}{k_BT}\right)^2 \frac{1}{[\exp(h\nu_j/k_BT)-1][1 - \exp(-h\nu_j/k_BT)]} \qquad (11.15)$$

Entropies and heat capacities can thus now be calculated using more elaborate models for the vibrational densities of states than the Einstein and Debye models discussed in Chapter 8. We emphasize that the results are only valid in the quasiharmonic approximation and can only be as good as the accuracy of the underlying force-field; calculation of such properties can thus be a very sensitive test of interatomic potentials.

Negative thermal expansion

Not all substances expand on heating. Some contract markedly over a wide range of temperature. Lattice dynamics simulations can provide insight into this puzzle and reveal the mechanisms operating at the atomic level that are responsible. Probably the most familiar example of negative thermal expansion (negative isobaric expansivity) from everyday life is the increase in density of liquid water between $0\,°C$ and $4\,°C$, which is crucial for the preservation of aquatic life during very cold weather. The phenomenon is however found more often in solids, and interest in the subject was renewed by the dramatic discovery of Sleight and co-workers in 1996 [4] that cubic zirconium tungstate, ZrW_2O_8, contracts on heating from below 15 K up to its decomposition temperature of ≈ 1500 K. Negative thermal expansion materials with practical applications (from astronomical telescope mirrors to cooking ware) include β-eucryptite ($LiAlSiO_4$), cordierite ($Mg_2Al_2Si_5O_{18}$), β-spodumene ($Li_2Al_2O_4.nSiO_2$) and the NZP ($NaZr_2P_3O_{12}$)–CTP ($Ca_{0.5}Ti_2P_3O_{12}$) family. A useful procedure is to mix materials having negative expansion with others having positive expansion so as to generate a mixture having a net expansion of approximately zero.

The contraction of solids on heating seems anomalous because it offends the intuitive concept that atoms will need more room to move as the vibrational amplitudes of the atoms increase. However, this argument is incomplete. Figure 11.9 plots schematically the variation of A with V at two temperatures, for both positive and negative thermal expansion. The volumes marked explicitly on the V-axis give the minima of each A vs. V isotherm. These are the equilibrium volumes at temperatures T_1 and T_2 respectively ($T_2 > T_1$) and zero pressure.

It is useful to relate the expansivity to the volume dependence of the entropy. This is readily seen by manipulating the expression for the isobaric expansivity, α,

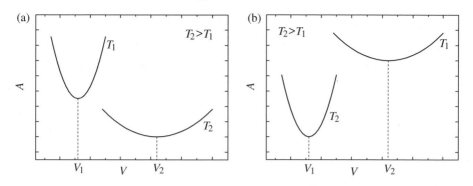

Figure 11.9 Schematic variation of Helmholtz energy A with volume V for (a) positive and (b) negative thermal expansion.

in eq. (11.13). The Maxwell relationship (Section 1.3), $(\partial V/\partial T)_p = -(\partial S/\partial p)_T$, gives

$$
\begin{aligned}
\alpha &= \frac{1}{V}\left(\frac{\partial V}{\partial T}\right)_p = -\frac{1}{V}\left(\frac{\partial S}{\partial p}\right)_T \\[2mm]
&= -\frac{1}{V}\left(\frac{\partial V}{\partial p}\right)_T\left(\frac{\partial S}{\partial V}\right)_T \\[2mm]
&= \kappa_T\left(\frac{\partial S}{\partial V}\right)_T; \quad \kappa_T = -\frac{1}{V}\left(\frac{\partial V}{\partial p}\right)_T
\end{aligned}
\tag{11.16}
$$

κ_T is the isothermal compressibility and is always positive (a thermodynamic stability condition; Section 5.1). So the change in volume on heating will always be in the direction of increasing entropy: a negative α thus indicates that the entropy increases when the substance is compressed isothermally. Entropy (disorder) would *normally* be expected to increase with volume, and usually α is indeed positive. *Only* in this *limited* sense is negative thermal expansion 'anomalous'. In an ideal gas, α is always positive; the origin of negative thermal expansion must therefore lie in the interactions between the particles. For example, the negative expansion of liquid water below 4 °C is associated with an increase of entropy on compression due to the break-up of tetrahedral H-bonding with increasing temperature, which over this temperature range more than compensates for other effects tending to decrease the entropy.

In most solids vibrations *parallel* to bond directions *decrease* in frequency as the volume increases and the entropy (eq. (11.14)) *increases* with volume; $(\partial S/\partial V)_T$ and the thermal expansion are positive. Negative thermal expansion is usually associated with more open structures where coordination numbers are low and vibrations *perpendicular* to bond directions can dominate the change in entropy with volume and thus the derivative $(\partial S/\partial V)_T$.

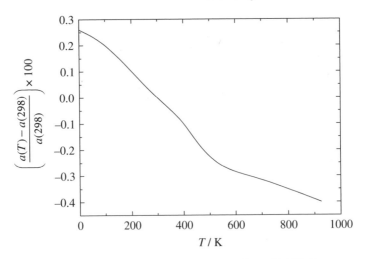

Figure 11.10 Negative thermal expansion of ZrW_2O_8.

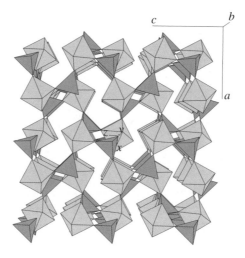

Figure 11.11 Crystal structure of ZrW_2O_8, showing the WO_4 tetrahedra and ZrO_6 octahedra.

Consider ZrW_2O_8. The expansivity is shown as a function of temperature in Figure 11.10, and the crystal structure is shown in Figure 11.11. The crystal is cubic, with a rather complex structure. WO_4 tetrahedra and ZrO_6 octahedra are linked so that each ZrO_6 unit shares its corners with six different WO_4 units, while each WO_4 unit shares only three of its corners with ZrO_6 units. The remaining oxygen in each WO_4 tetrahedron is formally singly coordinate. Gibbs energy minimizations [5] have reproduced the negative expansivity and indicate $(\partial S/\partial V)_T$ is negative for ZrW_2O_8. This is largely due to the presence of the two-coordinate bridging oxygens which form Zr–O–W linkages. The Zr–O–W *transverse* vibrations *increase* in frequency with increasing

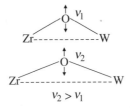

Figure 11.12 The tension effect in Zr–O–W linkages.

volume – compare the transverse vibrations of a violin string which increase in frequency when it is stretched). Called the **tension effect** (Figure 11.12), the increased frequencies lead to *lower* entropies at larger volumes, and so $(\partial S/\partial V)_T$ is negative.

Around 400 K there is a phase transition to a disordered but still cubic structure associated with the 'terminal' oxygen atom in a WO_4 tetrahedron which can migrate to another tetrahedron, thereby reversing the direction in which a pair of tetrahedra point. Nevertheless the same general atomic mechanisms are responsible for the negative thermal expansion up to the decomposition temperature.

Gibbs energy minimization has also predicted negative isobaric expansion coefficients for certain crystalline zeolite framework structures, which subsequently were confirmed experimentally [6]. Many solids show negative thermal expansion at very low temperatures, including even some alkali halides (Barron and White (Further reading)). Many other solids on heating expand in some directions and contract in others.

Configurational averaging – solid solutions and grossly non-stoichiometric oxides

The use of energy minimization can be extended to solid solutions and highly non-stoichiometric compounds. In principle the method is simple: we take a suitable thermodynamic average over the results of minimizations of different possible arrangements of the atoms. The overall procedure for a solution $A_{0.5}B_{0.5}$ is then as follows:

1. For a given unit cell size (supercell) generate different individual arrangements (configurations) k. For the alloy $A_{0.5}B_{0.5}$ these will comprise all arrangements with 50% of the atoms of type A and 50% type B.

2. Minimize the Gibbs energy G_k of each supercell at temperature T with respect to all the lattice vectors and atom positions.

3. Determine the thermodynamic properties of the system by taking a thermodynamic average. For example, the enthalpy is given by

$$H = \frac{\sum\limits_{k=1}^{K} H_k \exp(-G_k/k_B T)}{\sum\limits_{k=1}^{K} \exp(-G_k/k_B T)} \tag{11.17}$$

where overall there are K possible arrangements for the unit cell size chosen. H_k is the enthalpy for the relaxed structure of each possible arrangement within the supercell. The Gibbs energy is given by

$$G = -k_B T \ln \sum_{k=1}^{K} \exp(-G_k/k_B T) \qquad (11.18)$$

4. Check convergence with supercell size. This presents a major problem since small supercells – for which it is possible to carry out minimizations for all possible arrangements – exclude many possible arrangements by imposing an artificial short-range periodicity. The number of possible arrangements rises sharply with supercell size. In our example the total number of arrangements K is given by $(N_A + N_B)!/(N_A!N_B!)$ where N_A and N_B are the number of atoms of type A and B respectively. For a 50:50 composition, K equals 6 for a 4-atom supercell, 70 for a 8-atom supercell and 6×10^8 for a 64-atom supercell. Thus for even moderately sized supercells it is not feasible to carry out minimizations for other than a small fraction of arrangements. One strategy for solutions that are not too strongly non-ideal is to select a subset of configurations at random for a given supercell, and check convergence with the number of randomly chosen configurations. When restricted to K' configurations, eqs. (11.17) and (11.18) become

$$H = \frac{\displaystyle\sum_{k=1}^{K'} H_k \exp(-G_k/k_B T)}{\displaystyle\sum_{k=1}^{K'} \exp(-G_k/k_B T)} \qquad (11.19)$$

and

$$G = -k_B T \ln K - k_B T \ln \left(\sum_{k=1}^{K'} \exp(-G_k/k_B T)/K' \right) \qquad (11.20)$$

See also discussions of the related **cluster variation method** [7].

An example of this method is shown in Figure 11.13, in which the calculated enthalpy, entropy and Gibbs energy of mixing of MnO–MgO at 1000 K determined using a supercell containing 128 ions and 250 configurations is plotted as a function of composition [3]. The enthalpy of mixing is positive at all compositions. The entropy of mixing is slightly in excess of the ideal value; it is important to realize that this includes both configurational and vibrational contributions. The calculations indicate the vibrational contribution is typically about 10% of the total entropy of mixing; the configurational contribution is about 10% less than the ideal value reflecting the tendency of the cations of the same type to cluster together.

The same general approach can be applied to grossly non-stoichiometric oxides [8]. For an oxygen deficient perovskite such as $SrFeO_{2.5}$ this involves an explicit

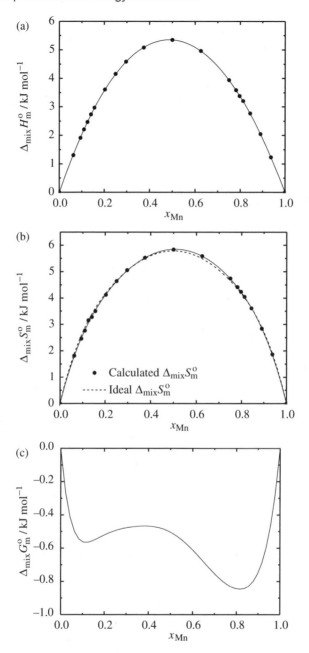

Figure 11.13 (a) Enthalpy, (b) entropy and (c) Gibbs energy of mixing of MnO–MgO at 1000 K, all calculated using the configurational averaging technique.

average, for a given cell size, over the different possible arrangements of the oxygen vacancies. In this example oxygen vacancy–vacancy interactions are considerable. These lead to the stabilization, at low temperature, of an orthorhombic structure, containing 4- and 6-coordinate Fe atoms. The order–disorder transition at high temperature appears to be associated with the generation of a large concentration of 5-coordinate Fe; the number of Fe atoms with coordination numbers lower than four is negligible. The concept of an ideal solution of oxygen vacancies in such systems is thus highly questionable. It is better to describe the disordered system in terms of disordered arrangements of square pyramids containing 5-coordinate Fe as well as the structural entities present in the ordered structure at low temperatures.

11.2 Monte Carlo and molecular dynamics

The energy minimization techniques we have discussed produce minimum energy configurations for the system of interest. In contrast, in a Monte Carlo simulation configurations are generated by making random changes to the positions of the atoms or molecules present and as the calculation proceeds statistical averages are calculated to obtain the thermodynamic properties of the system. In addition, molecular dynamics methods directly probe time-dependent behaviour.

In both Monte Carlo and molecular dynamics methods a box is set up containing the atoms or molecules of interest (typically of the order of thousands and up to 10^6 using modern computers). To simulate a liquid or solid the box is usually surrounded with replicas of the original box, thus avoiding an unwanted interface at the sides. This use of *periodic boundary conditions* is shown in Figure 11.14. Whenever a particle leaves the box through one of its faces, its image arrives through the opposite face so that the total number of particles remains constant.

Monte Carlo

If we wish to calculate a particular property Q of a system with a constant number of particles, temperature and volume (the canonical ensemble – usually referred to as *NVT*). Classical statistical mechanics shows that the average value of that property $\langle Q \rangle$ is given by

$$\langle Q \rangle = \int Q(Z)P(Z)\,\mathrm{d}Z \tag{11.21}$$

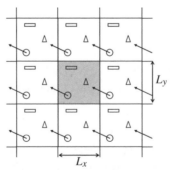

Figure 11.14 Periodic boundary conditions.

$$P(Z) = \frac{\exp(-U(Z)/k_B T)}{\int \exp(-U(Z)/k_B T)\, dZ} \tag{11.22}$$

where $P(Z)$ is the Boltzmann weighted probability, $U(Z)$ the internal energy and Z represents all possible states of the system of interest.

We might have thought we could find the values of these integrals by simply sampling different configurations of the system, i.e., just moving atoms at random. From the energies calculated at each move it would be possible to obtain estimates of $Q(Z)P(Z)$, and then take the average to get $\langle Q \rangle$. But this plan is seriously flawed! Since $P(Z)$ is proportional to the Boltzmann factor $\exp(-U(Z)/k_B T)$, *only* configurations with low energies will make a significant contribution to $P(Z)$. In this naïve approach we have just outlined states – both of high and low energy – that are generated with equal probability and *then* assigned a weight $\exp(-U(Z)/k_B T)$. Thus many of the configurations generated will have little significance, so efficient sampling of the states of the system has not been achieved.

What we do instead is to adopt the famous **Metropolis algorithm**, in which the generation of configurations is biased towards those that make the most significant contribution to the integral. The method generates states with a probability proportional to $\exp(-\Phi(Z)/k_B T)$ (equal to their Boltzmann probability) and then counts each of them equally.

So the Monte Carlo method generates configurations randomly and uses a special set of criteria (usually the Metropolis scheme) to decide whether or not to accept each new configuration. These criteria ensure that the probability of obtaining a given configuration is equal to its Boltzmann factor $\exp(-\Phi(Z)/k_B T)$. $\Phi(Z)$ is calculated as in molecular mechanics using a given set of interatomic/intermolecular potentials. Configurations with a low energy are thus generated with a higher probability than configurations with a higher energy. For each configuration that is accepted the values of the desired properties are calculated, and as the simulation proceeds the averages of these properties are obtained by simply averaging over the number M of values calculated, i.e.

$$\langle Q \rangle = \frac{1}{M} \sum_{i=1}^{M} Q(Z) \tag{11.23}$$

In a Monte Carlo simulation each new configuration of the system may be generated by randomly moving a single atom or molecule. Sometimes new configurations may also be obtained by moving several atoms or molecules, or by rotating about one of more bonds. $\Phi(Z)$ is then calculated for the new configuration. Then:

- If the energy of the new configuration is *lower* than the energy of its predecessor then the new configuration is accepted.

- If the energy of the new configuration is *higher* then the energy of its prede-cessor then the *Boltzmann factor of the energy difference* is calculated: $\exp(-\Delta\Phi/k_BT)$. A random number between 0 and 1 is then generated and com-pared with this Boltzmann factor. If the random number is *higher* than the Boltzmann factor then the move is rejected and the original configuration retained for the next iteration. If the random number is *lower* then the move is accepted and the new configuration becomes the next state. This procedure has the effect of permitting moves to states of higher energy. The smaller the uphill move, i.e. the smaller is ΔU, the greater is the probability that the move will be accepted. For more details see, for example, the book by Frenkel and Smit (Further reading).

It is also straightforward to carry out Monte Carlo simulations with a constant number of particles, temperature and pressure (the *NPT* ensemble). In such simula-tions, in addition to random moves of the atoms or molecules random changes in the volume of the simulation cell are also attempted, and in the Metropolis step $\Phi(Z) + pV$ replaces $\Phi(Z)$. Monte Carlo calculations, both *NVT* and *NPT*, have thus been extremely useful in establishing equations of state.

Monte Carlo simulations are also useful for the study of solid solutions, yielding information such as enthalpies of mixing and detailed information about the struc-ture of such solutions. In these explicit exchanges of the different types of atoms present in the alloy are attempted, thus sampling many different configurations or atomic arrangements (see e.g, Binder (Further Reading). For a useful summary of work on mineral solid solutions see Warren *et al.* [9]. Figure 11.15 shows the enthalpy of mixing of MnO and MgO calculated using Monte Carlo in this way [3]; compare Figure 11.13.

Much attention has been paid to Monte Carlo simulations of magnetic ordering, and its variation with temperature. Such models assume a particular form for the magnetic interactions, e.g. the Ising or Heisenberg Hamiltonian (see e.g. Binder

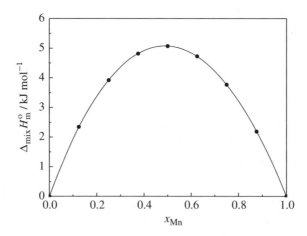

Figure 11.15 Enthalpy of mixing of MnO–MgO at 1000 K calculated using Monte Carlo.

(Further reading)). Monte Carlo simulations have thus played, for example, an important role in developing an understanding of behaviour approaching critical points, and provided valuable insights, for instance, into the fundamental physics responsible for the values of critical exponents.

Problems involving adsorption are also conveniently tackled using **grand-canonical Monte Carlo**. In these simulations the chemical potential, volume and temperature are kept constant; the number of particles may change during the simulation. The three basic moves in such a simulation are attempts to move an atom, to remove (annihilate) a particle, and to create a particle at a random position. Grand-canonical Monte Carlo has proved very useful for the calculation of isotherms for the adsorption of noble gases and hydrocarbons in zeolites, since the pressure can be directly calculated from the input chemical potential (see Frenkel and Smit, Further reading). Such calculations have also provided valuable insight into the underlying atomic mechanisms responsible for the selectivity of a given zeolite.

Molecular dynamics

Temperature effects are included *explicitly* in molecular dynamics simulations by including kinetic energy terms – the balls representing the atoms are now on the move! The principles are simple. In the microcanonical ensemble (*NVE*):

1. We generate the 'start-up' configuration – all particles in the box are assigned positions r_i and velocities v_i. Velocities are randomly distributed according to a Maxwellian distribution for some given temperature.

2. The force f_i on each particle i is calculated using the interatomic potentials.

3. We specify a time step, during which the forces are assumed to remain constant. Then r_i and v_i are updated. There are several schemes for this to overcome problems associated with finite rather than infinitesimal time steps. The force (and thus the acceleration) is assumed to remain constant throughout the time step Δt. For example, in the Leapfrog Verlet algorithm (e.g. Allen and Tildesley, Further reading),

$$v_i(t + \Delta t/2) = v_i(t - \Delta t/2) + \frac{f_i}{m_i} \Delta t$$

$$r_i(t + \Delta t) = r_i(t) + v_i(t + \Delta t/2)\Delta t$$

(11.24)

where m_i is the mass of atom i.

The choice of time step is crucial. It must be smaller than the time-scale of any important dynamical processes, so it must be at least an order of magnitude smaller than the typical period of atomic vibrations (10^{-12}–10^{-13} s). But too small a time step leads to very long computer times as the calculation samples phase space too slowly. Too large a time step leads to large truncation errors;

instabilities may arise as atoms approach too closely and the simulation may 'blow up'. A typical commonly used value is 10^{-15} s.

4. Step 3 is repeated many thousands of times, leading to equilibration when system properties have converged to equilibrium values and do not change significantly with further time steps. For example, the temperature of the simulation is calculated from the kinetic energies of all the atoms in the system:

$$\frac{3}{2}Nk_{B}T = \frac{1}{2}\sum_{i}m_{i}v_{i}^{2} \tag{11.25}$$

5. In the production stage the simulation is continued, recording velocities and coordinates of all the atoms at successive time steps.

Properties of interest, such as the mean square displacement of an atom or molecule from its starting position, $<r^2>$, may thus be determined. Since we know this as a function of time, we have a direct route to diffusion coefficients ($<r^2> = 6Dt$, where D is the diffusion constant).

Similar schemes to the above can be used in molecular dynamics simulations in other ensembles such as those at constant temperature or constant pressure (see Frenkel and Smit, and Allen and Tildesley (Further reading)). A molecular dynamics simulation is computationally much more intensive than an energy minimization. Typically with modern computers the 'real time' sampled in a simulation run for large cells is of the order of nanoseconds (10^6 time steps). Dynamical processes operating on longer time-scales will thus not be revealed.

An interesting application of molecular dynamics reveals the mechanisms responsible for the dramatic properties of fast-ionic ('superionic') conductors and the consequences for thermodynamic properties. The motion of the ions in these structures is sufficiently fast to probe the details of the underlying atomic mechanisms by following the movements of ions over a few picoseconds. For example, at 419 K, β-AgI undergoes a phase transition to the superionic α-AgI phase; the entropy change which accompanies the phase transition is large, equal in magnitude to approximately half the entropy of fusion of NaCl. Molecular dynamics simulations confirm the existence of a disordered cation sublattice in α-AgI – there is a body-centred arrangement of I$^-$ ions with two Ag$^+$ ions per unit cell distributed over 42 possible sites – and give much detailed information related to rapid diffusion of the Ag$^+$ ions.

The investigation of structure at high temperatures is particularly well suited for attack by molecular dynamics, as exemplified by the large number of studies on silicate glasses. Such studies start with either a crystalline structure or a random atomic distribution within the simulation cell, and the system is first run at a sufficiently high temperature in order to generate a simulated melt with no memory of the initial input configuration. Cooling then takes place, rescaling the velocities to the desired temperature for some appropriate number of time steps. Results for silica (SiO_2) glass suggest that five- and six-membered rings (containing five and six SiO_4 tetrahedra respectively in a closed loop) dominate the structure, although there are also some three-membered rings, which are thought to be responsible for the 606 cm^{-1} peak in the Raman spectrum. No edge-sharing tetrahedra are seen in bulk simulations, although they are seen in surface simulations, consistent with experiment. Molecular dynamics simulations have also provided an atomic level explanation for the decrease in ionic conductivity of alkali glasses when more than one alkali metal cation is present (the mixed alkali effect) based on an increased activation energy for hopping (e.g. Catlow, Further reading).

Several recent molecular dynamics simulations (e.g. [10] and references therein) have focussed on the wetting of interfaces (Section 6.1) and, for example, the behaviour of very small droplets at the nanoscale. Such simulations are able to relate the atomistic behaviour directly to relevant macroscopic parameters such as the contact angle and are able to show the dramatic effects at this length scale of addition of surfactant molecules or roughening of the surface.

As mentioned above, unlike lattice dynamics, it is very difficult to obtain *abso-lute* Gibbs and Helmholtz energies from Monte Carlo and molecular dynamics simulations. The reason here is that such calculation requires summation over regions of phase space of the system corresponding to higher energies (we need the ensemble average $<\exp(U/k_BT)>$, rather than $<U>$), and these are the regions which Monte Carlo and molecular dynamics simulations are designed to neglect. As a result, considerable attention has been paid to methods that give Gibbs and Helmholtz energy *differences*, which are often all that is required, e.g. for calculation of phase diagrams, Gibbs energies of binding and partition coefficients. Two powerful methods, to which we now turn, are thermodynamic perturbation and thermodynamic integration.

Thermodynamic perturbation

The *difference* in Helmholtz energy of two systems A and B is given by

$$A_A - A_B = -k_BT \ln Z_A + k_BT \ln Z_B = -k_BT \ln \frac{Z_A}{Z_B} \tag{11.26}$$

and so

$$A_A - A_B = -k_BT \ln(\sum \exp(-(U_A - U_B)/k_BT)) \tag{11.27}$$

remembering the definition of a partition function Z. In the course of a Monte Carlo or molecular dynamics calculation this can be calculated as an ensemble average:

$$A_A - A_B = -k_BT \ln \langle \exp(-(U_A - U_B)/k_BT) \rangle_B \tag{11.28}$$

The exponential in this equation involves the difference of two energies, rather than an energy itself, and as long as this is sufficiently small compared with $k_B T$, a typical simulation run is able to provide a good estimate of the difference in Helmholtz energy of A and B using eq. (11.28).

So overall we carry out a simulation of B, and as this simulation proceeds we evaluate this ensemble average using the potential functions appropriate for A. Alternatively we can carry out a simulation of A and as this proceeds evaluate the ensemble average using the potential functions appropriate for B. Suppose B corresponds to MgO and A to a mixture of MnO and MgO. The Helmholtz energy difference can be obtained from a simulation of pure MgO in which we not only calculate the value of the energy for each configuration as usual but also the energy it has when some of the Mg^{2+} ions are *temporarily* assigned the potential parameters for Mn^{2+}. Alternatively, and also as a very useful check, we can carry out a simulation of the mixture in which we not only calculate the value of the energy at each step as usual, but also the energy each individual configuration has when all the Mn^{2+} ions are temporarily assigned the potential parameters for Mg^{2+}. It is important to appreciate that the transformation of A to B need not correspond to any transformation that is *physically* realizable.[4]

If the energy difference is larger we can introduce a number of intermediate states between A and B by using a *coupling parameter* λ ($0 \leq \lambda \leq 1$) such that

$$U(\lambda) = \lambda U_A + (1 - \lambda)U_B \qquad (11.29)$$

Successive values of the coupling parameter should be chosen so that each ensemble average is performed over energy changes comparable with $k_B T$. The difference in Helmholtz energy between any two intermediate states is given by eq. (11.28) and the overall difference between A and B simply given by their sum. Again the intermediate states will often have no physical meaning.

Thermodynamic integration

It is also possible to show that

$$A_A - A_B = \int_0^1 \left\langle \frac{\partial U(\lambda)}{\partial \lambda} \right\rangle d\lambda \qquad (11.30)$$

In practice this integral is turned into a discrete summation over particular values of λ between 0 and 1, so we use

$$A_A - A_B \approx \sum_l \left\langle \frac{\partial U(\lambda)}{\partial \lambda} \right\rangle \Delta\lambda \qquad (11.31)$$

---

4 The computer code can thus easily fulfil the alchemists' dream of turning any element into another.

The difference between this result and the thermodynamic perturbation result is that in the former (eq. (11.28)) the average is taken of (finite) differences in energy while eq. (11.31) averages over a differentiated energy function; often the required derivative of the energy with respect to the coupling parameter can be obtained analytically and the averaging involved here is no more complicated than with eq. (11.28).

Considerable use has been made of the thermodynamic perturbation and thermodynamic integration methods in biochemical modelling, calculating the relative Gibbs energies of binding of inhibitors of biological macromolecules (e.g. proteins) with the aid of suitable thermodynamic cycles. Some applications to materials are described by Alfè *et al.* [11].

11.3 Quantum mechanical methods

One of the greatest challenges of condensed matter theory is to obtain accurate approximate solutions of the many-electron Schrödinger equation. Quantum chemistry and physics are complex and rapidly moving subjects needing volumes to themselves; see the texts listed under 'Further reading'. Space permits us only the briefest of overviews, and precludes, for example, discussion of periodic systems. Readers may well wish to skip this section, which necessarily requires some familiarity with the subject, and move on to the next section, where we discuss some representative recent applications for which, from a thermodynamic viewpoint, we need appreciate only that the ground state energy of the system is calculated using a quantum mechanical (also referred to as *ab initio* or first principles) method. *Ab initio* calculations are orders of magnitude more expensive in terms of computer time than the potential-based methods described earlier.

The basic problem is to solve the time-independent electronic Schrödinger equation. Since the mass of the electrons is so small compared to that of the nuclei, the dynamics of nuclei and electrons can normally be decoupled, and so in the **Born–Oppenheimer** approximation the many-electron wavefunction Ψ and corresponding energy may be obtained by solving the time-independent Schrödinger equation in which the nuclear positions are *fixed*. We thus solve

$$\hat{H}\Psi = E\Psi \tag{11.32}$$

where the non-relativistic Hamiltonian $\hat{H}$ in atomic units takes the form:

$$\hat{H} = -\frac{1}{2}\sum_i \nabla_i^2 - \sum_i \sum_\alpha \frac{Z_\alpha}{|r_i - d_\alpha|} + \sum_i \sum_{j>1} \frac{1}{|r_i - r_j|} + \sum_\alpha \sum_{\beta>\alpha} \frac{Z_\alpha Z_\beta}{|d_\beta - d_\alpha|} \tag{11.33}$$

in which r_i denotes the electron positions, d_α the nuclear positions and Z_α the nuclear charges. These four terms represent respectively the kinetic energy of the electrons, the electron–nuclear attractions, the electron–electron repulsions and the nuclear–nuclear repulsion (a constant since the nuclei are assumed to be frozen).

For all but the simplest systems the Schrödinger equation must be solved approximately. It is assumed that the true wavefunction, which is too complicated to be found directly, can be approximated by a simpler function. For some types of function it is then possible to solve the electronic Schrödinger equation numerically. Provided the assumption made regarding the form of the function is not too drastic, a good approximation will be obtained to the correct solution. Electronic structure theory consists of designing sensible approximations to the wavefunction, with an inevitable trade-off between accuracy and computational cost.

The energy of an approximate wavefunction Ψ is given by the expectation value of the Hamiltonian $\hat{H}$:

$$E = \frac{\int \Psi^* \hat{H} \Psi \, d\tau}{\int \Psi^* \Psi \, d\tau} \tag{11.34}$$

The best wavefunction is that which gives the lowest value of E – this is the famous **variational principle**.

The many-electron wavefunction must obey the Pauli Principle, i.e. possess the right permutation symmetry, such that it changes sign when any two electrons are exchanged, i.e.

$$\Psi(x_1, x_2, \ldots, x_i, \ldots, x_j, \ldots) = -\Psi(x_1, x_2, \ldots, x_j, \ldots, x_i, \ldots) \tag{11.35}$$

where $x_i = \{r_i, \sigma_i\}$ represents the space and spin coordinates of an electron. Due to the antisymmetry no two electrons can possess the same set of quantum numbers.

Hartree–Fock theory

The most usual starting point for approximate solutions to the electronic Schrödinger equation is to make the orbital approximation. In **Hartree–Fock** (HF) theory the many-electron wavefunction is taken to be the antisymmetrized product of one-electron wavefunctions (spin-orbitals):

$$\Psi(x_1, x_2, \ldots) = \mathfrak{A} \prod_{i=1}^{N} \psi_i(x_i) \tag{11.36}$$

Here $\mathfrak{A}$ is the antisymmetrizer which ensures that the Pauli Principle is obeyed. A convenient way of writing eq. (11.36) is as a Slater determinant:

$$\Psi(x_1, x_2, \ldots) = \frac{1}{\sqrt{N!}} \begin{vmatrix} \psi_1(x_1) & \psi_1(x_2) & \ldots & \psi_1(x_N) \\ \psi_2(x_1) & \psi_2(x_2) & \ldots & \psi_2(x_N) \\ \ldots & \ldots & \ldots & \ldots \\ \psi_N(x_1) & \ldots & \ldots & \psi_N(x_N) \end{vmatrix} \tag{11.37}$$

The spin-orbitals are the products of spatial and spin factors, i.e.

$$\psi_i(x_i) = \psi_i(r_i)\alpha(\sigma_i) \quad \text{or} \quad \psi_i(x_i) = \psi_i(r_i)\beta(\sigma_i) \tag{11.38}$$

where α and β are spin functions. The spatial orbitals $\psi_i(r_i)$ are generally approximately expanded in a basis set, as a linear combination either of atomic functions or plane waves:

$$\psi_i(r) = \sum_{j=1}^{n_{\text{basis}}} c_{ij}\phi_j(r) \tag{11.39}$$

Given the trial wavefunction – the Slater determinant eq. (11.37) – we then use the variational principle to minimize the energy – the expectation value of the Hamiltonian $\hat{H}$ – with respect to the orbital coefficients c_{ij} (eq. (11.39)). This leads after a fair amount of algebra to the self-consistent Hartree–Fock equations:

$$\hat{f}\psi_i = \varepsilon_i\psi_i \tag{11.40}$$

where ε_i are the orbital energies and $\hat{f}$, the Fock operator, is given by

$$\hat{f} = -\frac{1}{2}\nabla^2 + V_{\text{nuc}} + J - K \tag{11.41}$$

$$V_{\text{nuc}} = -\sum_{\alpha}\frac{Z_\alpha}{|r - d_\alpha|}, \quad J\psi_i(r) = \sum_j \int dr' \frac{|\psi_j(r')|^2}{|r - r'|}\psi_i(r),$$

$$K\psi_i(r) = \sum_j{}' \int dr' \frac{\psi_j^*(r')\psi_i(r')}{|r - r'|}\psi_j(r)$$

The terms on the right-hand side of eq. (11.41) denote the kinetic energy, the electron–nuclear potential energy, the Coulomb (J) and exchange (K) terms respectively. Together J and K describe an effective electron–electron interaction. The prime on the summation in the expression for K exchange term indicates summing only over pairs of electrons of the same spin. The Hartree–Fock equations (11.40) are solved iteratively since the Fock operator $\hat{f}$ itself depends on the orbitals ψ_i.

Hartree–Fock theory is relatively simple but does require the evaluation of a large number of six-dimensional integrals involving $1/|r - r'|$. In semi-empirical methods such as Hückel theory, tight-binding and MNDO (see e.g. Leach, Further reading), a large number of these integrals are simplified or neglected. Often some of the integrals are given empirical values, adjusted to reproduce some known thermodynamic properties of certain simple atoms, molecules or solids.

In Hartree–Fock theory, electrons interact only with the average positions of the other electrons, so this inevitably leads to an incorrect treatment of the electron–

electron repulsion. The **correlation** between the electronic motions caused by instantaneous electron–electron repulsion is neglected. Correlation energies are a small fraction of the total energy, but can be a large percentage of binding energies. There are many post Hartree–Fock methods that aim to rectify this. One method commonly used in molecular calculations and capable of high accuracy is the use of a linear combination of determinants rather than a single determinant. The main problem with such expansions is the very large number of determinants that are required. For solids more use is made of density functional theory, to which we now turn.

Density functional theory

Possibly the most popular method at present for calculating the electronic properties of large molecules and solids is density functional theory (DFT). The basis of density functional theory is a famous theorem due to Hohenberg and Kohn [12] which states that the ground-state properties of a many-electron system can be obtained by minimizing an energy functional $E[\rho]$ of the electron density $\rho(\mathbf{r})$. The minimum value of the functional is the exact ground-state energy obtained when $\rho(\mathbf{r})$ is the exact ground-state density.

Kohn and Sham later introduced the idea of an auxiliary non-interacting system with the same electron density as the real system. They were able to express the electron density of the interacting system in terms of the one-electron wavefunctions of the non-interacting system:

$$\rho(\mathbf{r}) = \sum_{i=1}^{N} |\psi_i(\mathbf{r})|^2 \tag{11.42}$$

The energy functional is then written in the form

$$
\begin{aligned}
E[\rho] &= T_S[\rho] + V_{\text{nuc}}[\rho] + J[\rho] + E_{\text{xc}}[\rho] \\
&= -\frac{1}{2} \sum_{i=1}^{N} \int \psi_i^*(\mathbf{r}) \nabla^2 \psi_i(\mathbf{r}) \, \mathrm{d}\mathbf{r} - \sum_{\alpha} \int \rho(\mathbf{r}) \frac{Z_\alpha}{|\mathbf{r} - \mathbf{d}_\alpha|} \, \mathrm{d}\mathbf{r} \\
&\quad + \frac{1}{2} \iint \frac{\rho(\mathbf{r})\rho(\mathbf{r}')}{|\mathbf{r} - \mathbf{r}'|} \, \mathrm{d}\mathbf{r} \, \mathrm{d}\mathbf{r}' + E_{\text{xc}}[\rho]
\end{aligned}
\tag{11.43}
$$

On the right-hand side of eq. (11.43) the terms in order are the kinetic energy of the non-interacting system with electron density $\rho(r)$, the electron–nuclear attraction, the Coulomb term and the exchange-correlation energy. Minimizing the total energy functional of eq. (11.43) gives rise to a self-consistent set of one-electron equations for the ψ_i, that can again be solved iteratively, and thus the ground state electron density and energy can be obtained.

The problem is that the exchange correlation functional E_{xc} is unknown. Approximate forms have to be used. The most well-known is the **local density approximation** (LDA) in which the expressions for a uniform electron gas are used. This approximation surprisingly also works well even when the electron density is clearly not uniform and highly inhomogeneous, such as in molecules and at surfaces.

More refined functionals (**generalized gradient approximation (GGA)**) use the gradient of the charge density as well as the charge density itself. There are also many hybrid schemes such as the popular B3LYP functional which combines some of the Hartree–Fock exchange with density functional exchange and correlation.

11.4 Applications of quantum mechanical methods

Carbon nitride

Molecular modelling is a particularly attractive way of predicting new compounds and novel forms of matter. Superhard materials are one such area which has received considerable attention due to their potential industrial applications. A useful rule of thumb is that materials with a bulk modulus K_T ($V(d^2U/dV^2)$) (in excess of ≈ 250 GPa) can be considered as superhard. The large bulk modulus of diamond (442 GPa), associated with the three-dimensional 'giant' network of strong covalent bonds, suggests that similarly hard materials might be formed from materials containing neighbours of C in the Periodic Table, such as B, C or N.

Nitrides of silicon or carbon have been studied extensively. A representative and interesting study is that of C_3N_4 by Teter and Hemley [13], who calculated E vs. V curves for five possible structures, using density functional theory and LDA. Four were diamond-like with 4-coordinated carbon: the α-, β-(β-Si$_3$N$_4$), cubic (high-pressure willemite-II Zn$_2$SiO$_4$), and pseudocubic (α-CdIn$_2$Se$_4$) forms. The fifth was a graphitic structure. At zero pressure the calculated lowest energy phase is the graphitic (Figure 11.16(a)), and the next lowest the α-form. A transition to a cubic C_3N_4 phase is predicted at 12 GPa. This cubic phase, shown in Figure 11.16(b), is predicted to have a very high bulk modulus as much as 50 GPa in excess of that of diamond (442 GPa). High thermal conductivity is also expected. Unfortunately a bulk synthesis of C_3N_4 remains elusive!

More recently, similar calculations [14] have been carried out for C_3P_4, for which in contrast to C_3N_4 the lowest energy phase at zero pressure was predicted to be the pseudocubic (Figure 11.16(c)). Structures such as the pseudocubic which are high in energy for C_3N_4 are low in energy for C_3P_4, reflecting the different structural preferences of N and P, as seen in their molecular chemistry.

Nanostructures

There has been tremendous interest in the study of atomic nanostructures over the last few years. At the atomic scale nanomanipulation is increasingly opening up a new world of nanosize clusters and structures, many of which have properties distinct both from those of the macroscopic solid materials and also from those of small molecules. The nanoscale is often the critical size at which properties start to change

(a)   (b)

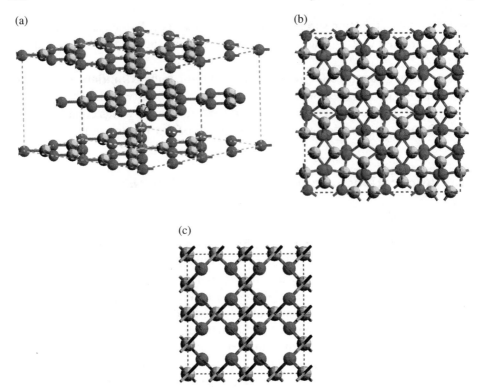

(c)

Figure 11.16 The (a) graphitic, (b) cubic and (c) pseudocubic forms of C_3N_4. Carbon atoms are dark and nitrogen atoms light.

from molecular to solid-like. Determination of the structure of nanomaterials is particularly important since it governs so many other properties, but is a difficult experimental task.

Recent first principles calculations for example have helped to understand monoatomic gold wires, which can be produced, for instance, with the tip of a scanning tunnelling microscope. Particularly puzzling was the spacing ($\approx$5 Å) between gold atoms in wires containing just a few atoms. This was almost twice the typical Au–Au distance; linear chains were predicted to break when the interatomic separation exceeds 3 Å. Density functional theory results [15] have resolved this apparent paradox by showing that the binding energy of the chain is larger for a zigzag rather than linear geometry. The calculations also revealed that the barrier to rotation of the zigzag chain was very small, and thus the transmission electron microscopy images would show only an average of the rotating atoms. Thus the 5 Å apparent separation corresponds to the distance between odd-numbered atoms in the wire and so is much larger than the real interatomic distance between adjacent gold atoms. This example illustrates nicely our increasing understanding of the interplay between energy (and thus chemical and physical behaviour), and details of structure and dynamics at the atomic level.

Lithium batteries

Modelling techniques have recently been used to study the electrodes in lithium ion battery systems. Such cells use typically a lithium anode (either as the pure metal, a compound, or most usually a graphite bronze) and the cathode is a transition metal oxide. Examples of cathode materials include $LiCoO_2$, $LiMn_2O_4$ and V_2O_5, all of which have structures with layers or cavities that allow intercalation of Li^+ ions. Organic polymers with good ionic conducting properties are used as the electrolyte. A key property is the voltage at which Li can be inserted in the cathode.

For example, lithium ion intercalation into the transition metal oxide cathode system V_6O_{13} has been studied by Braithwaite *et al.* [16]. Optimized structures were obtained using periodic density functional theory for $Li_xV_6O_{13}$ ($x = 0.0, 0.5$, 1.0, 2.0, 4.0), identifying sites for the Li ions in initial geometries on the basis of potential-based calculations. Li intercalation into V_6O_{13} leads to the selective reduction of V atoms that lie on the edges of cavities occupied by the Li. From the calculated *ab initio* energies of the series it is possible to work out the internal energy change accompanying the intercalation reaction:

$$Li_xV_2O_5 + yLi \rightarrow Li_{x+y}V_2O_5 \tag{11.44}$$

per intercalated lithium ion for a range of x. Neglecting entropic contributions to the Gibbs energy, average cell voltages can then be estimated using the well-known equation

$$\Delta G = -nFE_{cell} \tag{11.45}$$

Calculated voltages are somewhat lower than the experimental values, but the underestimation is consistent and trends are reproduced well.

The advantage of modelling here is that traditionally the determination of the voltage of a new material requires laborious synthesis and electrochemical measurement for each composition. These modelling studies provide a very cost-effective alternative. For example, Ceder *et al.* [17] have studied Li intercalation into $LiCoO_2$ and the effects of replacing Co with other metals. Most transition metal oxides were predicted to have a lower voltage than Li_xCoO_2 and they also predicted that including some aluminium raised the voltage, while also decreasing the density of the material. This has subsequently been confirmed experimentally.

Ab initio molecular dynamics

The overall scheme of *ab initio* molecular dynamics is similar to that of classical molecular dynamics described earlier; but instead of using interatomic potentials, the Schrödinger equation is solved to provide the energy and the forces acting on the particles. The computational cost is huge and most studies are limited to small simulation cells (< 100 atoms) and time-scales of a few picoseconds. Within density functional theory, a useful scheme is that of Car and Parrinello [18] in which electronic and nuclear dynamics are carried out simultaneously. The number of applications is growing rapidly. *Ab initio* molecular dynamics is contributing significantly to our understanding of liquid water, indicating for example that the anisotropy in the electron density is reduced in the liquid phase compared to the free molecule. The technique is particularly useful for problems where bonds are broken or created. Such properties are completely beyond the scope of the classical technique. It is also valuable where interatomic potentials are difficult to obtain, e.g. for the properties of iron under the conditions appropriate to the Earth's core (e.g. [19]). A further application of geological interest has been to the equation of state and elastic properties of the perovskite $MgSiO_3$, which, although unstable under ambient conditions, is now accepted to be the most abundant mineral in the Earth's lower mantle. With a simulation cell containing 80 atoms, under mantle conditions the structure is predicted to be an orthorhombic perovskite with space group *Pbnm*; the adoption of possible alternative symmetries at very high temperatures is still an area of active debate (compare e.g. [20] and [21]).

An interesting recent application [22] has involved the application of *ab initio* molecular dynamics (see below) together with thermodynamic integration to the problem of the temperature and composition of the Earth's core, both of which are major uncertainties in Earth Sciences. Determination of the difference in Gibbs energy between the solid and liquid phases gives the variation of the melting point of iron with pressure. By calculating the chemical potential of an impurity in both liquid and solid as a function of composition and imposing equality of chemical potential of all species in liquid and solid phases, it is possible to determine the partition of the impurity between solid and liquid. In this way Alfè *et al.* [22] show that the core could not have been formed from a binary mixture of Fe with S, Si or O and propose a ternary or quaternary mixture with 8–10% of S/Si in both liquid and solid and an additional 8% of oxygen in the liquid, suggesting a temperature at the boundary between the solid inner core and the liquid outer core of 5650 ± 600 K.

Surfaces and defects

Increasingly, computational techniques are able to provide detailed thermodynamic information about defect processes and the creation of extended defects such as interfaces. A striking example relating to *surfaces*, first established by static energy minimization but subsequently confirmed by detailed density functional theory calculations and then by neutron diffraction experiments, is the structure of the {0001} surface of the important ceramic material α-Al_2O_3 (Figure 11.17) under vacuo. These results were obtained by minimization of the static energy of a slab of the material with respect to the positions of the ions in the slab by letting the ions 'relax' from the positions they would occupy if the surface was a 'perfect' termination of the bulk. The surface lattice vectors were kept fixed. From the difference in energy between slab and bulk we can work out the surface energies before and after relaxation.

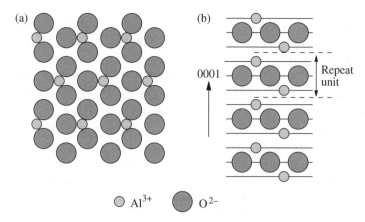

Figure 11.17 (a) The {0001} surface of α-Al_2O_3 and (b) the stacking sequence perpendicular to this surface.

The minimizations reveal dramatic atomic relaxations accompanying a large reduction in surface energy from that given by a 'perfect' termination of the bulk structure. As Table 11.2 makes clear, interatomic distances in the outermost layers can change by over 50% relative to the bulk values!

Surface relaxation thus has several effects. It modifies and reconstructs the surface atomic structure. Surface energies are reduced (possibly by as much as a factor of three in the above example – from 6.0 to 2.0 J m^{-2}). More generally, it can reorder the relative stability of different surfaces and thus have a profound effect on the crystal morphology.

Considerable attention has also been paid to modelling the thermodynamics of **defects**. This includes, for example, studies of the enthalpies of formation of vacancies or interstitial atoms and the association energies associated with the clustering of such defects. It is usually crucial to allow for the relaxation of the

Table 11.2 Comparison of theoretical and experimental surface relaxations (%) in the distances between layers in α-Al_2O_3 {0001}. Inner surface layers are numbered sequentially according to increasing distance from the surface (Figure 11.17(b)).

Layer	Energy minimization (using interatomic potentials) [23]	*ab initio* (DFT) [24]	Experiment [25]
(1) Al			
	−59	−86	−51
(2) O			
	2	3	16
(3) Al			
	−49	−54	−29
(4) Al			
	26	25	20
(5) O			

atoms surrounding a defect from the positions they occupy in the undefective structure. These studies can be very useful in determining the fundamental defects
present in a given material and thus in the interpretation of conductivity and diffusion experiments and diffraction data. Further examples of defect calculations are
of energies associated with the incorporation of trace elements and dopants, either
by substitution or in interstitial sites. Results for the incorporation of trace elements in the garnets have been presented in an earlier chapter.

Defect thermodynamics is by no means just of academic interest. For example,
graphite is used in nuclear reactors and irradiation produces defects such as vacancies. The defects can associate in exothermic processes, leading to dangerous
releases of energy implicated in, for example, the fire at the Windscale nuclear
reactor in the UK in 1957. This energy release is expected to involve the recombi-
nation of interstitials and vacancies and energies calculated using *ab initio* density
functional theory are playing an important contribution in unravelling the details
of the process [26] and revealing quite unexpected behaviour as the formation of
vacancy complexes in graphite over the large interlayer distances.

Quantum Monte Carlo

An exciting development of increasing importance offering the calculation of energies to very high accuracy and very high precision in the treatment of electron correlation energies is **quantum Monte Carlo** [27]. One such technique is **diffusion
Monte Carlo** which starts from the time-dependent Schrödinger equation:

$$i\frac{d\Psi}{dt} = -\frac{1}{2}\sum_j \nabla_j^2 \Psi + V\Psi \tag{11.46}$$

Replacing t by $-i\tau$ yields the imaginary-time Schrödinger equation

$$\frac{d\Psi}{d\tau} = \frac{1}{2}\sum_j \nabla_j^2 \Psi - V\Psi \tag{11.47}$$

which is analogous to the well-known diffusion equation:

$$\frac{dC}{dt} = D\nabla^2 C + \mathfrak{S}C \tag{11.48}$$

So eq. (11.47) can be viewed as a diffusion equation in the spatial coordinates of
the electrons with a diffusion coefficient D equal to $\frac{1}{2}$. The source and sink term $\mathfrak{S}$
is related to the potential energy V. In regions of space where V is attractive (negative) the concentration of diffusing 'material' (here the wavefunction) will accumulate and it will decrease where V is positive. It turns out that if we start from an
initial trial wavefunction and propagate it forward in time using eq. (11.47),

simulating a random walk, it will eventually at large τ be dominated by the ground-state wavefunction, thus yielding the ground-state energy. In order to generate an antisymmetric total wavefunction in keeping with the Pauli principle, the positions of the nodes (from e.g. a Hartree–Fock trial wavefunction) are kept fixed.

Impressive results have been obtained for cohesive (binding) energies of tetra-hedrally bonded semiconductors. Other examples include studies of the relative stabilities of the high-pressure monoatomic phases of solid hydrogen which have highlighted the importance made by zero-point energies. There is a trade-off between the electronic energy term, which favours low coordination numbers, and the zero-point energy, favouring higher coordination numbers. A transition to an atomic diamond-like phase is predicted at $\approx$300 GPa [28].

11.5 Discussion

Computational materials science is a rapidly evolving area and there remains much to be done. For many applications much more accurate energies are required – especially for large molecules or solids. For example, entropies and Gibbs ener-gies, excited states, processes involving bond-breaking and making and extremes of temperature and pressure all still present major challenges. Length scales and time-scales remain serious problems.

Structure prediction

A fundamental challenge so far largely unfaced is the prediction of structure *ab initio*, i.e. *just* from the molecular formula. In the examples considered above, and indeed almost throughout the literature, chemical intuition and the structures of related compounds are used to guess a set of suitable input structures, which all turn out to be local minima on the potential energy hypersurface. Calculations gave the relative energies of all of these candidate structures, but an ever-present danger is that a further structure which is even lower in energy has been missed! Given the number of variables involved this enormous challenge of global optimization remains for the future. There are several possible approaches. In **simulated annealing** the temperature of a molecular dynamics or Monte Carlo simulation is first increased to a very high value, making greater motion possible and more states accessible to the system; this is then followed by a rapid quench to low tempera-tures. Beginning to be explored is the use of **genetic algorithms**. The basic idea is to have a large 'population' of structures (starting configurations), which each pos-sess a set of 'genes'. These parent structures generate children having a mixture of the parent genes, i.e. the starting configurations 'evolve' according to a simple cost function based, for example, on target coordination numbers. Thanks to the cost function, low-energy structures are more likely to contribute to the next generation than ones high in energy. Limited mutation is usually allowed, i.e. structures are randomly changed to produce structural features outside the range in the current population. It is worth mentioning brave attempts such as those of Bush *et al.* [29]

and Woodley *et al.* [30]. For example, Bush *et al.* [29] solved the structure of the complex ternary oxide Li_3RuO_4, which had previously eluded structure determination, using a combination of genetic algorithm techniques and energy minimization.

There may be *no* single global energy minimum. Garzón *et al.* [31] have studied gold nanoclusters (Au_n, $n = 38, 55, 75$) using dynamic and genetic optimization methods. The search for minima was carried out using potential-based models, but the most stable structures were further studied using density functional calculations. For these three clusters no single ordered structure with a definite symmetry was obtained as the global minimum. Instead, several isomers with almost identical energies were found. Most of these have little symmetry and are effectively amorphous and glass-like.

References

[1]   G. W. Watson, P. Tschaufeser, A. Wall, R. A. Jackson and S. C. Parker, in *Computer Modelling in Inorganic Crystallography* (C. R. A. Catlow ed.). London: Academic Press, 1997, Chapter 3.

[2]   N. L. Allan, M. J. Dayer, D. T. Kulp and W. C. Mackrodt, *J. Mater. Chem.* 1991, **1**, 1035.

[3]   N. L. Allan, G. D. Barrera, J. A. Purton, C. E. Sims and M. B. Taylor, *Phys. Chem., Chem. Phys.* 2000, **2**, 1099.

[4]   T. A. Mary, J. S. O. Evans, T. Vogt and A. W. Sleight, *Science.* 1996, **272**, 90.

[5]   A. K. A. Pryde, K. D. Hammonds, M. T. Dove, V. Heine, J. D. Gale and M. C. Warren, *J. Phys. Condens. Matt.* 1996, **8**, 10973.

[6]   J. W. Couves, R. H. Jones, S. C. Parker, P. Tschaufeser and C. R. A. Catlow, *J. Phys. Condens. Matt.* 1993, **5**, L329.

[7]   J. M. Ziman, *Models of Disorder: the Theoretical Physics of Homogeneously Disordered Systems.* Cambridge: Cambridge University Press, 1979.

[8]   E. Bakken, N. L. Allan, T. H. K. Barron, C. E. Mohn, I. T. Todorov and S. Stølen, *Phys. Chem., Chem. Phys.* 2003, **5**, 2237.

[9]   M. C. Warren, M. T. Dove, E. R. Myers, A. Bosenick, E. J. Palin, C. I. Sainz-Diaz, B. S. Gutton and S. A. T. Redfern, *Min. Mag.* 2001, **65**, 221.

[10] M. Lundgren, N. L. Allan, T. Cosgrove and N. George, *Langmuir*, 2003, **19**, 7127.

[11] D. Alfè, G. A. de Wijs, G. Kresse and M. J. Gillan, *Int. J. Quant. Chem.* 2000, **77**, 871.

[12] P. Hohenberg and W. Kohn, *Phys. Rev.* 1964, **136**, B864.

[13] D. M. Teter and R. J. Hemley, *Science.* 1996, **271**, 53.

[14] N. L. Allan, P. W. May, J. M. Oliva and P. Ordejón, *Chem. Commun.* 2002, **21**, 2494.

[15] D. Sánchez-Portal, J. Junquera, P. Ordejón, A. Garcia, E. Artacho and J. M. Soler, *Phys. Rev. Lett.* 1999, **83**, 3884.

[16] J. S. Braithwaite, C. R. A. Catlow, J. H. Harding and J. D. Gale, *Phys. Chem., Chem. Phys.* 2001, **3**, 4052.

[17] G. Ceder, Y.-M. Chiang, D. R. Sadoway, M. K. Aydinol. Y.-I. Jang and B. Huang, *Nature*, 1998, **392**, 694.

[18] R. Car and M. Parinello, *Phys. Rev. Lett.* 1985, **55**, 2471.

[19] G. Steinle-Neumann, L. Stixrude, R. E. Cohen and O. Gulseren, *Nature*. 2001, **413**, 57.

[20] L. Stixrude (2001), in *Molecular Modelling Theory in the Geosciences* (R. T. Cygan and J. D. Kubicki eds.). Reviews in Mineralogy and Geochemistry, vol. 42, Geochemical Society, Mineralogical Society of America, Chapter 9.

[21] A. R. Oganov, J. P. Brodholt and G. D. Price, *Earth Planet Sci. Lett.* 2001, **184**, 555.

[22] D. Alfè, M. J. Gillan and G. D. Price, *Miner. Mag.* 2003, **67**, 113.

[23] W. C. Mackrodt, *J. Chem. Soc., Faraday Trans II.* 1989, **85**, 541.

[24] I. Manassidis and M. J. Gillan, *J. Am. Ceram. Soc.* 1994, **77**, 335.

[25] P. Guenard, P. Renaud, A. Barbier and M. Gautier-Soyer, *Mat. Res. Soc. Symp. Proc.* 1996, **437**, 15.

[26] R. H. Telling, C. P. Ewels, A. A. El-Barbary and M. I. Heggie, *Nature Materials.* 2003, **2**, 333.

[27] W. M. C. Foulkes, L. Mitas, R. J. Needs and G. Rajagopal, *Rev. Mod. Phys.* 2001, **73**, 33.

[28] V. Natoli, R. M. Martin and D. Ceperley, *Phys. Rev. Lett.* 1995, **74**, 1601.

[29] T. S. Bush, C. R. A. Catlow and P. D. Battle, *J. Mater. Chem.* 1995, **5**, 1269.

[30] S. M. Woodley, P. D. Battle, J. D. Gale and C. R. A. Catlow, *Phys. Chem., Chem. Phys.* 1999, **1**, 2535.

[31] I. L. Garzón, K. Michaelian, M. R. Beltrán, A. Posada-Amarillas, P. Ordejón, E. Artacho, D. Sánchez-Portal and J. M. Soler, *Phys. Rev. Lett.* 1998, **81**, 1600.

Further reading

M. P. Allen and D. J. Tildesley, *Computer Simulation of Liquids*. Oxford: Clarendon Press, 1987.

P. W. Atkins and J. de Paula, *Atkins' Physical Chemistry*, 7th edn. Oxford: Oxford University Press, 2002.

T. H. K. Barron and G. K. White, *Heat Capacity and Thermal Expansion at Low Temperatures*. New York: Kluwer Academic/Plenum, 1999.

C. R. A. Catlow (ed.), *Computer Modelling in Inorganic Crystallography*. London: Academic Press, 1997.

R. T. Cygan and J. D. Kubicki (eds.), *Molecular Modelling Theory in the Geosciences*. Reviews in Mineralogy and Geochemistry, vol. 42, Geochemical Society, Mineralogical Society of America, 2001.

P. Deák, T. Frauenheim, M. R. Pederson (eds.), *Computer Simulation of Materials at the Atomic Level*. Berlin: Wiley-VCH, 2000.

D. Frenkel and B. Smit, *Understanding Molecular Simulation, from Algorithms to Applications*, 2nd edn. London: Academic Press, 2002.

F. Jensen, *Introduction to Computational Chemistry*. Chichester: John Wiley, 1999.

D. P. Landau and K. Binder, *A Guide to Monte Carlo Simulations in Statistical Physics*. Cambridge: Cambridge University Press, 2000.

A. R. Leach, *Molecular Modelling, Principles and Applications*, 2nd edn. Upper Saddle River, NJ: Prentice Hall, 2001.

K. Ohno, K. Esfarjani and Y. Kawazoe, *Science: from Ab Initio to Monte Carlo Methods*. Springer Series in Solid-State Sciences. Berlin, Heidelberg: Springer-Verlag, 1999.

D. G. Pettifor, *Bonding and Structure of Molecules and Solids*. Oxford: Clarendon Press, 1995.

J. Simons, *An Introduction to Theoretical Chemistry*. Cambridge: Cambridge University Press, 2003.

J. Simons and J. Nichols, *Quantum Mechanics in Chemistry*. Oxford: Oxford University Press, 1997.

A. P. Sutton, *Electronic Structure of Materials*. Oxford: Clarendon Press, 1993.

Symbols and data

Symbol	Meaning
a_i	activity of component i
a_i^{α}	activity of component i in phase α
a_i^{H}	activity of component i with Henrian standard state
a_i^{R}	activity of component i with Raoultian standard state
A	Helmholtz energy
A_k	affinity
A_{s}	surface area
α	isobaric expansivity
$\boldsymbol{B}$	magnetic flux density
β	critical exponent
c_i	principal curvature of a surface
$\boldsymbol{c}$	velocity vector
c_i^{α}	concentration of component i in phase α
C	number of components
C_{ij}	elastic stiffness coefficients
C_{44}	shear modulus
C_p	heat capacity at constant pressure
$C_{p,\mathrm{m}}$	molar heat capacity at constant pressure
C_V	heat capacity at constant volume
$C_{V,\mathrm{m}}$	molar heat capacity at constant volume
C_{E}	electronic heat capacity

C_{dil}	dilational heat capacity
D	diffusion constant
E^o	standard potential
E	energy
$\boldsymbol{E}$	electric field strength
E_{cell}	cell voltage (electromotive force) (Chapter 11 only)
E_{xc}	exchange–correlation energy
ε_n	energy of nth state of an oscillator
ε_F	Fermi energy
ε_1^B	self-interaction coefficient
ε_0	vacuum permittivity
ε_i	orbital energy
f	force of elongation
$\hat{f}$	Fock operator
f_i	fugacity of component i, force acting on particle i (Chapter 11 only)
F	number of degrees of freedom (of system)
ϕ	dihedral angle
ϕ_i	volume fraction of component i
ϕ_j	atomic or plane wave basis function (Chapter 11 only)
Φ	potential energy (of system)
Φ_i	electric potential of particle i
g_i	degeneracy of energy state i
$g(\nu); \; g(\omega)$	vibrational density of state
G	Gibbs energy
γ	electronic heat capacity coefficient
$\gamma; \; \gamma^{\alpha\beta}$	surface energy; surface energy between phase α and β
γ_i	activity coefficient of component i
γ_i^H	activity coefficient of component i with Henrian standard state
γ_i^R	activity coefficient of component i with Raoultian standard state
γ_i^∞	activity coefficient of component i at infinite dilution
γ_G	Grüneisen parameter
Γ_i	adsorption of component i
$\Gamma_B^{(A)}$	relative adsorption of component B with respect to component A

$\hbar = h/2\pi$	Planck's constant
H	enthalpy
$\hat{H}$	Hamiltonian operator
J	Coulomb operator
j_i	surface activity of component i
k	wetting coefficient
k_B	Boltzman's constant
k_i	rate constant of reaction i
$k_{H,i}$	Henry's law constant for component i
k_r	bond force constant
k_θ	angle force constant
K	equilibrium constant; force constant; exchange operator
K_T	isothermal bulk modulus
$K_{T,0}$	isothermal bulk modulus at zero pressure
$K'_{T,0}$	pressure derivative of isothermal bulk modulus at zero pressure
κ_T	isothermal compressibility
L	Avogadro's number
l	length in direction of a force f
λ	wavelength; coupling parameter (Chapter 11 only)
m_i	mass of atom i
$\boldsymbol{m}$	magnetic moment
μ_i	chemical potential of component i
μ_i^o	standard chemical potential of component i
μ_i^α	chemical potential of component i in phase α
μ_i^*	chemical potential of component i in a given standard state
μ_i^H	chemical potential of component i with Henrian standard state
μ_i^R	chemical potential of component i with Raoultian standard state
n_i	number of mole of component i
N	total number of atoms
N_i	number of atoms i
N_{ij}	number of pair interactions ij
ν	frequency
p	pressure

p_c	critical pressure
p_i	partial pressure of component i
p^o	standard pressure (1 bar)
p_{ext}	external pressure
p_{tot}	total pressure
$\bar{p}$	p/p_c
$\boldsymbol{P}$	electric dipole moment
Ph	number of phases in a system
q	heat
q_{rev}	heat of a reversible process
$\boldsymbol{q}$	wave vector
q_i	electric charge of species i
θ	contact angle, fractional coverage
θ	bond (valence) angle
Θ_D	Debye temperature
Θ_E	Einstein temperature
r_i	principal radius of curvature of surface
r_{ij}	distance between atoms i and j
$\boldsymbol{r}$	position vector
R	gas constant
ρ	density; electron density (Chapter 11 only)
S	entropy
$\mathscr{S}$	total spin quantum number
ΔS_{sur}	entropy change of the surroundings of a system
ΔS_{tot}	entropy change of a system and its surroundings
σ_{ion}	ionic conductivity
σ_{tot}	total conductivity
$\sigma \, ; \, \sigma^{\alpha\beta}$	surface tension; surface tension between phase α and β
σ_i	spin coordinate (of electron i)
Σ	Gibbs dividing surface
t	time
t_{ion}	transference number of ion
T	temperature

T_c	critical temperature
T_C	Curie temperature
T_{fus}	melting temperature
T_F	Fermi temperature
T_g	glass transition temperature
T_K	Kauzman temperature
T_N	Néel temperature
T_{trs}	transition temperature
$\bar{T}$	T/T_c
τ	entropy term in the quasi-regular solution model
V	volume
V_c	critical volume
V_m	molar volume
V_{tot}	total volume of a system
V_{ij}	interaction energy between atoms i and j (two-body potential)
V_{ijk}	interaction energy between atoms i, j and k (three-body potential)
V_{ijkl}	interaction energy between atoms i, j, k and l (four-body potential)
V_n	torsion barrier
V_{nuc}	electron-nuclear potential energy
V_{vdw}	van der Waals energy
$\bar{V}$	V/V_c
w	work
w_{max}	maximum work
w_{pV}	pV-work
w_{non-e}	non-expansion work
$w_{non-e, max}$	maximum non-expansion work
ω	angular frequency
ω_{AB}	interaction energy $[u_{AB} - \frac{1}{2}(u_{AA} + u_{BB})]$
Ω	regular solution constant
Ω^{α}	regular solution constant for phase α
u_{ij}	interaction energy between i and j
u_n	atomic position
U	internal energy

x_i mole fraction of component i

X_{ij} pair fraction of pair ij

X_i ionic fraction of ion i

Y_i coordination equivalent fractions of component i

Z partition function; compressibility factor

Z_α atomic number (of nucleus α)

Z_c critical compressibility factor

z coordination number

ξ extent of reaction

ψ spatial wavefunction

Ψ many-electron wavefunction

Notation for extensive thermodynamic properties exemplified by enthalpy, *H*

H enthalpy

H_m molar enthalpy

H_m^o standard molar enthalpy

$\Delta_0^T H_m^o$ standard molar enthalpy at temperature T relative to zero K

$\overline{H}_i$ partial molar enthalpy

$\Delta_f H_m^o$ standard molar enthalpy of formation

$\Delta_{fus} H_m^o$ standard molar enthalpy of fusion

$\Delta_{vap} H_m^o$ standard molar enthalpy of vaporisation

$\Delta_{trs} H_m^o$ change in standard molar enthalpy of a phase transition

$\Delta_{latt} H$ lattice enthalpy

$\Delta_{atom} H$ enthalpy of atomization

$\Delta_{ion} H$ ionization enthalpy

$\Delta_{eg} H$ electron gain enthalpy

$\Delta_{diss} H$ enthalpy of dissociation

$\Delta_{ox} H$ enthalpy of oxidation

$\Delta_{f,ox} H_m^o$ standard molar enthalpy of formation (of a ternary oxide) from (binary) oxides

$\Delta_r H^o$ change in standard enthalpy of reaction

$\Delta_{mix} H_m$ molar enthalpy of mixing

$\Delta_{mix}^{id} H_m$ ideal molar enthalpy of mixing

$\Delta_{mix}^{exc} H_i$	excess molar enthalpy of mixing
$\Delta_{mix} \overline{H}_m$	partial molar enthalpy of mixing of component i
$\Delta_{vac} H$	enthalpy of vacancy formation
H^σ	surface enthalpy
H_k	enthalpy of an individual configuration k

Prefixes

pico	10^{-12}
nano	10^{-9}
micro	10^{-6}
milli	10^{-3}
kilo	10^3
mega	10^6
giga	10^9
tera	10^{12}

Fundamental constants

Gas constant	$R = 8.31451$ J K^{-1} mol^{-1}
Avogadro's number	$L = 6.02214 \times 10^{23}$ mol^{-1}
Boltzmann's constant	$k_B = 1.38066 \times 10^{-23}$ J K^{-1}
Faraday's constant	$F = 9.64853 \times 10^4$ C mol^{-1}
Elementary charge	$e = 1.602177 \times 10^{-19}$ C
Planck's constant	$h = 6.62620 \times 10^{-34}$ J s
	$\hbar = h/2\pi$

Pressure units

pascal	1 Pa	1 N m^{-2}
bar	1 bar	10^5 Pa
atmosphere	1 atm	101 325 Pa
torr	1 torr	133.32 Pa

Index